Studies in Logic
Logic and Argumentation
Volume 111

The Cognitive Dimension of Social Argumentation
Proceedings of the 4th European Conference on Argumentation
Volume III

Volume 101
The Logic of Partitions. With Two Major Applications
David Ellerman

Volume 102
Bounded Reasoning Volume 1: Classical Propositional Logic
Marcello D'Agostino, Dov Gabbay, Costanza Larese, Sanjay Modgil

Volume 103
The Fertile Debate. Affective Exploration of a Controversy
Claire Polo

Volume 104
Argument, Sex and Logic
Dov Gabbay, Gadi Rozenberg and Lydia Rivlin

Volume 105
Logic as a Tool. A Guide to Formal Logical Reasoning
Valentin Goranko

Volume 106
New Directions in Term Logic
George Englebretsen, ed

Volume 107
Non-commutative Algebras. Pseudo-BCK Algebreas versus m-pseudo-BCK Algebras
Afrodita Iorgulescu

Volume 108
Semitopology: decentralised collaborative action via topology, algebra, and logic
Murdoch J. Gabbay

Volume 109
The Cognitive Dimension of Social Argumentation. Proceedings of the 4th European Conference on Argumentation, Volume I. Fabio Paglieri, Alessandro Ansani and Marco Marini, eds.

Volume 110
The Cognitive Dimension of Social Argumentation. Proceedings of the 4th European Conference on Argumentation, Volume II. Fabio Paglieri, Alessandro Ansani and Marco Marini, eds.

Volume 111
The Cognitive Dimension of Social Argumentation. Proceedings of the 4th European Conference on Argumentation, Volume III. Fabio Paglieri, Alessandro Ansani and Marco Marini, eds.

Studies in Logic Series Editor
Dov Gabbay   dov.gabbay@kcl.ac.uk

The Cognitive Dimension of Social Argumentation
Proceedings of the 4th European Conference on Argumentation
Volume III

Edited by
Fabio Paglieri
Alessandro Ansani
and
Marco Marini

© Individual author and College Publications, 2024
All rights reserved.

ISBN 978-1-84890-473-6

College Publications
Scientific Director: Dov Gabbay
Managing Director: Jane Spurr

http://www.collegepublications.co.uk

Cover prepared by Laraine Welch

All rights reserved. No part of this publication may be reproduced, stored in a retrieval system or transmitted in any form, or by any means, electronic, mechanical, photocopying, recording or otherwise without prior permission, in writing, from the publisher.

The Cognitive Dimension of Social Argumentation:

Proceedings of the 4th European Conference on Argumentation

Volume 3

Edited by
Fabio Paglieri
Alessandro Ansani
Marco Marini

TABLE OF CONTENTS

INTRODUCTION V
FABIO PAGLIERI, ALESSANDRO ANSANI, MARCO MARINI

ARGUMENTATIVE SEQUENCE LABELLING USING 1
TRANSFER LEARNING
DAVIDE LIGA

HOW TO DESCRIBE ARGUMENTS IN FUNDAMENTAL 17
CRITERIA FOR GOOD ARGUMENTS?
CHRISTOPH LUMER

THE RHETORICAL FOUNDATIONS OF LOGIC: 31
RETHINKING JØRGENSEN'S DILEMMA
MAURIZIO MANZIN

REASONISM AND INFERENCISM IN THE THEORY OF 41
ARGUMENT
HUBERT MARRAUD

ARGUMENTATION QUALITY ASSESSMENT: AN ARGUMENT 53
MINING APPROACH
SANTIAGO MARRO, ELENA CABRIO, SERENA VILLATA

MODULATING EPISTEMIC COMMITMENT TO DISAGREEING 69
AND AGGRESSIVE CONTENTS EXPLOITING THE
EVIDENTIAL FUNCTION OF IMPLICIT COMMUNICATION
VIVIANA MASIA

"WE HAVE WEATHERED THE OMICRON STORM...": A CASE 85
STUDY ON THE ARGUMENTATIVE USE OF METAPHOR IN
IRELAND'S POLITICAL DISCOURSE ON COVID-19
DAVIDE MAZZI

FROM DEEP DISAGREEMENT TO RATIONALLY 97
IRRESOLVABLE DISAGREEMENT
GUIDO MELCHIOR

CONFLICTING CHARACTERIZATION FRAMES AND ARGUMENTATION IN THE PUBLIC CONTROVERSY SURROUNDING FASHION SUSTAINABILITY CHIARA MERCURI	111
PERSUASION WITHIN DELIBERATION: SYMBIOSIS OR PARASITISM? HENRI MÜTSCHELE	125
AN ARGUMENTATIVE RECONSTRUCTION OF MENCIUS'S VIRTUE THEORY ZI-HAN NIU & MING-HUI XIONG	139
WITTGENSTEIN'S HINGES AND THE LIMITS OF ARGUMENT PAULA OLMOS	151
THE THIRD PARTY IN SHARED DECISION-MAKING: THE ROLE OF PATIENT COMPANIONS IN DISCUSSIONS BETWEEN PATIENTS AND HEALTHCARE PROFESSIONALS ROOSMARYN PILGRAM & LOTTE VAN POPPEL	165
RHETORIC, ANTHROPOLOGICAL MODELS AND THE LAW FEDERICO PUPPO	181
LISTING AS AN ARGUMENTATIVE RESOURCE IN POLITICAL DISCOURSE MENNO H. REIJVEN, ALINA DURRANI & GONEN DORI-HACOHEN	189
CONSPIRACY AND NON-CONSPIRACY ARGUMENTS IN ANTI-VACCINE DISCOURSE THÉOPHILE ROBINEAU	201
EXPLORING THE ARGUMENTATIVE POTENTIAL OF DOUBT IN MEDICAL CONSULTATIONS MARIA GRAZIA ROSSI, DIMA MOHAMMED & SARAH BIGI	217

THE TYPES OF APPEAL TO RELIGIOUS AUTHORITIES. KAIROS AND INTERRELIGIOUS COMMUNICATION LUCIA SALVATO	231
ARGUMENTATION AND EPISTEMIC DISTRIBUTED VIGILANCE CRISTIÁN SANTIBÁÑEZ	245
ARGUMENTATION, CRITICAL THINKING, LIBERAL DEMOCRACY, AND NON-WESTERN STUDENTS MENASHE SCHWED	257
THE METHODOLOGY OF THE NEW RHETORIC (AND WHY IT STILL MATTERS) BLAKE D. SCOTT	267
A PROOF OF CONCEPT ON DIALOGUE GAMES FOR EXPLAINABLE ARTIFICIAL INTELLIGENCE ILIA STEPIN, ALEJANDRO CATALA & JOSE M. ALONSO-MORAL	277
FORENSIC ARGUMENTATION IN EDUCATION SERENA TOMASI	291
EXPANDING DEEP DISAGREEMENTS: INCOMPATIBLE NARRATIVES AS INTERPRETATIVE REPERTOIRES IN POLITICAL DISAGREEMENTS MEHMET ALÌ ÜZELGÜN	301
THE FALLACY OF POPULARITY JAN ALBERT VAN LAAR	315
COLLECTIVE ETHOS, AD HOMINEM, AND AUDIENCE JUDGMENT: THE CREDIBILITY OF THE INSTITUTIONAL "FACES" OF FACEBOOK JIANFENG WANG & CHRISTOPHER W. TINDALE	327
COGNITIVE BIAS IN SOCIAL ARGUMENTATION MARK WEINSTEIN	341

THE ROLE OF FRAMES IN PLATO'S LACHES   353
HARALD R. WOHLRAPP

THE ORIGINS AND CLASSIFICATION CRITERIA OF   363
CRITICAL QUESTIONS
SHIYANG YU & FRANK ZENKER

MULTI-CONTEXT POLYLOGUES: ANALYSIS OF   375
INTERTWINED LEGAL & PUBLIC ARGUMENTS
GÁBOR Á. ZEMPLÉN & JÁNOS TANÁCS

INTRODUCTION

FABIO PAGLIERI
Istituto di Scienze e Tecnologie della Cognizione, Consiglio Nazionale delle Ricerche (ISTC-CNR), Roma, Italy
fabio.paglieri@istc.cnr.it

ALESSANDRO ANSANI
Centre of Excellence in Music, Mind, Body and Brain, University of Jyväskylä, Finland
alessandro.a.ansani@jyu.fi

MARCO MARINI
Istituto di Scienze e Tecnologie della Cognizione, Consiglio Nazionale delle Ricerche (ISTC-CNR), Roma, Italy
marco.marini@istc.cnr.it

The European Conference on Argumentation (ECA) is an academic initiative launched in 2013 by a group of Europe-based argumentation scholars, with the aim of inaugurating a series of biennial international conferences on this thriving area of studies (for details, see https://ecargument.org/), to complement other large-scale events on similar or related topics: the conference of the International Society for the Study of Argumentation (ISSA), the conference of the Ontario Society for the Study of Argumentation (OSSA), the conference on Computational Models of Argument (COMMA), the Rhetoric in Society conference and, more recently, the events organized by the Argumentation Network of the Americas (ANA). The first edition of ECA took place in Lisbon (PT) in 2015, followed by a second one in Fribourg (CH) in 2017 and a third one in Groningen (NL) in 2019: then the Covid-19 pandemic struck and the next edition of ECA had to be postponed to 2022, when it took place in Rome, from September 28 to September 30.

The conference lasted two days and a half, with a very intense and diverse programme, including 3 keynote talks, 1 plenary panel, 16 long papers with invited commentators, and as many as 118 regular papers (14 of which were presented as part of 3 thematic panels). Most of these contributions are collected in written form in these three volumes, as follows:

- *Volume 1* includes 2 of the 3 keynotes presented at the conference, authored by Catarina Dutilh Novaes and Harvey Siegel, followed by 16 long papers, ordered alphabetically by first author's surname: the majority of those (9 out of 16) are accompanied by their respective commentary.
- *Volume 2* includes 30 regular papers, ordered alphabetically by first author's surname, from Aikin & Casey to Licato et al.
- *Volume 3* includes the remaining 30 regular papers, again ordered alphabetically by first author's surname, from Liga to Zemplén & Tanács.

Even though these proceedings cover only a selection of what was presented at the ECA conference in Rome, they provide a faithful approximation of the breadth and depth of the ongoing discussion in argumentation scholarship. They also attest the interdisciplinary character of this field: this has been the hallmark of argumentation studies since their inception, yet the disciplines brought to bear on this subject matter have steadily increased over the years; whereas philosophy and linguistics were always partners in this endeavor, nowadays they are supported also by computer science and experimental psychology, as well as communication and media studies in a broader sense. At the same time, the study of argument from a philosophical perspective is no longer regarded as a specialistic niche for philosophers with a grudge against deductive logic as a model of human reasoning, but it is taking back its place as a central concern for philosophical inquiry in general, as it was at the dawn of the discipline (Aristotle's work is an obvious example).

These are welcome developments in the natural evolution of argumentation studies, which the ECA initiative has always intended to promote and nourish: thus, we expect to see more of the same in future editions of ECA, starting with ECA 2025 in Warsaw, Poland, on "Argumentation in the digital society".

ARGUMENTATIVE SEQUENCE LABELLING USING TRANSFER LEARNING

DAVIDE LIGA
University of Luxembourg
davide.liga@uni.lu

Abstract

This work presents an approach for Argumentative Sequence Labelling using Transfer Learning. Specifically, a famous pre-trained neural architecture, BERT, has been employed using the Transfer Learning technique known as "fine-tuning" and employing two different data formats for sequence labelling (BIO and BILUO). The neural architecture has been fine-tuned on two famous corpora to recognize not only the boundaries of argumentative units, but also the specific types of argumentative component. The resulting model not only outperforms the results of previous models, but it is also easier to implement, since it does not require highly-engineered features. An evaluation at token-level is performed, as well as a preliminary error analysis.

1. Introduction

Transformers architectures like BERT, and more-generally pre-trained language models (PLMs) have been extremely popular in the last two years achieving several records in the State of the Art of Natural Language Understanding (Devlin et al., 2018). However, their use in the field of Argumentation Mining (Lawrence and Reed, 2020) has been relatively small so far. In this work, a Transfer Learning approach has been applied to the task of detecting argumentative spans. More precisely, we want to assess the ability of Transfer Learning to facilitate the task of labelling argumentative sequences. Transfer Learning methodologies, particularly the fine-tuning and contextual-embeddings techniques, have been recently used in many NLP tasks, achieving remarkable steps forward in artificial Natural Language Understanding. BERT and its derivations have been among the most successful natural language models employed for Transfer Learning in the last couple of years. The reason for the success of BERT is the fact that it is able to encode important information about language. To do so, BERT has been pre-trained on a large amount of data (mostly from wikipedia) performing tasks that are designed to force the neural architecture to learn language features. As explained in Devlin et al. (2018), one of these tasks is the Masked Language task: this task forces BERT's neural architecture to predict randomly masked tokens. Thanks to this simple idea and thanks to its attention-based mechanisms (Galassi et al., 2019), BERT is able to learn many features of human language, and this knowledge is incorporated in its neural architecture. As shown in Devlin et al. (2018), this knowledge can be then transferred (hence the expression Transfer Learning) to downstream tasks in two ways: the first option is to use BERT to output the embeddings produced by its neural architecture, using these embeddings as features for downstream tasks; the second option is to fine-tune the pre-trained neural architecture, namely performing new training epochs on it, while using downstream data.

Since fine-tuning has been surprisingly efficient with many NLP-related tasks, including sequence labelling tasks such as Named Entity Recognition (Devlin et al., 2018), we think it can be useful to assess BERT's performances on argumentative sequences employing the fine-tuning method to transfer learning from the pre-trained model to the downstream task of labelling argumentative sequences.

In Section 2, some related works will be mentioned. Section 3 will describe the datasets employed in this work. In Section 4, the proposed methodology will be presented. Section 5 will describe the achieved results, while Section 6 will offer a discussion about the results. Finally, Section 7 will conclude the work.

2. Related Works

While the tasks of sequence labelling and tagging are well known in NLP (Manning, 2011; Akbik et al., 2018), only few attempts have been performed in the field of Argumentation Mining, especially with regard to the labelling of argumentative spans. To the best of our knowledge, the first attempt to label argumentative sequences has been proposed by Stab (2017), where the modeling of argumentative sequences employs highly-engineered features including Structural, Syntactic, Lexical-Syntactic and Probabilistic elements, while the classification employs a CRF (Lafferty et al., 2001) implemented in CRFsuite (Okazaki, 2007) with an averaged perceptron (Collins and Duffy, 2002). In this study, Stab (2017) adapted the standard BIO format to the purpose of the Argumentative Sequence Labelling, using the labels Arg-B (for those tokens that are at beginning of an argumentative span), Arg-I (for all other argumentative tokens) and Arg-O (for tokens that are not within an argumentative span) with the Argument Annotated Essays Corpus (Stab, 2017). The resulting classification of these three labels achieved a macro averaged F1 score of 0.867.

Ajjour et al. (2017) carried out further experiments, showing that results can be improved by combining all the features provided by Stab (2017) with a Bi-LSTM (bidirectional long short-term memory) neural network, and showing that the Bi-LSTM outperforms SVM and CRF classifiers. Importantly, the classification consider not only the Argument Annotated Essays Corpus but also other two famous corpora: the Webis-Editorials-16 corpus (Al Khatib et al., 2016) and the Argument Annotated UserGenerated Web Discourse corpus (Habernal and Gurevych, 2017).

The above-mentioned studies employs highly-engineered features and their models (the CRF and the BiLTSM) are designed for the labelling of argumentative spans in general (without considering the differences between argumentative components, which is considered as a separate task). Moreover, the two models only consider the BIO format, despite the fact that other formats showed better learning performances in some cases, e.g. the BILUO format, which considers also the last token of spans and the spans with just one token (Ratinov and Roth, 2009).

The main novelties of our paper, are: (a) the assessment of the performances of fine-tuning as a Transfer Learning methodology for Argumentative Sequence Labelling on two famous datasets: the Argument Annotated Essays Corpus (Stab and Gurevych, 2017) and the Argument Annotated User-Generated Web Discourse corpus (Habernal and Gurevych, 2017); (b) the assessment of the performances of two different data formats for the sequence labelling: BIO and BILUO; (c) the division of the experiment in two separate sub-tasks to assess whether Argumentative Sequence Labelling can be performed not only to detect argumentative vs non-argumentative spans, but also to detect the spans of different argumentative components, e.g. premises, claims (somehow combining sequence labelling and text classification). The research questions addressed in this paper are:

1. Can Transfer Learning outperform previous Argumentative Sequence Labelling scores without employing highly-engineered features?
2. Can different labelling formats (e.g. BIO, BILUO) affect these scores significantly?
3. Can we use the same methodology to discriminate argumentative components at a more granular level (distinguishing, for example, premises and claims)?

3. Data

Essays. This corpus is composed of 402 persuasive essays written by students and annotated by three experts. Particularly, the annotation considers three types of argumentative units (premises, claims and major claims) considering all the other spans as non-argumentative spans. The authors split the corpus into 322 essays for training and 80 essays for test. This split has been preserved also in the present paper.

Web Discourse. This corpus consists of 340 comments written in online newspapers, forums and blogs. In this case, the annotation has been performed by considering a five different types of argumentative unit, which are similar to the famous Toulmin argument model (Toulmin et al., 1958): claim, premise, rebuttal, backing and refutation. Since the dataset does not provide any training-test split, we followed the classic 80/20 split, also proposed by Ajjour et al. (2017).

These two datasets have a very different nature: the first one is composed of well-structured arguments written by students that were asked to write about specific topics, the second one shows less predictable ways of expressing arguments, which is frequent in comments from the Internet.

4. Methodology

We divided our experiments into two parts. The first one reproduces the same task of sequence labelling discussed in Stab (2017) and Ajjour et al. (2017). The second one is instead an extension which increases the complexity of the labelling task. Table IX, in the Appendix A, describes this complexity showing the number of targeted token-level labels considered in the two tasks.

Task 1 focuses on the labelling of sequences as belonging to an argument or not: in this case, each tokens has been labelled as belonging or not to an argumentative span. We created a script to convert each corpus into BIO and BILUO format. For both corpora, the BIO format is represented by three labels (B-ARG, I-ARG and O), while the BILUO format is represented by five labels B-ARG, I-ARG, L-ARG, U-ARG and O (however, no U-ARG have been found in the two corpora). Task 2 presents exactly the same neural architecture, but we trained it for a more complex task: labelling each tokens as belonging to specific argumentative components. This means that each token is considered as being or not part of a specific argumentative component, such as premise, claim, and so on. In this case, we converted each corpus into a BIO and BILUO using the argumentative components specifically belonging to them. This means that we applied the prefix (B-, I- and L-) to three labels in the case of the Essays corpus (Claim, Premise, MajorClaim) and to five labels in the case of the Web Discourse corpus (Claim, Premise, Rebuttal, Backing, Refutation). In other words, Task 2 is more complex because the neural architecture tries to classify more types of sequences. To understand this complexity, we can consider the prefixes of the chosen format (B-, I- in the case of the BIO format; B-, I-, L-, U- in the case of the BILUO format) and multiply the number of prefixes by the number of argumentative components in the two corpora (three in the case of the Essays corpus, five in the case of the Web Discourse corpus), plus the label O.

Before feeding our model with the textual data, we used spaCy[1] to automatically separate each document into different sentences. We noticed that this process of separation improves significantly the learning results. Regarding the employed neural architecture, we implemented Google's BERT[2], a famous attention-based (Galassi et al., 2019; Spliethöver et al., 2019) Transformer (Devlin et al., 2018). More specifically, we used the pre-trained BERT base-uncased (we will sometimes refer to it as "BERT base") which is a neural architecture consisting of 12 encoder layers, 768 hidden units and 12 attention heads and it is pre-trained on a large amount of data (including Wikipedia), resulting in 110M parameters. To use BERT, each sentence of the corpora has been tokenized using wordpiece (Wu et al., 2016) tokenization as required by BERT (Devlin et al., 2018). Moreover, we truncated and padded all sentences to a fixed length, and we trained our model in 4 epochs, with a batch size of 32 and a learning rate of 5e-5.

Finally, to assess the ability of our model to understand argumentative spans, we evaluated the results of the classification at token-level, considering Precision, Recall and F1 scores for the correct classification of each tokens.

5. Results

This section reports the results for the two parts of the experiment. The pre-trained neural architecture is exactly the same in both in Task 1 and Task 2 and we compared it with a baseline, to show whether a simple pre-trained BERT model can improve previous results. More specifically, we considered as baseline the previous scores achieved by the BiLSTM model proposed by Ajjour et al. (2017) on the same two corpora. While Ajjour et al. (2017) employed just the BIO format, we employed both the BIO and BILUO format. Table I reports the average macro F1 score of our BERT model for the Essay and Web-Discourse corpora considering both the BIO and BILUO formats.

Importantly, our simple BERT implementation is able to reach and outperforms the results of the highly-engineered BiLSTM baseline (which answers our first research question). When considering the Essays corpus in the BIO format, our model almost reaches the Baseline at 87.43; when considering the BILUO format, our model slightly outperforms the

---

[1] https://spacy.io/
[2] github.com/google-research/bert

baseline at 88.97. However, the situation is different when considering the Web Discourse: using the BILUO format, our model performs slightly worse than the baseline at 52.66, while using the BIO format, our model outperforms the baseline with a more evident improvement at 60.26. This is a confirmation that the chosen format can affect the ability of neural architectures to learn (as has been showed for example in Ratinov and Roth (2009)), which is also an answer for our second research question.

5.1. Task 1: Argumentative Span Detection

Table II is a more complete report, showing the F1 scores per class for the two datasets. This report provides a better understanding of the ability of the model to classify tokens correctly. For example, we can see that using the BIO format BERT is capable of discriminating correctly between tokens that are at the beginning of an argumentative span with an F1 score of 0.87 while it can recognize inner tokens with a F1 score of 0.92. Interestingly, when using the BILUO format, performances improve. In this case, the support for I-ARG is split (1,264 tokens are considered as L-ARG tokens, concluding an argumentative span), however BERT seems capable to recognize the difference between B-, I- and L- tokens with an average macro F1 of .89 (i.e., .8897 as showed in Table I).

Table I. Comparison with the baseline. Mean F1 scores (macro) for the token labelling considering the BIO format (3 classes: B-ARG, I-ARG and O) and the BILUO format (5 classes: B-ARG, I-ARG, L-ARG, U-ARG and O). In bold, the macro F1 scores that outperformed the baseline.

Corpus	$BERT_{base}$ BIO	$BERT_{base}$ BILUO	Baseline BiLSTM BIO
Essays	.87	**.89**	.88
Web Discourse	**.60**	.53	.55

Table II. Results for the token-level classification of the Task 1 for the two datasets: detecting argumentative spans on the two corpora using BIO and BILUO formats.

Task 1 Essays corpus				
BIO format	P	R	F1	support
B-ARG	.86	.88	.87	1,266
I-ARG	.91	.94	.92	18,750
O	.86	.81	.83	9,412
macro avg:			.87	
weighted avg:			.89	
BILUO format	P	R	F1	support
B-ARG	.87	.88	.88	1,266
I-ARG	.91	.94	.93	17,484

L-ARG	.90	.92	.91	1,266
O	.88	.82	.85	9,412
macro avg:			.89	
weighted avg:			.90	

Task 1				
Web Discourse corpus				
BIO format	P	R	F1	support
B-ARG	.42	.60	.49	201
I-ARG	.59	.66	.63	7,615
O	.72	.66	.69	10,325
macro avg:			.60	
weighted avg:			.66	
BILUO format	P	R	F1	support
B-ARG	.44	.49	.46	201
I-ARG	.60	.52	.56	7,414
L-ARG	.43	.42	.43	201
O	.68	.74	.71	10,325
macro avg:			.54	
weighted avg:			.64	

5.2. Task 2: Argumentative Component Span Detection

Not surprisingly, results for the Task 2 show lower scores. Table IV reports the achieved scores for the two datasets considering BIO and BILUO formats. In the case of the Essays corpus, the labelling involved 3 types of argumentative components, while in the case of the Web Discourse corpus, we attempted two strategies of classification: firstly, we trained our sequence labelling involving all 5 types of argumentative components of the dataset, which roughly follow the Toulmin's argument model. However, since results were quite unsatisfactory, we simplified the task by considering 2 classes only (on the one side the class Claim, on the other side a class Argumentative, including the other 4 argumentative classes). Importantly, we attempted two ways for classifying the spans of the argumentative components of the Essays corpus. In the first method, we simply considered all three classes already mentioned (Claim, Major Claim and Premise). In the second method, we considered Claim and Major Claim as the same label, to see if the classifier could recognize the difference between tokens belonging to claims (both major claims and normal claims) and tokens belonging to premises.

The left side of Table III shows the results of the token classification on the Essays corpus considering the BIO and BILUO format for all the three types of argumentative components of the Essays corpus (Claim, Major Claim, Premise). In right side of Table III, instead, we considered Claim and Major Claim as belonging to the same category (a general Claim category). Interestingly, the average macro F1 score improves when considering Major Claim and Claim as a unique class: it increases from .65-.66 (for BIO and BILUO formats, respectively) to .72-.70. Moreover, in both cases, the BIO format outperforms the BILUO format. It is important to notice that even if the support for some classes is relatively small, results seems encouraging (especially in the when considering just two classes). These scores demonstrate that BERT is capable of not only recognizing argumentative spans, but also distinguishing between different types of argumentative spans.

In Table IV, the token classification results for the Web Discourse corpus are presented. In this case, the model attempted to classify 11 token-level labels in the BIO format and 16 token-level labels in the BILUO format. The average F1 score is quite low for both formats, indicating that the model struggles with classification. This is likely due to the low support for some classes, meaning there may not be enough instances for the model to learn from. The labels with the smallest supports achieve the lowest F1 scores. The weighted F1 score provides an alternative description of this unbalanced scenario, showing a slightly more encouraging .55/.56 for BILUO and BIO formats, respectively. In this scenario as well, the BIO format shows better performance.

In Table V, we simplify the scenario by proposing a classification of 5 token-level labels in the BIO format and 7 token-level labels in the BILUO format. In this scenario, the goal is to distinguish Claims from Argumentative spans (with the latter including all argumentative spans that are not claims). We found that this approach resulted in better classification performance, with a weighted F1 score of .70/.72 for the BIO and BILUO formats, respectively.

Table III. Task 2 for the Essays corpus, using BIO and BILUO formats and considering Claim, MajorClaim, Premise (left), and Claim and Premise (right).

Task 2				
Essays corpus *(considering three classes)*				
BIO format	P	R	F1	support
B-CLAIM	.42	.49	.45	304
B-MAJORCLAIM B-PREMISE	.69	.68	.69	153
	.73	.69	.71	809
I-CLAIM	.43	.49	.46	3,920
I-MAJORCLAIM I-PREMISE	.72	.71	.72	1,970
	.81	.79	.80	12,860
O	.85	.82	.83	9,412
macro avg:		.66		
weighted avg:		.75		
BILUO format	P	R	F1	support

	P	R	F1	support
B-CLAIM	.40	.49	.44	304
B-MAJORCLAIM	.68	.67	.68	153
B-PREMISE	.73	.69	.71	809
I-CLAIM	.38	.51	.44	3,616
I-MAJORCLAIM	.73	.66	.69	1,817
I-PREMISE	.80	.77	.78	12,051
L-CLAIM	.43	.51	.46	304
L-MAJORCLAIM	.73	.67	.70	153
L-PREMISE	.75	.73	.74	809
O	.87	.81	.84	9,412
macro avg:	.65			
weighted avg:	.74			

Task 2
Essays corpus *(considering two classes)*

BIO format	P	R	F1	support
B-CLAIM	.62	.62	.62	457
B-PREMISE	.71	.73	.72	809
I-CLAIM	.61	.62	.61	5,890
I-PREMISE	.78	.81	.79	12,860
O	.86	.80	.83	9,412
macro avg:			.72	
weighted avg:			.76	

BILUO format	P	R	F1	support
B-CLAIM	.57	.62	.60	457
B-PREMISE	.72	.70	.71	809
I-CLAIM	.58	.62	.60	5,435
I-PREMISE	.79	.77	.78	12,049
L-CLAIM	.59	.64	.61	457
L-PREMISE	.76	.72	.74	809
O	.84	.83	.84	9,412
macro avg:			.70	
weighted avg:			.76	

Table IV. Task 2 for the Web Discourse corpus, using BIO (left) and BILUO (right) formats and considering Claim, Premise, Rebuttal, Backing, Refutation.

Task 2 Web Discourse corpus (five classes)				
BIO format	P	R	F1	support
B-CLAIM	.71	.14	.23	36
B-PREMISE	.30	.42	.35	106
B-BACKING	.45	.12	.19	43
B-REBUTTAL	.00	.00	.00	12
B-REFUTATION	.00	.00	.00	4
I-CLAIM	.40	.42	.41	680
I-PREMISE	.44	.45	.45	4,247
I-BACKING	.32	.26	.29	2,089
I-REBUTTAL	.32	.15	.20	453
I-REFUTATION	.00	.00	.00	146
O	.68	.71	.70	10,325
macro avg:			.26	
weighted avg:			.56	

Task 2 Web Discourse corpus (five classes)				
BILUO format	P	R	F1	support
B-CLAIM	.43	.08	.14	36
B-PREMISE	.33	.35	.34	106
B-BACKING	.50	.12	.19	43
B-REBUTTAL	.00	.00	.00	12
B-REFUTATION	.00	.00	.00	4
I-CLAIM	.35	.34	.35	644
I-PREMISE	.43	.38	.40	4,141
I-BACKING	.31	.23	.27	2,046
I-REBUTTAL	.11	.08	.10	441
I-REFUTATION	.00	.00	.00	142
L-CLAIM	.33	.31	.32	36
L-PREMISE	.33	.36	.34	106
L-BACKING	.00	.00	.00	43
L-REBUTTAL	.00	.00	.00	12
L-REFUTATION	.00	.00	.00	4
O	.67	.76	.71	10,325
macro avg:			.20	
weighted avg:			.55	

Table V. Task 2 for the Web Discourse corpus, using BIO (left) and BILUO (right) formats and considering two classes: Claim and Argumentative ("Arg").

Task 2 Web Discourse corpus (two classes)				
BIO format	P	R	F1	support
B-CLAIM	.48	.31	.37	36
B-ARG	.30	.41	.34	106
I-CLAIM	.51	.37	.43	674
I-ARG	.44	.45	.44	3,941
O	.80	.80	.80	12,350
macro avg:	.48			
weighted avg:	.70			

Task 2 WebDiscourse corpus (two classes)				
BILUO format	P	R	F1	support
B-CLAIM	.50	.22	.31	36
B-ARG	.38	.31	.34	106
I-CLAIM	.51	.34	.41	643
I-ARG	.52	.36	.43	3,856
L-CLAIM	.45	.29	.35	36
L-ARG	.34	.24	.28	85
O	.79	.89	.84	12,350
macro avg:			.42	
weighted avg:			.72	

6. Discussion on the results

The research questions mentioned before can now be addressed: results show, in fact, that a simple BERT model can reach and even outperform the previous models. This is important, because previous records employed highly engineered models, while we are using a simple pre-trained model without changing its neural architecture. We also assessed that the choice of BIO and BILUO can indeed affect results. However, it seems that the Web Discourse corpus is the one that is most affected by this change. In this regard, we are interested in determining whether this is due to the composition of the argumentative data coming from the web (which are less well-structured than the Essays corpus, or structurally more variable). Further studies are needed to investigate this aspect.

The poor results on the Web Discourse corpus should be viewed as an ongoing challenge in the field of Argument Mining, since they show that we still need to figure out how to deal with inconsistent or misshapen natural language arguments. Even if our weighted F1 scores are encouraging, the macro F1 scores show that there are important limitations and that although the label "O" receives relatively high scores, the other labels struggle to achieve satisfactory results, proving that the detection of argumentative components in everyday language is a very challenging target.

Finally, answering to the third research question, we showed that the sequence labelling achieves encouraging results also at more granular levels, discriminating among different kinds of argumentative components.

6.1. Preliminary Error Analysis

We are currently performing an Error Analysis which shows that BERT can recognize patterns of language commonly employed in natural arguments (e.g. the use of connectors such as "In my view," or "Finally,") and also the beginning and the conclusion of argumentative spans are detected with precision (please, see Appendix B). However, missing information about the context unavoidably affects results: for example, some sentences which seems argumentative (especially in other contexts) but are not (w.r.t. the topic of the discussion) might generate false positives (see the first false positive reported in Appendix B). Other false positives might be generated by connectors such as "because" (see the second false positive reported in Appendix B). Finally, there are cases in which the argumentative sentence is detected but the match is not perfect (BERT wrongly adds or misses argumentative spans).

7. Conclusion

This study outperformed previous results in the State of the Art of Argumentative Sequence Labelling, showing that BERT can reach and outperform previous benchmarks on the Argument Annotated Essays corpus and on the Argument Annotated User-Generated Web Discourse corpus. More precisely, we divided the work in two Tasks: in the first one, we focused on the recognition (at token-level) of argumentative spans vs non-argumentative spans, while in the second task we focused on a more fine-grained classification of the tokens as belonging to specific argumentative components.

Importantly, we showed that the choice of the labelling format (e.g., BIO, BILUO) can affect scores, although the extent of such influence seems related to the underlying data employed. In this regard, further research is required to understand what kind of format are more performing and how these performances are related to the underlying argumentative data. Furthermore, we showed that BERT is able not only to recognize sequences of argumentative tokens (considering argumentative vs non-argumentative), but also to recognize what kind of argumentative components are involved (premise, claim, rebuttal, etc.).

In future, the performances of other PLMs might be assessed. In any case, we think that this work can be a starting point for future research employing more complex Transfer Learning architecture for Argumentative Sequence Labelling.

References

Yamen Ajjour, Wei-Fan Chen, Johannes Kiesel, Henning Wachsmuth, and Benno Stein. 2017. Unit segmentation of argumentative texts. In Proceedings of the 4th Workshop on Argument Mining, pages 118–128.

Alan Akbik, Duncan Blythe, and Roland Vollgraf. 2018. Contextual string embeddings for sequence labeling. In Proceedings of the 27th International Conference on Computational Linguistics, pages 1638–1649.

Khalid Al Khatib, Henning Wachsmuth, Johannes Kiesel, Matthias Hagen, and Benno Stein. 2016. A news editorial corpus for mining argumentation strategies. In Proceedings of COLING 2016, the 26th International Conference on Computational Linguistics: Technical Papers, pages 3433–3443.

Michael Collins and Nigel Duffy. 2002. Convolution kernels for natural language. In Advances in neural information processing systems, pages 625–632.

Jacob Devlin, Ming-Wei Chang, Kenton Lee, and Kristina Toutanova. 2018. Bert: Pre-training of deep bidirectional transformers for language understanding. arXiv preprint arXiv:1810.04805.

Andrea Galassi, Marco Lippi, and Paolo Torroni. 2019. Attention, please! a critical review of neural attention models in natural language processing. arXiv preprint arXiv:1902.02181.

Ivan Habernal and Iryna Gurevych. 2017. Argumentation mining in user-generated web discourse. Computational Linguistics, 43(1):125–179.

John Lafferty, Andrew McCallum, and Fernando CN Pereira. 2001. Conditional random fields: Probabilistic models for segmenting and labeling sequence data.

John Lawrence and Chris Reed. 2020. Argument mining: A survey. Computational Linguistics, 45(4):765–818. Christopher D Manning. 2011. Part-of-speech tagging from 97% to 100%: is it time for some linguistics? In International conference on intelligent text processing and computational linguistics, pages 171–189. Springer.

Naoaki Okazaki. 2007. Crfsuite: a fast implementation of conditional random fields (crfs).

Lev Ratinov and Dan Roth. 2009. Design challenges and misconceptions in named entity recognition. In Proceedings of the Thirteenth Conference on Computational Natural Language Learning (CoNLL-2009), pages 147–155, Boulder, Colorado, June. Association for Computational Linguistics.

Maximilian Spliethöver, Jonas Klaff, and Hendrik Heuer. 2019. Is it worth the attention? a comparative evaluation of attention layers for argument unit segmentation. arXiv preprint arXiv:1906.10068.

Christian Stab and Iryna Gurevych. 2017. Parsing argumentation structures in persuasive essays. Computational Linguistics, 43(3):619–659.

Christian Matthias Edwin Stab. 2017. Argumentative writing support by means of natural language processing. Ph.D. thesis, Technische Universita ̈t Darmstadt.

Stephen Toulmin et al. 1958. The uses of argument.

Yonghui Wu, Mike Schuster, Zhifeng Chen, Quoc V Le, Mohammad Norouzi, Wolfgang Macherey, Maxim Krikun, Yuan Cao, Qin Gao, Klaus Macherey, et al. 2016. Google's neural machine translation system: Bridging the gap between human and machine translation. arXiv preprint arXiv:1609.08144.

Appendix A. Description of the span classes

Table IX. Description of the span classes for the two parts of the experiments, depending on corpus and format. The dagger (†) refers to the fact that no U- tokens have been actually found. The asterisk (*) refers to the fact that the classes Claim and Major Claim can be joint into a unique class, producing 7 total token-level labels instead of 10.

Task 1 (Argumentative span detection)			
Corpora	Prefixes	Token-level classes	Classes
Essays (BIO)	B- I-	B-ARG I-ARG O	3
Essays (BILUO)	B- I- L- U-	B-ARG I-ARG L-ARG U-ARG† O	4†
Web Discourse (BIO)	B- I-	B-ARG I-ARG O	3
Web Discourse (BILUO)	B- I- L- U-	B-ARG I-ARG L-ARG U-ARG† O	4†

Task 2 (Argumentative component span detection)			
Corpora	Prefixes	Token-level classes	Classes
Essays (BIO)	B- I-	B-CLAIM, B-MAJORCLAIM, B-PREMISE, I-CLAIM, I-MAJORCLAIM, I-PREMISE O	7
Essays (BILUO)	B- I- L- U-	B-CLAIM, B-MAJORCLAIM*, B-PREMISE, I-CLAIM, I-MAJORCLAIM*, I-PREMISE L- CLAIM, L-MAJORCLAIM*, L-PREMISE U-CLAIM†, U-MAJORCLAIM*†, U-PREMISE† O	10† 7*†
Web Discourse (BIO)	B- I-	B-CLAIM, B-PREMISE, B-BACKING, B- REBUTTAL, B-REFUTATION I-CLAIM, B-PREMISE, I-BACKING, I-REBUTTAL, I-REFUTATION O	11
Web Discourse (BILUO)	B- I- L- U-	B-CLAIM, B-PREMISE, B-BACKING, B-REBUTTAL, B-REFUTATION I-CLAIM, I-PREMISE, I-BACKING, I- REBUTTAL, I-REFUTATION L-CLAIM, L-PREMISE, L-BACKING, L- REBUTTAL, L-REFUTATION U-CLAIM†, U-PREMISE†, U-BACKING†, U-REBUTTAL†, U-REFUTATION† O	16†

Appendix B. Error Analysis

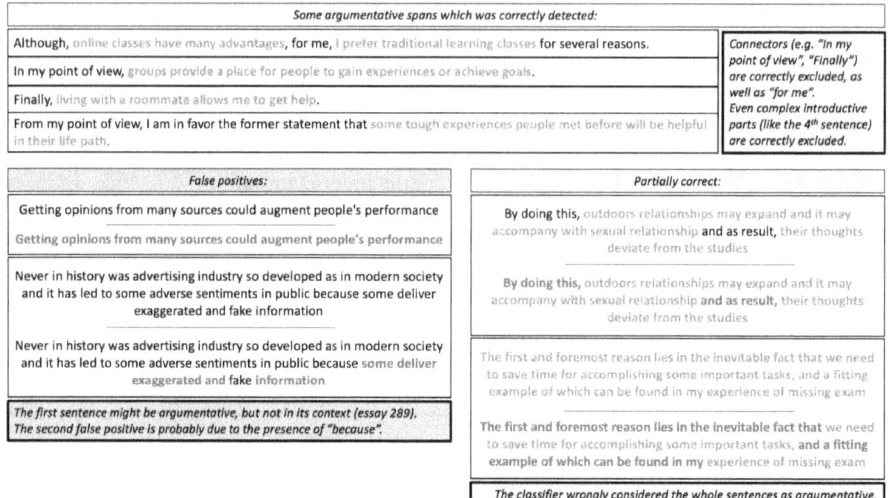

Figure 1. A preliminary error analysis on the Essay corpus for task 1. We selected 8 results: 4 correct, 2 partially correct, 2 false positives. Regarding false positives and partially correct sentences, both the true spans and the predicted span are reported: true spans are on the top, while their relative predictions are immediately below them. The red color is just used within predictions, to show errors (both false positives and false negatives). Please, compare predictions with the sentences above them. While false positive are those predictions where the classifier wrongly detected a non-existent argumentative span, partially correct sentences are those sentences where the match between true spans (on the top) and their relative predictions (below them) is not perfect, which means that the classifier either added or missed an argumentative span.

HOW TO DESCRIBE ARGUMENTS IN FUNDAMENTAL CRITERIA FOR GOOD ARGUMENTS?

CHRISTOPH LUMER
University of Siena, Italy
lumer@unisi.it

Abstract
How, in particular how abstractly, should arguments be described in fundamental argumentation-theoretical criteria for epistemically good arguments? In order to answer this question, it is first clarified which conditions must be fulfilled by epistemically oriented, adequate argumentation theories. Then, five differently abstract ways of describing arguments are presented, including argument schemes and characterisations of argument classes. Next, it is shown that theories of argumentation that formulate their fundamental criteria for good arguments in terms of argument schemes cannot be epistemically adequate, whereas theories that use characterisations of argument classes—as epistemological theories of argumentation do—can be. This is a strong argument against argument schema approaches in argumentation theory and in favour of epistemological ones.

The topic of this paper is: How should arguments be described in argumentation theory, at the most fundamental level, in criteria for good arguments? In particular: How abstract should they be described? By 'good arguments' I mean those that are epistemically good, that is, that guide the recognition of the acceptability of the thesis (where "acceptability" means: truth, verisimilitude or probable truth). I compare above all two types of descriptions of arguments in particular: *argument schemes*, on the one hand, and more abstract descriptions of argument types, namely *characterisations of argument classes*, on the other.

1. What are good arguments?

We all need true or—if certain truth is not attainable—*acceptable* (i.e. probably true or truth-like) information about the world in order to be able to orientate ourselves in it, to be able to achieve our goals, to be able to develop meaningful goals in the first place, etc. Our beliefs about such information must also be rationally justified. The advantage of *rational justification* is that it increases the proportion of true beliefs among one's own beliefs. (i) It does so because the first part of the justification consists in the fact that one has acquired the opinion on an epistemically correct path, which alone increases the proportion of true and acceptable ones among one's own opinions. (ii) Futhermore, the second part of the justification consists in a memory of the essential stages of the first part (the path of cognition); thus, incorrectly executed or weak justifications in the first part can eventually be recognised later and replaced by correct or stronger ones, which once more increases that proportion. (Lumer, 1990: 30-38; 2005a: 215.)

Epistemically oriented argumentation is a means of obtaining such rationally justified acceptable beliefs. Here, "good arguments" always means such epistemically functional arguments.1 Arguments produce rationally justified acceptable beliefs by guiding the addressee in recognising the acceptability of the thesis: The individual arguments are based on general epistemologically justified criteria for the acceptability of the thesis—which I call "epistemological principles". Such principles state that theses that fulfil such and such conditions are acceptable. The deductive epistemological principle, for example, states: 'If any true propositions p_1 to p_n logically imply another proposition q, then the latter proposition, q, is also true.' Other epistemological principles come from probability theory or rational decision theory. One of the practical epistemological principles, which underlie practical arguments for personal value judgements, e.g. is the definition of the concept of 'prospect desirability': 'The prospect desirability of event e for the subject s on the data base d is defined as e's intrinsic desirability (for s) plus the sum of the intrinsic desirabilities (for s) of all intrinsically non-neutral consequences of e weighted by their probability weight.' (Lumer, 2014: 8.) 2 The various epistemological principles are the bases of different argument types—the just quoted principles are the basis for deductive and (a special type of) practical arguments. Epistemically good arguments now consist in going

---

1 Epistemological argumentation theorists (e.g. Battersby, Biro, Feldman, Goldman, Lumer, Siegel, Weinstein) and theorists in their vicinity (e.g. Ennis, Freeman, Pinto) advocate an epistemic orientation of arguments, but also informal logicians (Blair, Johnson, Govier) or Walton.
2 Further but more specific epistemological principles: Lumer, 1990: 187; 225; 228; 235-237; 246; 276; 327; 328; 331; 429-430; 2005a: 222; 232; 2014: 8-10.

through the individual conditions of such an epistemological principle, and assert explicitly (or at least implicitly) for each one of these conditions that it is fulfilled. Deductive arguments therefore assert the truth of various propositions, i.e. those p_1 to p_n; and they then assert that these propositions imply the thesis q. If these assertions are true, i.e. if the conditions of the epistemological principle for the respective thesis are fulfilled, then, firstly, the thesis is also acceptable according to the epistemological principle; hence, the acceptability of the proposition is thus guaranteed by the argument (alethic, objective aspect of the argument). Secondly, these arguments also guide cognition (epistemic, subjective aspect of the argument): If the fulfilment of a condition is asserted in the argument and if the argument is adequate for convincing the respective addressee (e.g. he is already convinced of the premisses), then the addressee can directly check whether this is the case. In the case of deductive argumentations, he thus checks whether the propositions p_1 to p_n are true and whether they imply q logically. If the addressee during this examination always (correctly) arrives at positive results, he has established that all conditions of the epistemological principle for the acceptability of the thesis are fulfilled. If he now also at least implicitly knows the underlying epistemological principle, he will therefore conclude that the thesis is acceptable and believe this; he thus has recognised the acceptability of thesis q. Arguments that can in principle guide one in recognising the acceptability of the thesis in this way I call "*argumentatively valid*". For the following considerations, it is essential from what has just been said that epistemically good, argumentatively valid arguments are each based on an epistemological principle, which in turn is justified epistemologically in the broad sense: The principle helps to guarantee the thesis' acceptability, and it serves as a checklist for the addressee when he recogises the thesis' acceptability.

2. Adequacy conditions for epistemically oriented theories of argumentation

The descriptions of arguments in criteria for good arguments, to be examined here, are parts of epistemically oriented normative argumentation theories. Such theories of argumentation are instrumental, they must fulfil certain conditions, conditions of adequacy. Such conditions of adequacy are:

AC1: *Instrument delivery, decisiveness and constructive guidance:* 1. An instrumental theory must develop instruments in the first place, in this case criteria for good arguments. (These criteria are meta-instruments, namely instruments for creating and evaluating instruments, in this case epistemically functional arguments.) 2. And these criteria for good arguments should be so elaborated and *precise* that they allow a clear decision about the quality of an argument and 3. are an aid in the design of arguments, if possible they can function as a *design guide*.

AC2: *Epistemic effectiveness:* The minimum requirement for instrumental theories is: the instrument they develop at least produces the desired output. An epistemologically oriented argumentation theory must therefore classify arguments as good that lead to justified, acceptable beliefs and only arguments that do so. I call this "*epistemic effectiveness*".

AC3: *Completeness:* A good argumentation theory, first, for each type of *proposition* for which one can somehow argue should have a type of argument in its programme with which one can argue for such propositions. Second, it should cover all good (epistemically effective) *types of arguments* with which one can argue for the same proposition. I call this condition "*completeness*" (cf. Hansen, 2011: 745).

AC4: *Efficiency:* Instruments should be designed in such a way that the output aimed at with the instrument is realised with as little effort as possible. Efficiency here refers to the effort involved, for example, in handling the instrument (cf. also: Hansen, 2011: 745).

AC5: *Practical justification of the argumentation criteria:* A full justification of an instrument and in particular of argument schemes should take all possible relevant (good or bad) aspects of the object into account, weigh them against each other and show that the instrument in question is the best instrument for achieving the desired output. Furthermore, the output itself should also be practically justified as good and important. (On prudential practical justifications: Lumer, 2014.) Only via this kind of reasoning will it be possible to identify or to construct the best instrument.

3. Ways of describing arguments

Now I will distinguish various ways of describing arguments. *Characterising descriptions* state the *properties* of an object. Besides this, in the case of linguistic objects, i.e. in particular arguments, there are *reproducing descriptions* that *quote* the linguistic object or pieces of it or its structure: e.g.: 'The argument 'p_1, p_2, therefore q".

Descriptions of arguments vary in their level of abstraction. The descriptions are *more abstract* in that they omit more details, either completely or replace them with variables, which then stand for such various possible detailed descriptions. (The *objects* themselves, i.e. the meant arguments, however, remain the same in all cases and thus also equally abstract; only the description becomes more abstract). With increasing abstractness, the number of differentiated argument types also decreases; at the lowest, least abstract level, no argument *types* are differentiated at all, but only individual arguments. I distinguish the following *levels of abstraction* in ascending order:

AD1: *(Reproducing) concrete argument (descriptions):* Example: 'All humans are mortal. Socrates is a human being. Therefore: Socrates is mortal.'

AD2: *(Reproducing) argument schemes:* Example: 'All human beings are mortal. y is a human being. So: y is mortal.'

AD3: *(Reproducing) analytical argument forms:* Example: 'All Φ are X. x is Φ. Therefore: x is X.'

AD4: *Characterisations of an argument class:* Example: 'Arguments of the deductive class have the general argument form 'p_1. ... p_n. Therefore, q.'. (The premises) $p_1, ..., p_n$ are all true. $p_1, ..., p_n$ together logically imply q.'

AD5: *(Reproducing) general argument form (more precisely: general form of an elementary argument):* Example: 'p_1. ... p_n. Therefore, q.' [3]

The examples given each outline complete descriptions of criteria for a type of good argument. Accordingly, the description modes AD4 and AD5 are different, even though AD4 contains the description mode AD5 (for describing the domain of definition): AD4 is less abstract, contains additional details that are missing in the description mode AD5.[4] The description modes AD1 and AD5 are not used in any theory as fundamental descriptions of good arguments; they are only listed here to illustrate the spectrum of degrees of abstraction.

AD1. Concrete argument descriptions simply cite an argument. They are not useful for formulating fundamental criteria for good arguments, because these criteria would have to list all good arguments. – I skip argument schemes for a moment and proceed to the third level of abstraction.

AD3. Analytical argument forms are reproducing argument descriptions that are developed from concrete argument descriptions in this way:

1. Substitution of individual constants: In analytic argument forms, no individual constants occur; they are usually replaced by free individual variables.

2. Replacement of predicate and operator constants: In analytic argument forms, with a few exceptions, predicates and function terms do not occur. The *exceptions* are: in probabilistic argument forms: probability concepts, logical and arithmetic operators; in practical argument forms: desirability operators ('the desirability of x for y'), arithmetic operators, probability

---

3 I got many helpful inspirations for defining this list from: Hansen 2011; 2020.
4 One could also add the general argument form AD5 to the description forms AD1 to AD3 to describe the domain of definition; this would even be formally more correct. But to my knowledge, no one has done this yet; therefore, I have also omitted this addition here in order not to distract from the essential.

operators and concepts of concomitance relations such as 'x is consequence of y'. The other predicates and function terms from the concrete argument description are replaced by free predicate or function term variables.

3. *Limit for the substitution of analytic operators:* The substitution of logical operators and of the exception terms mentioned under 2 does not go so far that the resulting form consists *only* of free variables for propositions. (For this would then already be the general argument form.)

AD2. In terms of the degree of abstraction, *argument schemes* lie between concrete argument descriptions and analytical argument forms. They are generated from concrete argument descriptions with the same steps as analytical argument forms, but the substitutions do not go as far as with analytical argument forms. Argument schemes thus contain at least one free variable for an individual term, a predicate, an operator or a proposition—this distinguishes them from concrete argument descriptions—; and they contain at least one individual term, a non-analytic predicate, a non-analytic operator or a proposition—this distinguishes argument schemes from analytic argument forms. The "argumentation schemes" developed by Walton, Reed & Macagno (2008) as well as those of Kienpointner (1992) are also argument schemes in the sense just defined.

AD4. Characterisations of argument classes are more abstract and more general than analytical argument forms. Their reproducing description of the arguments is usually maximally abstract; it only implies that the arguments have the general argument form. Subsequently, however, the set of objects captured in this way is narrowed down considerably by giving them an additional characterising description. An example is: '(The argument) a has the general argument form 'p_1. ... p_n. Therefore, q.'. The premises $p_1, ..., p_n$ are all true. $p_1, ..., p_n$ together logically imply q.' – The idea behind class formation is to group once again a more or less large set of structurally related analytic argument forms under one description. According to epistemological approaches, the structural relationship consists in the fact that these analytical argument forms are based on the same epistemological principle; all argument forms that are based on the same epistemological principle appertain to one class; and all arguments that are based on the same epistemological principle make up the class. More precisely, the characterisation of a class refers in each case to an epistemological principle that is not cited directly, but indirectly: The characterisation states that arguments of this class assert the individual conditions of the respective epistemic principle as fulfilled for the respective thesis. Thus, the characterisation 'The premises $p_1, ..., p_n$ are all true. $p_1, ..., p_n$ together logically imply q.' indirectly cites the deductive epistemological principle: 'If any true propositions p_1 to p_n logically imply another proposition q, then the latter proposition, q, is also true'.

AD5. In *general argument forms*, finally, all judgements of the concrete argument description are replaced by free variables for judgements.

4. The inadequacy of the argument scheme approach

Consequences can now be drawn from what has been worked out so far, critical ones in this section, constructive ones in the next sections. The criticism is directed against the argument schemes approach in argumentation theory. Walton's theory of argument schemes will be used as an example. However, the reasons against this theory are so general and fundamental that they also apply to the other theories that describe the positive criteria for arguments in terms of argument schemes.

My *critical thesis* is:

T1: *Inadequacy of the argument scheme approach:* Argumentation theories that formulate the fundamental criteria for epistemically good arguments by means of argument schemes (AD2) are inadequate.

Next, I will give some reasons for this thesis.

4.1. Problems of number

PAS1: Too many schemes: If one disregards the more exotic analytic argument forms and considers only the more common valid inference forms, one arrives at several hundred to perhaps 1,000. This is many despite the arbitrary cap, too many to remember and too many for a theoretically sophisticated account, which also explains why these analytic forms are valid. Epistemological approaches in argumentation theory solve this problem by raising the level of abstraction of description in the criteria for epistemically good arguments from analytical argument forms to the characterisation of (up to a handful of) argument classes. The schema approach, instead, goes—theoretically, not in the actual procedure of its protagonists—in the opposite direction: it replaces variables in an analytical argument form with constants—in the case of the Argument from Expert Opinion there were three predicate constants ('expert', 'subject domain', 'asserts'). For each predicate variable, one could substitute at least 1000 predicate constants. With three such replacements, this quickly turns the original up to 1000 analytical argument forms into 1000^4, i.e. 1 trillion argument schemes.

4.2. Problems of validity

Authors of schema approaches in argumentation theory usually collect these schemes empirically; and they do not provide a theoretical, in the broad sense epistemological, justification as to why the schemes are supposed to be inferentially valid in each case.

The inferential validity of the arguments, however, results from the epistemological principles underlying the argument, the conditions of which are asserted as fulfilled in the argument and which are justified by the appropriate theories, namely logic, probability theory and decision theory. Now, the argument schemes approach, the combination of a lack of theoretical justification of the possible inferential validity and a lack of (explicit or implicit) recourse to epistemolgial principles in the description of the good argument types leads to problems with regard to the inferential validity of the argument schemes. I will just list these *problems of an argument schemes approach* to criteria for good arguments.[5] Also PAS2-PAS5 are fairly general problems of argument schemes approaches.

PAS2: Arbitrariness in the selection of good schemes: No justification is discernible as to why particular schemes of existing arguments were included in the list of good schemes.

PAS3: Lack of a structural understanding of the schemes' form and functioning: There is no explanation as to why it should be rational to accept the conclusion after accepting the premises. In this way, the addressees cannot acquire knowledge.

PAS4: Invalid schemes: Three quarters of the argument schemes in Walton, Reed and Macagno's compendium (Walton et al., 2008: 308-346), if their variables are correctly substituted, are argumentatively invalid in the sense of not guaranteeing their theses' acceptability.

PAS5: Missing decisiveness: Because of the relatively precise description, the schemes in Walton, Reed and Macagno's Compendium certainly fulfil the function of guiding construction (AQ1.2), but because of the reference to the critical questions (which, incidentally, are often missing) and to the outcome of the dialogue, their schemes clearly fail the criterion of decisiveness (AQ1.1).

Because of these problems, the known lists of argument schemes violate all adequacy conditions except AC1.1 (delivery of an instrument at all) and AC1.3 (design guide).

5. The possibility and adequacy of the argument class characterisation approach

The previous criticism of the argument scheme approach in describing epistemically good arguments in fundamental criteria for valid arguments has always pointed, at least implicitly, to a better, epistemological alternative, i.e. characterisations of argument classes that refer to epistemological principles in the description of arguments. In this section,

---

5 More details can be found in: Lumer, 2016: 6-17; 2022: 230-283.

this approach is presented in more detail and it is shown that it fulfils the adequacy conditions. This proves the following constructive thesis:

Constructive thesis:

T2: *Adequacy of the approach of characterising argument classes:* (Properly constructed) argumentation theories that formulate the fundamental criteria for epistemically good arguments as characterisations of argument classes with the help of effective epistemological principles—the arguments represent the individual conditions of the epistemological principle specified for the thesis as fulfilled—are adequate.

What does a characterisation of an argument class with the help of an epistemological principle look like in more detail? The following is an extract from my own proposals.

The core of a simplified *characterisation of the class of valid deductive arguments* is:

"**DA0:** *Domain:* The argument consists of 1. a single statement, the thesis, 2. an indicator of argument (like 'therefore', 'for this reason'), and 3. a set of further statements, the premises. ('q. p_1. ... p_n.') [...]

DA2: *Guarantee of truth:* 1. The premises' propositions are true, 2. and they logically imply the thesis' proposition." (Lumer, 2005: 221)

The reproducing description in this characterisation, i.e. DA0, only indicates the general argument form. The characterisation in DA2 instead does not reproduce analytical argument forms, but characterises a class of good, valid arguments from the domain of definition by reference to the deductive epistemological principle, which is indirectly cited here: The components of DA2.2 imply that the two conditions of the deductive epistemological principle (the premises are true and logically imply the thesis) are satisfied for the thesis.

With the help of other epistemological principles, e.g. practical or probabilistic epistemological principles (see also fn. 2), criteria for further classes of valid arguments can then be developed, each containing characterisations of the arguments in that class (see Lumer, 1990: 187-189; 237-244; 258-260; 277-279; 362-366; 2005a: 221; 235-236; 2011: 1149-1151; 2014: 11-14). Such characterisations of classes of valid arguments then also capture *those* of the schemes developed by the proponents of the argument scheme approach that can be reconstructed as epistemologically justified.

How and why do these types of characterisations of argument classes fulfil the above adequacy conditions.

AC1: *Instrument delivery, decisiveness and constructive guidance:* Characterisations of argumentation classes such as those described are meta-instruments that describe the properties of the arguments to be constructed and evaluated with their help so precisely that unambiguous judgements can be made about them.

AC2: Epistemic effectiveness: By asserting that the conditions of an effective epistemological principle specified for the respective thesis are fulfilled, arguments constructed according to this approach guide the sufficiently competent addressee in verifying compliance with the conditions of this principle of cognition and thus bring him to realise the acceptability of the thesis.

AC3: Completeness: Despite the plurality of epistemological principles mentioned, presumably not all effective epistemological principles that can be used for arguments have yet been discovered. However, the outlined approach to normative argument description is, of course, open to the integration of new effective epistemological principles and the extension by additional argument classes. *Within an argument class*, however, all arguments based on this epistemological principle are covered because of the abstract characterisation of the class via the epistemological principle itself. In this respect, the approach is complete.

AC4: Efficiency: Because of the strict orientation towards the epistemological principle, the *arguments* constructed with the presented approach are as complex as necessary, but not more complex. And because of the working off of the epistemological principle, the arguments constructed according to this approach are also easily seen through by the addressee, are clear in their epistemic value. The application of an *argument characterisation* itself in the construction and evaluation of arguments is prima facie more difficult than the application of argument schemes because of the abstractness of the characterisation. Secunda facie, however, this advantage of the schema approaches only exists because they only cover a tiny number of the simplest arguments. If, instead, the schema approaches are applied to any arguments, then in most cases they provide no answer at all. And should the schema approaches—in the fictitious case—be complete at some point, then the enormous search effort then required will destroy this previous advantage of practicability. Moreover, at least Walton's approach, with its critical questions and reference to the dialectic of their answers, contains many long loose ends that make the evaluation of argumentation considerably more difficult and often impossible, so that the simple appearance of the schemes themselves concerns precisely only the first part of these argumentation criteria. (In the final section, it is outlined how the approach of characterising argument classes can be didactically simplified.)

AC5: Practical justification of the argumentation criteria and optimality: The importance of the standard output aimed at with the approach outlined here, namely to achieve rationally justified cognition, was justified at the beginning; and it was shown how the arguments oriented towards the epistemological principles can also realise this standard output (see AC2). It has also just been shown that the arguments themselves constructed according to this approach and the characterisations of argument classes fulfil their functions reasonably well and efficiently, measured against the complexity of the task. So far, arguments of no other approach to the epistemic normative description of arguments reliably realise the standard output; thus, they are not epistemic instruments at all. This makes the comparison (necessary for a complete practical justification of an instrument) of whether the approach outlined here also develops the best or near-best epistemic instruments very simple: if it is the only instrument (so far) to fulfil this function, then it is automatically also the best.

6. Singularity of the approach of characterising argument classes via recourse to epistemological principles

My exclusiveness thesis is:

T3: *Exclusiveness of the approach of characterising argument classes via epistemological principles:* Only argumentation theories that formulate the fundamental criteria for epistemically good arguments by means of characterisations of argument classes via a recourse to epistemological principles are adequate.

This thesis goes beyond the constructive thesis. It actually consists of two sub-theses: *T3.1:* Only by recourse to epistemological principles is the epistemic adequacy of epistemically normative argument descriptions attainable. *T3.2:* Argumentation theories that refer to epistemological principles are only adequate if they use characterisations of argument classes (with the help of these epistemological principles) in their fundamental normative descriptions of good arguments. I will now justify these two sub-theses in this order.

In my justification of T3.1, I focus on the crucial point, namely that only by referring to epistemic principles can the adequacy condition of epistemic efficacy (AC2: epistemically good arguments lead to rationally justified belief) be fulfilled. 1. The acceptability of a thesis is based on the fact that it fulfils an epistemological principle. If an argument is to guarantee the acceptability of a thesis, it must therefore claim that the conditions of an effective epistemological principle specified for the thesis in question are fulfilled, i.e. it must at least implicitly refer to the epistemological principle. 2. In order to enable an addressee to check the acceptability of the thesis on the basis of an epistemological principle that is (at least implicitly) known to him, the argument must be constructed in such a way that its premises judge the conditions of this epistemological principle specified for the respective thesis to be fulfilled. In order for the argument to be constructed accordingly by a user of argumentation theory, the normative descriptions of this argumentation theory, which are, as it were, the construction manuals for good arguments, must be oriented towards the epistemological principles, namely prescribe that the arguments assert the conditions of this epistemological principle specified for the thesis as fulfilled. 3. If the addressees want to go beyond an *intuitive* insight into the validity of an argument, then they must also know the epistemological principle explicitly and consciously apply it as a checklist when examining the argument. However, this examination can only lead to a positive result if the argument is actually oriented towards the checklist. 4. The recourse to the epistemological principle provides a justification as to why this argument guarantees the acceptability of the thesis.

What has just been said does not yet rule out the possibility that the fundamental normative descriptions of good arguments provided in the theory do not have the form of reproducing descriptions, that they are, for example, argument forms. The second sub-thesis T3.2 counters that the corresponding theory would then not be adequate or that, if it were nevertheless adequate, it would still use a characterisation of argumentation classes as a fundamental description. 1. For one thing, in order to achieve adequacy (i.e. so that the arguments authorised by the theory are really epistemically good), there would have to be a general *primary* characterising description in the justification of the argument forms selected as good, based on epistemological principles, of what the secondary reproducing descriptions propagated by the theory as normative descriptions should look like. This primary description, however, would then be a characterisation of the argument class; and it would be the actual fundamental criterion, which, however, would only occur in the justification. 2. For another thing, one would not achieve completeness without a characterising description, i.e. solely by listing the reproducing descriptions, e.g. analytical argument forms, because there are always quite exotic forms that are newly discovered, i.e. that are not included in the list, and whose validity must then be checked on the basis of a more general criterion.

After these general considerations as to why only characterisations of argument classes on the basis of epistemological principles can be adequate fundamental normative descriptions of epistemically good arguments and why other ways of describing them cannot be adequate, this will be further elaborated here more concretely and briefly— argument schemes have already been criticised in section 4—by considering two more of the ways of describing arguments listed in section 3.

Why not lists of *analytical argument forms* (AD3)? A look at logic may help here. In logic, too, the criteria for valid inferences are not described in the form of a list of valid forms of inference, but by more abstract criteria such as the exclusion of interpretations in which the premises are true but the conclusion is false. Reasons for this more abstract approach are, just as in epistemological argumentation theory, among others, firstly, that this is theoretically informative and explains the meaning of a valid inference or an argumentatively valid argument. Secondly, enumerating even only all non-exotic valid analytic forms would be far too long and, thirdly, would still require an independent proof of validity.

Why not *general argument forms* (AD5)—as one of the anonymous referees insisted—? Well, the general argument form '$p_1. ... p_n$. Therefore, q.', describes only the domain of definition. And this, of course, does not circumscribe the *good* arguments, but also includes all bad arguments, fallacies and, even worse, mainly nonsensical sequences of judgement, where no one would dream of regarding this as an argument. Therefore, the general argument form as the form of describing the fundamental criterion for good arguments is a non-starter, while the additional characterisations of argument classes provide the necessary restrictions from the mere domain of definition to the good arguments.

Having completed the justification of the three main theses of this article, it only remains to make a small remark on didactics. The formulation of criteria for valid arguments via the characterisation of argument classes is rather abstract and may also require a high level of abstraction from the user when interpreting arguments. Therefore, a schemes approach in (normative) argumentation theory may be easier for beginners at first. But even in the medium term it leads nowhere, for several reasons. The learner has too many schemes to learn; these nevertheless fail to capture many arguments and thus do not help her in assessing them: many schemes actually advanced by the theory will be wrong and thus also lead to a false assessment of the arguments' validity and to the adoption of false beliefs; and she does not develop an understanding of the meaning and functioning of arguments. The characterisations of argument classes via epistemological principles (as described above), on the other hand, provide all this. There is nothing to be said against using argument schemes for the presentation of criteria for valid arguments in beginners' classes. But these schemes must, for one thing, themselves be developed with the help of epistemologically justified criteria for valid arguments, such as the above characterisations of argument classes. They cannot themselves be the fundamental criteria for valid arguments. Furthermore, they are only suitable for teaching *beginners*; the teaching of argumentation already at high schools and even more so at the university should go beyond this level and also teach the theoretically deepened fundamental criteria.

References

Hansen, H. V. (2011). Using Argument Schemes as a Method of Informal Logic. In F. van Eemeren, B. Garssen, D. Godden, & G. Mitchell (Eds.), *Proceedings of the 7th Conference of the International Society for the Study of Argumentation* (pp. 738-749). Amsterdam: Rozenberg, Sic Sat.

Hansen, H. V. (2020). Argument Scheme Theory. In C. Dutilh Novaes, H. Jansen, J. A. van Laar, & B. Verheij (Eds.), *Reason to Dissent. Proceedings of the 3rd European Conference on Argumentation* (vol. 2, pp. 341-355). London: College Publications.

Kienpointner, M. (1992). *Alltagslogik. Struktur und Funktion von Argumentationsmustern*. Stuttgart: Frommann-Holzboog.

Lumer, C. (1990). *Praktische Argumentationstheorie. Theoretische Grundlagen, praktische Begründung und Regeln wichtiger Argumentationsarten*. Braunschweig: Vieweg.

Lumer, C. (2005a). The Epistemological Theory of Argument – How and Why? *Informal Logic 25(3)*, 213-243.

Lumer, C. (2005b). The Epistemological Approach to Argumentation – A Map. *Informal Logic 25(3)*, 189-212.

Lumer, C. (2011a). Argument Schemes – An Epistemological Approach. In F. Zenker (Ed.), *Argumentation. Cognition and Community. Proceedings of the 9th International Conference of the Ontario Society for the Study of Argumentation* (32 pp.). Windsor, Canada: University of Windsor. URL = <http://scholar.uwindsor.ca/cgi/viewcontent.cgi?article=1016&context=ossaarchive>.

Lumer, C. (2011b). Probabilistic Arguments in the Epistemological Approach to Argumentation. In F. van Eemeren, B. Garssen, D. Godden, & G. Mitchell (Eds.), *Proceedings of the 7th Conference of the International Society for the Study of Argumentation* (pp. 1141-1154). Amsterdam: Rozenberg, Sic Sat.

Lumer, C. (2014). Practical Arguments for Prudential Justifications of Actions. In D. Mohammed, & M. Lewiński (Eds.), *Virtues of Argumentation. Proceedings of the 10th International OSSA Conference* (16 pp.). Windsor, Canada: University of Windsor. URL = <http://scholar.uwindsor.ca/cgi/viewcontent.cgi?article=2077&context=ossaarchive>.

Lumer, C. (2016). Walton's Argumentation Schemes. In P. Bondy, & L. Benaquista (Eds.), *Argumentation, Objectivity, and Bias. Proceedings of the 11th International Conference of the Ontario Society for the Study of Argumentation* (pp. 1-20). Windsor, Canada: University of Windsor. URL = <http://scholar.uwindsor.ca/cgi/viewcontent.cgi?article=2286&context=ossaarchive>.

Lumer, C. (2022). An Epistemological Appraisal of Walton's Argument Schemes. *Informal Logic 42(1)*, 203-290. URL = <https://informallogic.ca/index.php/informal_logic/article/view/7224>.

Perelman, C., Olbrechts-Tyteca, L. (1958). *La nouvelle rhétorique. Traité de l'argumentation*. 2 vols. Paris: Presses Universitaires de France.

Walton, D., Reed, C., & Macagno, F. (2008). *Argumentation Schemes*. Cambridge: Cambridge U.P.

THE RHETORICAL FOUNDATIONS OF LOGIC: RETHINKING JØRGENSEN'S DILEMMA

MAURIZIO MANZIN
University of Trento
maurizio.manzin@unitn.it

Abstract

In analytical legal philosophy Jørgensen's Dilemma (JD) has long been considered a watershed between logic and ethics or morals. According to a common interpretation of JD (i) only apophantic sentences can count as premises for a logical inference, and (ii) only declarative sentences (in the indicative mood) are apophantic. Thus (iii) normative sentences (in the imperative mood) cannot count as premises for a logical inference. In the case of law (which largely manages normative sentences) we should then admit that it cannot be logic. But in the same place where Jørgensen establishes his "puzzle" he indirectly suggests that any definition of logic is in itself rhetorical. If so, we could explore the legal rhetorical situation (namely the judicial one) – context, language, audience, speaker, argumentative moves etc. – without fear of being inevitably non-logic. In the following lines I will first stress the relevance of common JD's interpretations in legal philosophy (sect. 1), then I will face some preliminary possible objections on the fact that: JD were actually a dilemma (sect. 2); or rather a mere reformulation of Hume's is-ought problem (sect. 3); not so or no longer relevant for argumentative studies (sect. 4); anyway, surpassed by contemporary other-than-deductive logics (sect. 5). Secondly, I will analyze some parts of JD's text in Jørgensen (1937) in order to bring to light its remarkable rhetorical foundations (sect. 6), eventually addressing the issue of legal reasoning as possibly "cured" by a rhetorically-oriented interpretation of JD.

1. The enduring relevance of JD for legal philosophy

This is only a brief introduction (more will be addressed about JD relevance in sect. 3-5). For the moment I only want to roughly note that in analytical accounts of legal argumentation as well as in many handbooks and essays used in the faculties of law and law schools (and even in law firms), JD is still considered a milestone in legal thought. JD's formulation, almost always interpreted in a formalistic way, counts as a divide between logic and other kinds of knowledge, admitting or not that normative sentences can stand for premises of a logical inference (as supposed in deontic logic). For example, Guastini (2001) authoritatively states that when looking at JD there could be (at least) six reasonable answers to the question: is legal reasoning a *logical* one? (i) Yes, because the first horn of JD is wrong: normative sentences *can* be true or false just like the apophantic ones (this is a typical meta-ethical cognitivistic account, as in the natural law theory). (ii) No, because norms are some kind of facts and between facts only causal relations are possible, and not logical connections .(iii) Yes, provided that normative sentences can be considered from the point of view of the fulfillment of the behavior they command or forbid: in this case they function *as it were* apophantic sentences. (iv) Yes, because although norms are not apophantic, sentences describing norms (which are syntactically identical to the first ones) *do* function as premises in a deontic inference. (v) Yes partially, because every normative sentence has a performative part (a command or a prohibition) and an apophantic one (the description of the mandatory or forbidden behavior) – and the sentence about the fulfillment of the behavior can be true or false. (vi) Yes, because the domain of logic is wider than the one of truth, and despite normative sentences are not predictable as true or false, nevertheless they can draw logical inferences (Guastini 2001).

As it can be easily observed, a major Italian theorist of legal reasoning (quotations are taken from one of the most influential sources for Italian legal scholars: the *Treccani online Encyclopedia of Social Sciences*) currently uses JD's formulation as a fundamental starting point for considerations on what is legal reasoning. No need to say that, therefore, the question of what JD *actually* implies for logic – in our case legal logic – and for the relation between logic and argumentation is of the utmost importance. For this reason JD has been taken here into account aiming at showing that at a deeper glance it reveals a crucial connection with rhetoric.

In order to justify my thesis, before consulting the very pages of the text of *Imperatives and Logic* (hereinafter IL), I will try to answer some preliminary questions related to the status of JD, its relevance and actuality in argumentative studies.

2.   Is JD a "dilemma"?

The issue is not a trivial one. If we pick up the article published in *Erkenntnis* by the Danish logician in 1937-1938 we could discover that the word "dilemma" has never been used by him. He always speaks of a "puzzle". The difference is significant: while "puzzle" denotes only a seemingly rational difficulty (a mystery you have to solve, a game you have to think about carefully in order to do it, a picture cut up in small pieces you have to reconstruct), "dilemma" deals with a situation in which you have two opposite alternatives to choose from: either one or the other. And usually the pragmatics of the term implicates a *difficult* choice and an *unpleasant* situation (see e.g. OED sub voce "dilemma"). But in no place of IL Jørgensen talks about dramatic alternatives, despite the fact that – when pointing at a puzzle – he in fact puts forward two lines of reasoning prima facie incompatible among each other (see IL, 288, 290 and 296).

I would tend to think that such use of the term "puzzle" instead of "dilemma" makes us somehow sense that the opposition between the two lines of reasoning does not lead to a cul-de-sac: this is precisely the reason why the situation would not be not so tragic. The choice between two alternatives is dramatic only if they draw a contradiction (and this can happen only if sentences opposed to each other are in the form of *p* vs *non-p*); but if JD is not actually a dilemma – and thus it is not dramatic – then we are only facing a kind of trouble in which our mind is *puzzled*. And it could be the case that looking carefully to the small pieces in which the picture has been cut up we will be able to find out the way to solve the problem.

But if Jørgensen would not be responsible for the term "dilemma" which made him famous, where does such idea come from? We can imagine that the choice of a term like "dilemma" implies a negative feeling about the possibility to mind the gap between the alternatives on the floor. According to G.H. von Wright (1999), Alf Ross was the very first who talked about a "Jørgensen's dilemma" (Ross 1941) making us believe that Jørgensen wrote something like: *aut* (i) logic is verofunctional and normative sentences are not apophantic (thus normative discourse is irrational), *aut* (ii) logic admits values different from the verofunctional ones and normative sentences are not apophantic (thus normative discourse is rational). On one side there is no doubt that Jørgensen argued this dualistic point of view in IL, while on the other it would seem that he did not intend it as a ruthless alternative between logic and non-logic. Managing Jørgensen's authentic expressions as a dilemmatic formula is therefore from Ross a kind of text forcing motivated probably by a legal-realistic approach, eager to get rid of matters that have too much to do with formal logic. In other words, rather than a "JD" we should speak of a "Ross' Dilemma".

3. Is JD a mere reformulation of Hume's *is-ought* problem?

As notoriously stated by David Hume in 1739: "For as this *ought*, or *ought not*, expresses some new relation or affirmation, it's necessary that it should be observed and explained; and at the same time that a reason should be given, for what seems altogether inconceivable, how this new relation can be a deduction from others, which are entirely different from" (Hume 1978: sect. 3. 1. 1). These words conclude a very famous passage in which Hume is *puzzled* by the idea of a possible deduction of sentences about moral behavior (we ought or we ought not do something) from sentences describing some state of affairs in the world (the things are or are not such and such).

Anyone knows that these concerns from Hume have been considered like a strict logical rule which later on deserved the name of "Hume's law". But once again, as in the case of JD, the very formulation through which Hume raised the problem of *is-ought* is more articulated. Rather than drawing a contradiction, the Scottish philosopher wonders if "a reason [c]ould be given" to justify a connection between sentences about a state of affairs and a behavior. And, yes, he admits that – in the form of a deduction – a similar kind of relation "seems [...] inconceivable" to him. But he also says that this "relation or affirmation" has to be observed and explained, so that we could not unduly suppose that a sort of "new" relation, which cannot be a (formal) deduction, could connect the *is* with the *ought*.

I do not want to affirm that Hume *personally* believed that such a relation does or could exist, but only that his empiricist attitude inclined him not to theoretically exclude it at all. My soft interpretation of Hume's words fits with the intuitiveness of the not 'illogical' conclusion that in practical judgments like "it is raining, then you should take an umbrella" one is not expressing two sentences that have nothing to reasonably do with each other.

The *is-ought* issue is also known as the problem of the "mixed premises", and though in his IL Jørgensen curiously does not quote Hume (but more significantly Poincaré) he raises the question in similar terms, given that he explicitly declares what Hume only alludes to in a very elliptic form – that in the end we are facing the rational difficulty of excluding mixed premises from logical inferences.

This is the reason why I maintain that JD is not a mere reformulation of *is-ought* problem but rather a heuristically richer variant of it, as I hope to better explain in the following sections.

4. Is JD relevant for argumentative studies?

At this point of my discourse I would argue that JD *is* relevant (i) if intended as a puzzle and not as a dilemma, and (ii) if intended as an heuristically richer variant of *is-ought* problem and not as a mere reformulation of it. Its relevance consists mainly in the fact that, interpreted this way, JD allows us to avoid all those too narrow accounts on logic that cut off the arguments and inferences we could build in informal domains (like the ones of the real world), where it is impossible and in the end inappropriate to establish axioms before starting the reasoning.

This is of special interest for legal scholars dealing with judicial reasoning, because in trials' decisions the so-called major premise of the alleged judge's syllogism (actually an enthymeme) is given by sentences – namely statutes and decisions – which are not axioms and which *have not to be* axioms to appropriately fit with situations in the real life (they are pour cause written in a non-formalized language).

It would be of course a platitude to observe that argumentative studies sink their very roots in the refusal of a strict divide between verofunctional logic as the only way to be rational and any other kind of verbal activity. The fact is that from this point of view we should reject JD as a 'logicist' position which, in line with the logical positivism, tends to keep intact the separation between the so-called "two cultures" (Snow 2012). But in my view a more sensitive interpretation of the authentic JD text would provide some useful insights in order to realize that Jørgensen did not want to merely re-affirm the primacy of verofunctional logic but, in a more critic and fruitful way, he wanted to address the question of how normative sentences, framed or not a in a deontic inference, could to some extent satisfy the demands of rationality.

In other words, rather than a door barred against non-verofunctional sentences (i.e. a limit for argumentative studies) or a challenge for deontic logicians, JD should be appreciated as a further resource for scholars of argumentation, legal ones included.

Then the question is *how* it could be considered a good argumentative resource. And I hypothesize that it depends on the extent to which JD is capable of highlighting the rhetorical structure of definitions like the nominal ones employed to establish axioms. In particular I will try to show how Jørgensen *himself* uses rhetorical practices in order to build his argument on what is logic about.

Even more, I think that the way he puts forward his "puzzle" – being implicitly rhetorical – makes it suitable for opening the door I have just mentioned to ascertain the rhetorical structure of logic and overcome the separation between the "two cultures".

5. Is JD an outdated issue?

Either the concept of logic or the one of inference that the epigones of the Wiener Kreis had in mind in their time have nowadays abundantly changed – that goes without saying. From this point of view a study on JD might seem of some interest only to scholars of the history of science. As for the others, well, we have achieved a lot of argumentative and non-deductive theories since the second half of the last century: why, then, waste our time brushing up an antiquated account on logic?

To reply to this objection I limit myself to quote two authors very distant (in every possible sense) from each other: Aristotle and Martin Heidegger. According to the former, the task of philosophy is to rethink again and again the same problems, starting every time from the beginning: *palin ex arches*, as he wrote. According to the latter, returning to the same issue does not necessary imply a repetition, given that *das Selbe* (the same) is not *das Gleiche* (the identical). For these two very good reasons I would suppose that taking back in hand a manifesto of logicism like JD and trying a reading of it as an invitation to discover its hidden rhetorical foundations could be a philosophical undertaking worth attempting.

It is indeed not negligible to note that *the same* Jørgensen in his IL asserts that we should "enlarge our concept of inference" (IL 290), meaning this way that the question is not the one of applicating or not the formal deduction to normative sentences (in order to logically manage the value judgments), but to find a connection between mixed premises *wider* than the formal deduction.

I cannot believe that an endeavor like such could be considered outdated.

6. Rhetorical insights in JD's text

After having briefly addressed some major objections about JD, I want now to indicate and discuss a number of *loci* in JD's text where there is room enough to conjecture a rhetorical activity by Jørgensen aiming at "puzzling" the supporters of the great divide between alethic judgments and the practical ones.

First of all I would like to focus on the surely most famous – and cited – assertion of JD as reported in IL (290):

> According to a generally accepted definition of logical inference only sentences which are capable of being true or false can function as premisses or conclusions in an inference; nevertheless it seems evident that a conclusion in the imperative

mood may be drawn from two premisses one of which or both of which are in the imperative mood.

This quotation is almost always used to show the alleged dilemma through which Jørgensen challenges what normally counts as logic. But what is remarkable to me is not so much the just saying 'general sense' of the proposition; more interesting is the fact that Jørgensen employs arguments like (i) "general acceptance" and (ii) "seemingly evidence" in order to justify the premise and the conclusion of his puzzle. The premise is that *on the base of a shared opinion* only verofunctional sentences can work in an inference; the conclusion is that, nevertheless, the contrary *seems evident*. Here it is the Jørgensen's puzzle. But to figure it as a puzzle it has been necessary for the author to resort to *the opinion of a large audience* (of educated people? of laymen? of logicians? – it is not specified), and to a quality of the propositions consisting of *seeming evident*.

Both the circumstances make us suppose that the common understanding of logic as intended by Jørgensen could be related to the typical "rhetorical situation" as famously described by Lloyd Bitzer and characterized by an exigence, an audience and a set of constraints (Bitzer 1968). The exigence is the puzzle, then we have an audience who "generally" shares a definition of logic, and the constraints are the ones of the verofunctional sentences and the deductive inference.

The choice of terms like (i) "general acceptance" and (ii) "seeming evidence" have not to be interpreted as casual or inaccurate, because there are other places of IL in which Jørgensen uses these terms.

For instance when he talks about inference (IL 288):

If we take the word "inference" in the generally accepted sense in which it is defined

where it is clearly admitted a *diffused persuasion* about the meaning of the word.

The same when he talks about the terms "true" and "false" (IL 289):

in any sense in which these words are used in logic

where Jørgensen alludes to a pragmatic condition (i.e. the linguistic use) as a ground for the development of his reasoning on logic and the imperatives.

As for the concept of "seeming evidence" we can observe the use of rhetorical linguistic markers like the verb *to seem* (occurring 16 times in the article) or adjectives like *reasonable* and *probable* (IL 295), and the adverbs *apparently* (IL 288) and *seemingly* (IL 290). All these forms of expression remind us the Aristotelian notion of *eikos*: what happens in a generality of cases (although it might be otherwise) – for Aristotle the ground of the rhetorical truth.

In a similar way, the appeal to widely diffused opinions quoted above brings us back again to the pages of Aristotle's *Rhetoric* where the *eikos* is connected with the authority of the opinions (*endoxa*), either because they are very popular or because they are supported by experts – and it would be difficult to deny that a scientific article like Jørgensen's one, which was also the transcript of a lecture given in front of prestigious colleagues, did imply the confrontation with an audience of experts. In all the passages of IL in which Jørgensen stresses a belief (that only sentences in the indicative mood can function as premises for logic; that a conclusion drawn by sentences one of which or both are in the imperative mood, seems to be logic; that imperative sentences are "older" than the indicative ones; that an inference is so and so, etc.) he always makes reference to common opinions and never to axioms. Thus it is clear enough that the basis of his reasoning on logic as intended by neo-empiricist logicians is not formal in itself, but it is part of a discourse built in a rhetorical way, where different opinions are taken into account in order to evaluate and select them, to find out the more persuasive ("probable") ones and to finally "enlarge our concept of inference".

In other words, Jørgensen's attempt to enlarge what counts as logic is carried out not only by *what* he says but more radically through *the way* he says that. From this point of view, JD ends up being the indication of a method.

Rhetoric, thrown out of the door of logic, returns through the window.

7. JD as a "cure" for legal argumentation

The word *pharmakon* designates in Greek both medicine and poison (Serra 2021). Similarly, the two snakes coiled on Hermes' caduceus symbolize the double power of poison to cure or kill. It is a matter of quantity.

If we take JD *out* of the rhetorical situation in which it has been conceived by his author its quantity and purity would be undoubtedly excessive: in this case JD will be toxic for other-than-deductive logics. To put it simply, JD would remain a dilemma whose two horns are dramatically irreconcilable: on one side *the* logic with its formal deductions and its sentences in the indicative mood that can only be true or false; on the other all discourses which could also be otherwise (*eikos*) and, among them, the legal ones where judgements are based on imperatives and opinions.

But if we take JD *within* the rhetorical situation it ceases to be a poison – ceases to be a razor blade that separates logic from law – and makes it possible (and rational) to draw conclusions which "seem evident" from sentences expressing "generally accepted" opinions, commands and prohibitions.

As I tried to show in the previous section discussing the issues (i) and (ii) of Jørgensen's formulation, the rhetorical construction of JD let us suppose that (the definition of) logic – and even (of) formal logic – is the product of rhetorical operations. According to Aristotle, such operations imply the use of techniques (*psychai*) which encompass *all* degrees of human cognition: reasonings (*logos*), feelings (*pathos*) and moral beliefs (*ethos*). This is also the reason why the metaphor of *pharmakon* (drug) seems to me particularly apt, if we take into consideration what the WHO defines as health: "a state of complete physical, mental and social well-being" (Constitution of World Health Organization: preamble). A holistic approach like such, in which drugs and medical practices are finalized to re-establish the complex balance of body, mind and environment – a balance which is never a definitive state, but the provisional outcome of continuous 'homeostatic' treatments – is exactly that of rhetoric, where the truth achieved through a balance of general or expert opinions, shared emotions and common values is never the ultimate one.

The *completeness* of the rhetorical approach, which does not exclude non-formal arguments from logic, or pathetic and ethical from rationality – and thus fits better with the "fractaliy" of real life (Tindale 2020, Manzin 2020) – could be the 'cure' for legal argumentation, namely the judicial one. The syllogistic model of argument imagined by the Enlightenment's authors under the fascination of formal sciences, if applied to legal decisions, is in fact not a 'healthy' one, since it pushes legal logic towards the dramatic alternative of JD as a razor blade which separates reason from the mere power of legislative and judicial authority.

We should then bring back legal reasoning to the rhetorical situation where, to quote Bitzer once again, we have a real exigency or problem, an audience to persuade and a set of constraints. In rhetorical legal situation there are "requirements of justice" (Dworkin 1977) to be fulfilled and concrete problems to solve, a judge (or more) and (sometimes) a jury to be persuaded, and a number of constraints represented in iure by statutes, previous judgements, legal doctrine, interpretative standards, and in facto by the procedural rules.

No need to say that issues like *general acceptance* and *probability* – just as (i) and (ii) in Jørgensen's construction of his puzzle – play a crucial role in trials and lawsuits when interpreting legal norms and looking for evidence "beyond any reasonable doubt".

My attempt to bring to light the rhetorical foundations of logic in JD's formulation aims at going precisely in this direction – the one of 'healthier' practices in legal judgements capable of managing the complexity of the cognitive tools to be applied in order to decide what is just in the case.

References

Bitzer, L.F. (1968). The Rhetorical Situation, Philosophy & Rhetoric, 1(1), 1-14.
Constitution of World Health Organization: https://www.who.int/about/governance/constitution
Dworkin, R. (1977). Taking Rights Seriously. Harvard: Harvard University Press.
Guastini, R. (2001). Giuridico, ragionamento. In Enciclopedia delle Scienze Sociali. Supplemento. Treccani.it (https://www.treccani.it/enciclopedia/ragionamento-giuridico_%28Enciclopedia-delle-scienze-sociali%29/).
Hume, D. (1978). A Treatise on Human Nature. Oxford: Oxford University Press.
Jørgensen, J. (1937). Imperatives and Logic, Erkenntnis, 7(1), 288-296.
Manzin, M. (2020). "Identity-Based" and "Diversity-Based" Evidence Between Linear and Fractal Rationality. OSSA Conference Archive / OSSA 12 (pp. 1-8). Windsor: Scholarship at Windsor (https://scholar.uwindsor.ca/ossaarchive/OSSA12/Thursday/13).
Ross, A. (1941). Imperatives and Logic. Theoria, 7, 53-71.
Serra, M. (2021). At the origins of an analogy: discourse as pharmakon. Rivista Italiana di Filosofia del Linguggio, 15(1), 4-17 (http://www.rifl.unical.it/index.php/rifl/article/view/638).
Snow, P. C. (2012). The Two Cultures. Cambridge: Cambridge University Press.
Tindale, C.W. (2020). Strange Fish: Belief and the Roots of Disagreement. In C. Dutilh Novaes, J. Henrike, A. van Laar Jan & B. Verheij (Eds.), Reason to Dissent. Proceedings of the 3rd European Conference on Argumentation, 1 (pp. 513-526). London: College Publications.
von Wright, G.H. (1999). Deontic Logic: A Personal View. Ratio Juris, 12(1), 26-38.

REASONISM AND INFERENCISM IN THE THEORY OF ARGUMENT[1]

HUBERT MARRAUD
Universidad Autónoma de Madrid (Spain)
hubert.marraud@uam.es

Abstract

Reasonism and Inferencism are two major conceptions of the Theory of Argument which differ both in their understanding of argumentation and in their definition of a logically good argument. For the inferencist, to argue is to present something as a logical consequence of something else, while for reasonist it is to present something as a reason for something else. Accordingly, for the inferencist a good argument is one in which the conclusion follows from the premises, and for the reasonist it is one that gives a good reason. The primary difference from which all other differences result is that reasons are weighted notions, while inferences are non-weighted. Since there can and usually will be good reasons for and against something, that P is a (good) reason for C does not authorize one to conclude C without further ado. On the contrary, that C follows or is logically inferred from P authorizes one to infer C from P, even if that permission is tentative and revisable. I will analyze the differences between these two approaches to the Theory of Argument and argue that inferencism, the predominant position, is a flawed theory of argument.

1. Introduction

The Theory of Argument is that part of Argumentation Theory that deals with arguments. According to a widespread characterization, arguments are products of the action or process of arguing. Although this characterization may be misleading and even incorrect (Goddu 2011), viewing the Theory of Argument as a part of Argumentation Theory entails analyzing arguments as components of those practices, or, at least, doing so in the context of those practices: "The theory of argument is a component of the theory of argumentation, in much the same way that

---

[1] This research has been funded by Project "Argumentative practices and the pragmatics of reasons" (Parg-Praz), reference number PGC2018-095941-B-I00.

argument is a component of the practice of argumentation." (Johnson 2000, 31).

Although in argumentation theory and critical thinking 'reasons' and 'premises' are often used as if they were interchangeable, reasons for believing something, doing something or adopting an attitude, and premises of an inference are quite different things. Indeed, reasonism and inferencism[2] are two major conceptions in the Theory of Argument, which differ both in their understanding of arguing and in their definition of a logically good argument. For the inferencist, to argue is to present something as a logical consequence of something else, while for the reasonist it is to present something as a reason for something else. Thus, the inferentialist reading of 'A so B' is B follows from A, and its reasonist reading is A is a reason for B Accordingly, for the inferencist a good argument is one in which the conclusion follows from the premises, and for the reasonist it is one that gives a good reason (which may mean either a *pro tanto* reason, or a conclusive, all things considered, reason). I will analyze the differences between these two approaches to the Theory of Argument and argue that inferencism, the predominant position, is a flawed theory of argument. Finally, I will argue that the unit of argumentative analysis is not the argument (premises plus conclusion), but the discourse.

2. The standard inferentist approach to argument

Inferentism is by far the predominant approach in the Theory of Argument. I will begin by stating the basic principles of this conception of the Theory of Argument.

　　　IP1. The Theory of Argument (or Logic in its broadest sense) deals with logical inferences: relations between statements that enable the inference of a statement (the conclusion) from others (the premises).

When there is a relation of logical inference between a set of statements and a statement, it is said that this statement is a logical consequence of that set, or that it is logically inferred or follows from it. Note that in its second occurrence in IP1 'infer' refers to a process by which, consciously, someone arrives at a conclusion from a set of data. This is important to prevent confusion, because 'follows from' and 'is a logical consequence of' denote a binary relation between a statement and set of statements,

---

2 I use this term, to avoid confusion with inferentialism, a view popularized by authors such as Robert Brandom. Inferentialism explains how our verbal or mental acts acquire content in terms of their embedding in a set of social practices normatively governed by inferential rules.

whereas 'infers from' denotes a ternary relation involving an agent, a set of statements and a statement.

> IP2 A good argument, logico sensu, is one in which the conclusion follows or can be inferred from the premises.

IP1 and IP2 define the inferentist approach to the Theory of Argument, as opposed to the reasonist approach, which I will describe later. Using 'valid' to qualify a logically good argument, the above postulates lead naturally to

> AP1. The validity of arguments depends exclusively on the relationship between their premises and their conclusion.

A consequence of AP1 is that validity becomes an intrinsic relational property of arguments; that is, a property determined by the relationship between the parts of an argument and is therefore contextually invariant.

Validity is the quintessential logical property, so we can ask whether all logical properties are intrinsic and contextually invariant. An argument is sound, as it is well known, if and only if it is valid and its premises are true. If we admit that the same statement can be true in some situations and false in others (unlike Quine's eternal sentences), it seems that soundness would not count as a logical property under the above criteria. Although that may be a reason to exclude soundness from the class of logical properties (especially from an aprioristic conception of logic), I think that when logic is viewed as a theory of arguments it is preferable to broaden the characterization of logical properties, to include properties of arguments that are completely determined by the properties of their parts and their mutual relations (Marraud 2021). I therefore reformulate AP1 as follows:

> AP2. The logical properties of arguments are those that are completely determined by the properties of their parts and their mutual relations.

A conception of logic that adheres to principles such as AP1 and AP2 is atomistic, a category to be understood here in opposition to holism, which is the thesis that the logical properties of arguments are affected by contextual factors that are not parts of the argument (Marraud 2020). Although atomistic inferentism is the standard theory, revocable inferences give rise to a holistic variant, as we shall see later.

The notion of logical inference is often defined in terms of inference rules, formal or non-formal, and then IP2 can be formulated as follows:

> IP2' A good argument, *logico sensu*, is one in which the conclusion follows or can be inferred from the premises by correctly applying a valid rule of inference.

Inferentialism then leads to generalism, the claim that the very possibility of assessing the logical quality of an argument depends on the provision of a suitable supply of inference rules, principles or licenses, something denied by particularism (Marraud 2020).

3. The reasonist approach to argument

Reasonism, the alternative to Inferentism, is based on two basic intuitions about arguments:

> RP1. To argue is to present, for examination, something to someone as a reason for something else,
> and
> RP2. A good argument *logico sensu* is, depending on purpose of the appraisal, either (a) one that presents a pro tanto reason, i.e., a reason that has genuine weight, but may be outweighed by other reasons, or (b) one that presents a conclusive reason, i.e., a reason that entitles us to assert the conclusion.

These two intuitions constitute the core of the reasonist approach to argument and argumentation. Joseph Wenzel condenses these intuitions into a question when he says that the key logical question is "should we accept this thesis for the reasons adduced in support of it?" (Wenzel 2006[1990]:16).

I will argue that RP1 and RP2 are incompatible with atomistic inferentism, and if so, since they are two basic intuitions, atomistic inferentialism is a flawed theory of arguments.

A standard definition states that a reason is a consideration that favors a position or claim on an issue, and a conclusive reason is one that entitles one to assert that claim. Reasonism defines validity in terms of conclusiveness: an argument is valid if and only if it supports the claim with a conclusive reason. The key to determining whether inferentism is compatible with reasonism is whether, when we rightly assert a claim on the basis of a given reason, that claim follows or is inferred from this reason (or from the statements expressing it). How do we draw conclusions from reasons? Reasons, unlike inferences, are paradigmatic weighted notions: establishing that a reason is conclusive requires its joint consideration with other concurrent reasons and considerations (conditions and modifiers -vid. Marraud, 2020), which vary from situation to situation (Lord & Maguire, 2017, 4). Thus, the same argument presents a conclusive reason in some situations, but not in others. Whether an argument is conclusive does not depend only on the properties of its parts and the relationships between them. This blatantly contradicts AP1 if conclusiveness is identified with validity; otherwise, and, according to the atomistic criteria AP2, conclusiveness would not be a logical property of arguments, in open contradiction with RP2.

To sum up, the main difference between reasons and logical inferences is that reasons are weighted and logical inferences are non-weighted. To reconcile inferentist tenets with basic reasonist intuitions, two different strategies have been designed, which I call 'conductivism' and 'revisionism'. However, neither succeeds, as I will show next.

4. Conductivism

Conductivism consists in treating counter-considerations, taking this term from Trudy Govier (2010, 375), as premises of a conclusive argument, and not as premises of opposing arguments.

Conductivism begins by admitting that premises can be positively or negatively relevant to the conclusion.

> A statement A is positively relevant to another statement B if and only if the truth of A counts in favor of the truth of B. This means that A provides some evidence for B, or some reason to believe that B is true.
>
> A statement A is negatively relevant to another statement B if and only if the truth of A counts against the truth of B. This means that if A is true, it provides some evidence or reason to think that B is not true. (Govier 2010: 148, 149).

Let us consider a simple real-life example.

> The director of the Robert Reid Cabral children's hospital, Clemente Terrero, acknowledged this Wednesday that although the country continues with dengue cases within the expected parameters and within the security zone, "we are seeing an acceleration of the disease in recent weeks."
>
> Dominican Today, September 22, 2022. Retrieved from https://dominicantoday.com/dr/local/2022/09/22/director-of-the-robert-reid-although-we-are-in-a-security-zone-there-has-been-an-acceleration-of-dengue/

Reid Cabral is arguing that the epidemiological situation in the Dominican Republic is worrisome because, although the country continues with dengue cases within the expected parameters and within the security zone, there is an acceleration of dengue in recent weeks. In a conductivist analysis, he offers an argument that combines a negative premise,

> (NP1) The country continues with dengue cases within the expected parameters and within the security zone,

and a positive premise,

> (PP1) We are seeing an acceleration of the disease in recent weeks,

The conclusion the epidemiological situation in the Dominican Republic is worrisome follows from these two premises if the positive premise outweighs the negative one. The metaphor of the weight of an argument can have very different meanings, so let us clarify that what is meant here by 'outweigh' is simply that the inference relation is not monotonic: it may happen that C follows from P, but not from P and Q, or even that not C follows from P and Q. Thus, the conclusion is not reached weighing up pro and con arguments, but by means of a peculiar type of inference that combines positive and negative premises. I shall call such inferences 'conductive inferences', again drawing inspiration from Govier (2010), but without intending to use that name in the same sense as she does.

On the very plausible assumption that A is the same argument as B if and only if A and B have the same constituents playing the same roles, the parts of an argument are those elements that differentiate it from any other argument. Conductivism multiplies the number of arguments, distinguishing, for example, between the arguments:

> A1. The epidemiological situation in the Dominican Republic is worrisome because, although the country continues with dengue cases within the expected parameters and within the security zone, we are seeing an acceleration of the disease in recent weeks.

and

> A2. The epidemiological situation in the Dominican Republic is worrisome because we are seeing an acceleration of the disease in recent weeks.

Govier, for instance, seems to suggest that A1 is stronger than A2 when she says that "Acknowledging counter-considerations does not necessarily weaken an argument." (op.cit., 356). From a holistic point of view, there is only one argument here, A2, and the additional consideration mentioned in A1 are but one of the many contextual factors relevant to its evaluation. If so, it does not make sense to attribute different weight to A1, and A2, although some modes of presentation may be more convincing than others.

The conductivist can avoid the proliferation of arguments by saying that these additional considerations are hidden premises, as it is often done. In this case, A2 would be an incomplete formulation of A1. But, since there is usually an indeterminate number of considerations relevant to evaluate an argument, the result of such a maneuver is that one can never make explicit all the premises of an argument, and consequently argument identity becomes fuzzy.

5. Revisionism

Revisionism turns counter-considerations into exceptions by introducing conditions of exception or rebuttal (Toulmin 2003 [1958], 93-95). Exceptions are unexpected events that block the passage from the premises to the conclusion and suspend the argument. I will call inferences subject to exceptions 'revocable inferences'. The idea is that any revocable inference is warranted only in the absence of some exceptional conditions, which would undercut the inference. Revisionism is usually particularist because exceptions are conceived as exceptions to a rule of inference (Toulmin, Rieke & Janik 1984, .99; Toulmin 2003 [1958], 94).

The previous example will help us to grasp the difference between conductivism and revisionism. On the conductive view, the conclusion C, the epidemiological situation in the Dominican Republic is worrisome, follows from the premises PP1 and NP1, since the positive premise outweighs the negative one. The revisionist sees two consecutive arguments where the conductivist saw only one:

A2. The epidemiological situation in the Dominican Republic is worrisome because we are seeing an acceleration of the disease in recent weeks.

A3. The epidemiological situation in the Dominican Republic is not worrisome because the country continues with dengue cases within the expected parameters and within the security zone.

The exceptions revokes A3. This prevents the conclusion that the epidemiological situation in the Dominican Republic is not worrisome, but by itself does not lead to any further conclusion. The conclusion C is provided by A2, insofar as no exception occurs. PP1 acts both as an exception with respect to A3 and as premises in A2. Thus, the conclusion results from argument A2 after discarding A3 as invalid. In other words, the fact that we are seeing an acceleration of the disease in recent weeks is a reason to be worried about the epidemiological situation in the Dominican Republic, while the fact that the country continues with dengue cases within the expected parameters and within the security zone is only an apparent reason not to be worried.

A major difference between conductivism and revisionism is that negative premises are parts of the argument while conditions are not. There is a difference between inferring C from P unless E and inferring C from P and not E. Admittedly, it seems more natural to say that that the country continues with dengue cases within the expected parameters and within the security zone is a relevant factor when evaluating A2 than to say that it is part of a different argument, A1, supporting the same conclusion. Hence, revisionism is an attempt to save inferentism by sacrificing atomism. Revisionism recognizes the existence of contextual factors relevant to evaluating an argument, which are not part of it. Revisionism thus rejects principles AP1 and AP2, resulting in holistic inferentism.

However defeasible inferences are not weighted notions. Using the familiar example of default logics, from Tweety is a bird you can conclude that Tweety flies, as long as you cannot infer from the available information that Tweety is a penguin. When you can infer from the available information that Tweety is a penguin, you cannot infer Tweety flies from Tweety is a bird. Thus, Tweety is not a penguin is a condition for inferring Tweety flies from Tweety is a bird. There is no comparison or choice between two arguments with opposite conclusions here, and therefore there is no weighing.

I understand by 'weighting' the process of comparing the weight of opposing reasons or arguments to reach a conclusion. Weighting occurs wherever there is a conflict of reasons. That is, weighting is in order when one has two arguments that present pro tanto reasons for two incompatible conclusions. Revisionism assumes that there are no genuine conflicts of reasons, and that a conflict of reasons is an indication of the presence of some exception to one of the opposing arguments. Exceptions come in two ways. First, the premise of one of the opposing arguments (A2) is at the same time an exception to the other argument (A3). Since all the relevant facts about the spread of dengue occur simultaneously, some explanation is required as to why the positive premise is an exception to A3 and the negative premise is not an exception to A2. Second, it may be an independent exception: for example, Dengue Case Count Trackers underestimate the number of cases. The belief that there will always be a hidden exception that prevents a conflict of reasons is simply an act of faith. Therefore, the revisionist explanation seems unnatural and arbitrary.

Weighting is contextual in a sense that revocable inferences are not. If E is a rebuttal condition to the inference rule 'if P then C', then the inference of C from P will be valid in all those situations in which E does not occur and invalid in all those in which it does. In this sense, revocable inferences are contextually dependent. But weighting is also contextual in the sense that the weighting of A2 and A3 in each situation only provides a solution for that particular situation. It is possible to conceive situations in which A2 has more weight than A3, and vice versa.

If those reasons are being considered in relation to diabetics, the conclusive argument might be A3, and if they are in relation to people taking anticoagulants, it might be A4. However, one can hardly interpret 'many people are diabetics' as an exception to A4, or 'many people are treated with anticoagulants' as an exception to A3. Of course, epicycles can always be constructed, and it could be argued that 'unless many people are treated with anticoagulants and daily intake of raw garlic increases the risk of bleeding' is a condition of rebuttal to A3. Just as one can always add hidden premises, one can always make the rebuttal conditions more and more complex.

For inferential generalism, reasoning is the application of rules, strict or with exceptions, to draw a conclusion from a set of data. When two arguments are weighed against each other, the conclusion, according to the most accepted position in legal argumentation theory, does not result from the application of any rule, and therefore weighting is a kind of particularist argumentation. In fact, when the clash of reasons can be resolved by applying a rule of priority, one cannot properly speak of a conflict of reasons.

The reasonist explanation, on the other hand, is simple and intuitive. We have reasons to believe that it is advisable to eat raw garlic every day, collected in A3, and reasons to believe that it is not advisable, collected in A4. After weighing them in a given situation, we came to the conclusion that it is advisable to eat raw garlic or that it is not, or we did not come to any conclusion at all. Thus, the logical properties of an argument are contextual and depend on its comparison to other concurrent arguments and on the presence or absence of contextual factors (called 'modifiers') that affect its relative weight.

To sum up, I believe that neither conductivism nor revisionism satisfactorily account for how we reach a conclusion in cases such as raw garlic. Conductivism multiplies arguments *praeter necessitatem* and blurs their identity, and revisionism cannot account for the weighting of arguments.

6. Different senses of 'argument' and 'conclusion'

Any theorist of argumentation who mentions the strength or weight of arguments is very careful to warn that it is a misleading metaphor. For example, Carl Wellman, who coined the term 'conductive inference', later recovered by Govier, writes:

> This suggests too mechanical a process as well as the possibility of everyone reading off the result in the same way. Rather one should think of the weighing in terms of the model of determining the weight of objects by hefting them in one's hands. This way of thinking about weighing brings out the comparative aspect and the conclusion that one is more than the other without suggesting any automatic procedure that would dispense with individual judgment or any introduction of units of weight. (1971: 58).

Wellman's caveats fall short, because metaphors of argument's strength or weight also suggest that argumentation is composed of discrete units that it is possible to isolate and compare with each other. Those units are, naturally, the arguments.

In argumentation theory the confusion of two senses of 'argument' causes quite a few headaches and obscures the deep differences between inferentism and reasonism. In both formal and informal logic an argument is a compound of premises and conclusion; but in linguistics and pragmadialectics an argument is a linguistic segment that, in the context in which it occurs, has a specific argumentative orientation.[3] Here are two examples of this second usage of 'argument'.

Be, for example, the chain:

---

[3] Juthé (2019) elaborates on the different senses of argument and argumentation, which he associates with the product-centered (informal logic) and process-centered (pragma-dialectical) approaches.

You drive too fast; you risk having an accident.
[...] Some semantic theorists think, and even write, that what we have here is a kind of reasoning that proceeds from a premise 'you drive too fast' to a conclusion 'you risk having an accident'. But such an interpretation seems to me to be completely absurd, since the word 'too much', which appears in the antecedent, can only be interpreted in relation to the consequent. What is driving 'too fast' if not driving at a speed at which there is a risk of producing undesirable consequences? [...] In other words, the very content of the argument cannot be understood except by the fact that it is directed to a conclusion. Outside such a chain, uttered or implied, it means nothing. (Ducrot 2004, 38-39; my translation).

The utterances advanced in the Argumentation are reasons, or, as we prefer call them, arguments relating to a standpoint. It is their function that makes arguments and standpoints different from other utterances [...] In the communication between language users, with a standpoint, a point of view is expressed that entails a certain position in a dispute; with an argument, an effort is made to defend that position. (Van Eemeren & Grootendorst 2016 [1992], 31).

Correlatively, 'conclusion' also has two meanings. In logic, the conclusion is one of the parts of the argument, but when we speak of the conclusion of a report, an article, or a debate we do not use 'conclusion' in its logical sense, since a report, an article or a debate is not an argument. The conclusion of a report, an article or a debate is the conclusion reached after examining the reasons given in the report, article, or debate. Thus, the term 'conclusion' is used both to refer to a part of an argument and to refer to the decision reached in a process of scrutiny and weighing of reasons.

7. Inferentist and Reasonist sense of 'conclusion'

Inferentists use 'argument' and 'conclusion' in their logical sense, so that the conclusion of a report, an article or a debate is the conclusion of one of the arguments offered in it, and the report, article or debate itself can be conceived as the process of selecting the strongest argument. This is the image suggested by almost every way of speaking of the weighting and strength of arguments. The task of the logician is then to provide criteria for determining which argument should be chosen. Abstract argumentation systems are very revealing of this way of thinking:

An abstract argumentation system is a collection of "defeasible proofs", called arguments, that is partially ordered by a relation expressing the difference in conclusive force. [...] Incompatibility and difference in conclusive force cause defeat among arguments. The aim of the theory is to find out which arguments eventually emerge undefeated. These arguments are considered to be in force. (Vreeswijk 1997, 225).

However, the conclusion of a debate may not coincide with that of any of the arguments considered. In a deliberation, for example, the probable consequences of various alternative courses of action are considered and discussed to choose the most desirable one. Whether an action is likely to have positive or negative consequences is a reason for choosing or rejecting it, respectively. The strength of the argument 'If A is done, B will probably happen, therefore it is advisable to do A' depends on the probability that doing A will have those effects, and on the intensity and beneficence of those effects. However, the conclusion of the deliberation cannot be 'It is advisable to do A', since the deliberation has a claim to determinacy or resolutive effectiveness (Vega 2020, 172), but rather 'We ought to do A' or 'Let us do A', which does not coincide, therefore, with the conclusion of any of the arguments examined in its course.

For reasonists the conclusion of an argument is defined in terms of the effect of the consideration adduced on the outcome of a discussion. The reasonist analysis of argumentation puts inter-argumentative relations (expressed by connectors such as 'also', 'but', etc.) before intra-argumentative relations (expressed by connectors such as 'therefore' or 'because'). Argumentation is no longer conceived as a competition of arguments, but as a network of interconnected reasons that participants weave interactively to reach a conclusion.

Analytically a dialogue (i.e., an exchange of reasons) begins with a question "Why…" and continues with a reason "Because…". If the dialogue does not end there, with the acceptance of the given reason, it can be continued with the presentation of a new reason, more or less explicitly related to the former one. From this moment on, each additional reason is presented as a reason that is chained, coincidental, opposite, or analogous to some previous reason. In this way the participants weave together an argumentative web that constitutes the object of study of the dialectic of arguments. Even when some participant tries to answer directly the initial question, her contribution is interpreted by the other participants in connection with the reasons already adduced. I call "macro-argument" this fabric of reasons. (Leal & Marraud 2022: 287).

We can extract from that macro-argument a consideration and the position it favors, and the result is an argument in logic's sense. But that should not make us forget that that consideration only favors that position in the context of the exchange in which it is adduced. What the logician must determine is not what the winning argument is, but what the overall orientation of the macro-argument is. Reasonism, in short, allows us to reconcile our behavior as theorists of argumentation with our behavior as practitioners of argumentation.

References

Ducrot, Oswald 2004. Argumentation rhétorique et argumentation linguistique. In M. Doury & S. Moirand, eds., *L'Argumentation aujourd'hui*, 17-34. Paris: Presses Sorbonne Nouvelle.

Eemeren, Frans H. van & Grootendorst, Rob 2016 [1992]. *Argumentation, Communication and Fallacies. A Pragma-dialectical Perspective*. New York: Routledge.

Goddu, Geoffrey C. 2011. Is 'argument' subject to the product/process ambiguity? *Informal Logic* 31 (2), 75-88.

Govier, Trudy 2010. *A Practical Study of Argument*, 7th ed., Belmont, CA: Wadsworth.

Johnson, Ralph H. 2000. *Manifest Rationality: A Pragmatic Theory of Argument*. Mahwah NJ: Lawrence Erlbaum.

Juthé, André 2019. Reconstructing Complex Pro/Con Argumentation. *Argumentation* 33, 413–454. https://doi.org/10.1007/s10503-018-9467-

Leal, Fernando M. & Marraud, Hubert 2022. *How Philosophers Argue. An Adversarial Collaboration on the Russell-Copleston Debate*. Cham: Springer.

Lord, Errol & Maguire, Barry, eds. 2016. *Weighing Reasons*. New York: Oxford University Press.

Marraud, Hubert 2020. Holism of Reasons and its Consequences for Argumentation Theory. In C. Dutilh Novaes et al., *Reasons to Dissent. Proceedings of the 3rd ECA Conference*, 167-180. London: College Publications.

Marraud, Hubert 2021. Cuatro modelos de argumento / Four models of argument. *Quadripartita Ratio: Revista de Retórica y Argumentación*, 6(11), 17-40.

Toulmin, Stephen E. 2003 [1958]. *The Uses of Argument*. New York: Cambridge University Press.

Toulmin, Stephen. E., Rieke, Richard y Janik, Allan 1984. *An Introduction to Reasoning*, 2nd edition. New York: McMillan.

Vega Reñón, Luis 2020. "Deliberando sobre la deliberación. Una revisión". *Lógoi. Revista de Filosofía* núm. 38, año 22, 166-200.

Vreeswijk, Gerard A.W. 1977. Abstract Argumentation Systems. *Artificial Intelligence* 90, 225-279.

Wellman, Carl. 1971. *Challenge and Response: Justification in Ethics*. Carbondale IL: Southern Illinois University Press.

Wenzel, Joseph. 2006 [1990]. Three Perspectives on Argument. Rhetoric, Dialectic, Logic. In R. Trapp, & J.H. Schuetz, eds., *Perspectives on Argumentation: Essays in Honor of Wayne Brockriede*, 9-26. New York: Idebate Press.

ARGUMENTATION QUALITY ASSESSMENT: AN ARGUMENT MINING APPROACH

SANTIAGO MARRO
Université CôteD'Azur, CNRS, Inria, I3S
smarro@unice.fr

ELENA CABRIO
Université CôteD'Azur, CNRS, Inria, I3S

SERENA VILLATA
Université CôteD'Azur, CNRS, Inria, I3S

Abstract

Argumentation is used by people both internally, by evaluating arguments and counterarguments to make a decision, and externally, e.g., by exchanging arguments to reach an agreement or to promote a position. A major component of the argumentation process concerns the assessment of a set of arguments and of their conclusions in order to establish their justification status, and therefore compute their acceptability degree. The assessment of the justification status of the statements supported by arguments allows the agent to decide what to believe and what to do. Argumentation semantics provide formal criteria to determine which sets of arguments (i.e., extensions) can be regarded as collectively acceptable (Baroni, Caminada, and Giacomin 2011). However, the assessment of the arguments acceptability is only a (basic) part of the complex assessment tasks required in argumentative processes in many everyday life applications, e.g., in medicine and education.

Assessing argumentation is a crucial issue in the context of artificial argumentation, encompassing various aspects such as identifying real natural language arguments and their relations in text, computing the justification status of abstract arguments, and gradually evaluating arguments. While some approaches have tackled the automatic assessment of natural language arguments (Wachsmuth et al. 2017, 2020), this issue remains largely unresolved.

In this paper, we address this open issue and we answer the following research question: what are the basic quality dimensions to characterize natural language argumentation and how to automatically assess them?

More precisely, we propose an Argument Mining (AM) approach to identify and classify natural language arguments along with quality dimensions.

In this work, we decide to characterize argument quality along with three quality dimensions for natural language argumentation, i.e., cogency, rhetoric, and reasonableness. The assessment of cogency involves determining the acceptability and sufficiency of the premises that support an argument's conclusion, while rhetoric identifies the use of rhetorical strategies such as ethos, logos, and pathos in the argument's conclusion. Additionally, reasonableness rates whether the argument effectively rebuts counterarguments, assessing the dialectical quality dimension of the argumentation.

Our interest focuses on the education scenario, where students are asked to interact with our AM system to assess the quality of their persuasive essays with respect to these three quality dimensions. To train our AM model, we annotated an existing dataset of 402 student persuasive essays (Stab and Gurevych 2017) with these quality dimensions.

We then propose a new deep learning AM method based on a transformer architecture, exploiting the structure of the argumentation graph through graph embeddings. Our approach automates the evaluation process proposed by Stapleton and Wu (2015) in social science by utilizing a scoring rubric for persuasive writing that combines the assessment of argumentative structural elements and reasoning quality. The obtained results are satisfactory and outperform standard baselines and similar approaches in the literature.

1. Introduction

A major component of the argumentation process concerns the assessment of a set of arguments and of their conclusions to establish their justification status, and therefore compute their acceptability degree (Baroni et al., 2011). Both qualitative and quantitative approaches have been proposed in the literature to assess the acceptance of an argument. However, the assessment of the arguments acceptability is only a (basic) part of the complex assessment tasks required in argumentative processes in many everyday life applications and contexts, e.g., in medicine and education.

The issue of assessing an argumentation is particularly critical when considering the different aspects of artificial argumentation, from the identification of real natural language arguments and their relations in text, to the computation of the justification status of abstract arguments (Baroni et al., 2011), to the gradual assessment of arguments (Amgoud et al., 2022) based, e.g., on the trustworthiness of the argument proponents (da Costa Pereira et al., 2011) or on the value promoted by the argument (Bench-Capon, 2003). Despite some approaches addressing the automatic assessment of natural language arguments (Wachsmuth et al., 2017), this issue remains largely unexplored and unsolved. In this paper, we address this open issue, and we answer the following research questions: *(i)* what are the basic quality dimensions to characterize natural language argumentation? and *(ii)* how to automatically assess these quality dimensions on natural language argumentative text?

More specifically, we propose an argument mining (Cabrio and Villata, 2018; Lawrence and Reed, 2020; Lauscher et al., 2022) approach to identify and classify natural language arguments along with quality dimensions. We first define and annotate three prominent quality dimensions for natural language argumentation, i.e., *cogency, rhetoric,* and *reasonableness*, on an existing dataset of student persuasive essays (Stab and Gurevych, 2017). We then train a neural network model classifier empowered with properties from the argument graphs to address the task. Our core contribution is twofold:

- We enrich a linguistic resource of persuasive essays (1908 arguments) with a new annotation layer, i.e., the quality dimensions of *cogency, rhetoric,* and *reasonableness*.
- We propose a new model architecture, exploiting the structure of the argument graph through graph embeddings. To the best of our knowledge, this is the first method that combines the graph structure of the argumentation with the textual content to assess the argumentation quality.

This paper is motivated by the lack of natural language argumentation resources annotated with quality dimensions, and the need for effective methods to address this task. Our contribution offers a novel resource and method to advance the field.

2. Related Work

Recent approaches in Argument(ation) Mining (AM) tackle specific argument qualities features, such as argument relevancy (Wachsmuth, 2017), convincing arguments (Habernal, 2016) and overall argument quality (Toledo, 2019).

Defining the characteristics of a good and successful argument is a hard task. First, we must address the several text rating procedures proposed in the literature. Different factors, such as the aim of the assessment, the freedom given to the raters, and the number of texts to be analyzed should be considered when evaluating the quality of argumentative texts. Following (Coertjens et al., 2017), rating procedures can be classified in two dimensions: Holistic vs. analytic and absolute vs. comparative. Holistic rating entails evaluating texts as a complete entity, while analytic rating involves assessing multiple text features. In absolute ratings, each text is assessed based on a predefined criteria or description, while in comparative ratings, texts are compared to each other to determine their score. In this study, our objective is to assess the quality of argumentative texts in persuasive essays through the application of a consistent and absolute analytic rating system. The aim is to ensure that the evaluation is based solely on the essay's content, rather than the subjective bias of the evaluator, and that the assessment results are consistent across all raters.

A commonly used rating method is rubrics. In an analytic rubric, text features are predetermined, but the weight assigned to each feature may not be predetermined. (Coertjens et al. 2017) found that evaluators may assign different weights to predetermined text features, potentially leading to variations in assessments of a single text among evaluators.

To tackle this issue, (Stapleton and Wu, 2015) describe the weight of the separate text features in a rubric as fixed. In this rubric, the authors stated that a strong argumentative text is composed of two important elements. *(i)* an argumentative text must be constructed considering all elements contributing to a good *quality of argumentation* and *(ii)* attention must be paid to the *quality of the content* of the text.

Different approaches have been proposed to assess both points from a logical, rhetorical, and dialectical point of view. Wachsmuth (2017), derive a taxonomy of argumentation quality that systematically decomposes quality assessment based on the interactions of 15 widely accepted quality dimensions. The three main characteristics are *Cogency*, *Effectiveness* and *Reasonableness*. As a follow up, Wachsmuth (2020), investigate how effectively each dimension can be automatically assessed, modelling features such as content, style, length, and subjectivity. This text-only assessment yields moderate learning success for most of the evaluated dimensions.

Following the work by (Stapleton and Wu. 2015) and (Wachsmuth et al. 2017), we argue that it is important to evaluate both the quality of argumentation and the quality of the content to provide a complete assessment of the argumentative texts. We, therefore, advance the state of the art of natural language argument quality assessment by investigating three main quality properties of *persuasive essays* (i.e., cogency, reasonableness, and rhetorical strategy) using the rubric provided by (Stapleton and Wu. 2015). Additionally, we propose a novel approach to evaluate argument reasonableness by integrating cogency properties with the argumentation graph structure.

3. Quality dimensions of persuasive essays

To annotate the quality dimensions on persuasive essays, we rely on the corpus built by (Stab and Gurevych, 2017), containing 402 persuasive essays annotated with the argument components (i.e., evidence, claims and major claims) and relations (i.e., support or attack). We add a new annotation layer by manually labelling for each argument in the essays the following three quality attributes: *cogency, reasonableness,* and *argumentation rhetoric.*

3.1. Annotation guidelines

Given that our goal is to assess persuasive essays written by students, we rely on an absolute analytic quality evaluation process proposed by (Stapleton and Wu. 2015). The authors propose a scoring rubric for persuasive writing that integrates the assessment of both argumentative structural elements and reasoning quality. This rubric contemplates several characteristics of the standard definition of Cogency and Reasonableness, such as Relevancy, Acceptability, and Soundness as well as the presence of counterarguments and rebuttals. Tables I, II and III show the analytic scoring rubrics proposed by (Stapleton and Wu. 2015). A scale of 0, 10, 15, 20, 25 is given to assess the Cogency and Reasonableness of a given argument.

Definition 1. Cogency *An argument should be seen as cogent if it has individually acceptable premises that are relevant to the argument's conclusion and that are sufficient to draw the conclusion* (Wachsmuth et al., 2017).

Table I. Analytic Scoring Rubric for assessing Cogency (Stapleton and Wu, 2015).

Score 25	Score 20	Score 15	Score 10	Score 0
a. Provides multiple reasons for the claim(s), and b. All reasons are sound/acceptable and free of irrelevancies.	a. Provides multiple reasons for the claim(s), and b. Most reasons are sound/acceptable and free of irrelevancies, but one or two are weak.	a. Provides one to two reasons for the claim(s), and b. Some reasons are sound/acceptable, but some are weak or irrelevant.	a. Provides only one reason for the claim(s), or b. The reason provided is weak or irrelevant.	a. No reasons are provided for the claim(s); or b. None of the reasons are relevant to/support the claim(s).

Following Table I, we define the *acceptable* premises as the ones that are worthy of being believed, and the *relevant* one as those that contribute to the acceptance or rejection of the argument's conclusion. These criteria are considered in point (b) (Table I) whilst the structural information about the argument graph is addressed in point (a). Examples 1, 2, 3 show the cogency annotation on three different persuasive essays from (Stab and Gurevych, 2017).

Example 1. We should attach more importance to cooperation during primary education. [**Through cooperation, children can learn about interpersonal skills which are significant in the future life of all students**]1.[*What we acquired from team work is not only how to achieve the same goal with others but more importantly, how to get along with others*]1. [*During the process of cooperation, children can learn about how to listen to opinions of others, how to communicate with others, how to think comprehensively, and even how to compromise with other team members when conflicts occurred*]2. [*All of these skills help them to get on well with other people and will benefit them for the whole life*]3.

Example 2. Animals should live in natural habitats instead of zoos. [**it is our responsibility to create a natural and safe environment for animals to live in**]1, [*Given the fact that human beings are responsible for the heavy pollution and severe damage to the natural habitats of many wild animals*].1[it is the right of wild species to live in a environment away from human beings].2

Example 3. Television devastate families ties. [**Most of people do not have a plan for make a limitation or schedule for watching television**]1.

The first sentence represents the major claim, while the claim to be assessed is in bold and the premises supporting it are in italics. Example 1 is annotated with cogency score 25, given that the author presents multiple premises which are acceptable and relevant to draw a conclusion. Example 2 shows a cogency score of 15, given that the author presents two premises that are relevant to the topic but not sufficient to draw the conclusion. Finally, Example 3 is annotated with a cogency score 0, given that the author does not presents a premise to support the claim.

Table II. Rubric for Reasonableness Counterargument (Stapleton and Wu, 2015).

Score 25	*Score 20*	*Score 15*	*Score 10*	*Score 0*
a. Provides multiple reasons for the counterargument claim(s), and b. All reasons for the alternative view(s) are sound/acceptable and free of irrelevancies.	a. Provides multiple reasons for the counterargument claim(s), and b. Most reasons for the alternative view(s) are sound/acceptable and free of irrelevancies, but one or two are weak.	a. Provides one to two reasons for the counterargument claim(s), and b. Some reasons for the alternative view(s) are sound/acceptable, but some are weak or irrelevant.	a. Provides only one reason for the counterargument claim(s), or b. The reason for the alternative view is weak or irrelevant.	a. No reasons are provided for the counterargument claim(s); or b. None of the reasons are relevant to/support the counterargument claim(s)/alternative views.

Definition 2. Reasonableness *An argumentation should be seen as reasonable if it contributes to the resolution of the given issue in a sufficient way that is acceptable to the target audience* (Wachsmuth et al., 2017).

The Analytic Scoring Rubric for Reasonableness (Table II and Table III) integrates these concepts and follows the idea of evaluating the argumentation graph with a focus on the counterarguments and their respective rebuttals. (Stapleton and Wu. 2015) separates the evaluation of Reasonableness for two different argumentative components, the counterarguments, and the rebuttals.

Table III. Analytic Scoring Rubric for assessing Reasonableness (Stapleton and Wu, 2015).

Score 25	Score 20	Score 15
a. Refutes/points out the weakness of all the counterarguments, and	a. Refutes/points out the weakness of all the counterarguments, and	a. Refutes/points out the weakness of all the counterarguments, and
b. All rebuttals are sound/acceptable, and	b. Most rebuttals are sound/acceptable, but one or two are weak.	b. Some rebuttals are sound/acceptable, but some are weak
c. The reasoning quality of all the rebuttals are stronger than that of the counterarguments.	c. The reasoning quality of most rebuttals are stronger than that of the counterarguments, while one or two are equal to that of the counterarguments.	c. The reasoning quality of some rebuttals are stronger than that of the counterarguments, while some are weaker to that of the counterarguments.

Score 10	Score 0
a. Refutes/points out the weakness of some counterarguments, or	a. No rebuttals are provided; or
b. Few of the rebuttals are sound/acceptable; most of them are weak, or	b. None of the rebuttals can refute the counterargument
c. The reasoning quality of most rebuttals are weaker than that of the counterarguments.	

The rubric score given to evaluate the reasonableness of the *counterarguments* (Table II) stipulates an analysis on the cogency of the counterargument, providing the same definitions given in Table I. Similarly to *cogency* and *reasonableness counterargument*, an evaluation on the soundness and acceptability of the text is required for the evaluation of the *rebuttals* (Table III). However, it differs from the others when evaluating if *(i)* the rebuttal refutes the counterarguments and *(ii)* does so with a stronger reasoning quality.

To evaluate the Reasonableness of an argument, one must analyze its counterarguments and related rebuttals. In Figure 1, the reasonableness quality dimension is assessed following Tables II and III. Counterargument Claim E receives a Score 0 for Reasonableness Counterargument as no supporting reasons or premises are provided. In contrast, for the rebuttal, Claim F provides a sound premise and stronger reasoning quality than the counterargument, resulting in a Score 25 for Reasonableness Rebuttal.

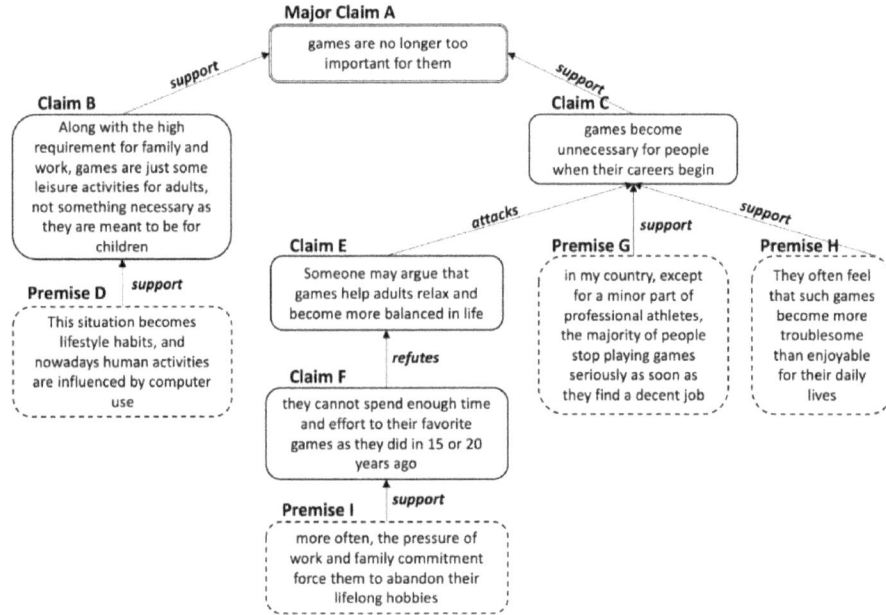

Figure 1. Example of an argument graph of a persuasive essay (Stab and Gurevych, 2017).

Argumentation Rhetoric. Annotators were asked to evaluate at the argument level which rhetoric strategy the argument is following among *ethos*, *logos*, and *pathos* (Aristotle, 2004). Examples 6, 7, and 8 show the rhetorical strategy annotation on three different arguments from (Stab and Gurevych, 2017).

Example 4. The advanced medical care brings with it more benefits than disadvantages. [**The main advantage of high-tech medical care is that people are better taken care so that they have a good health**][1]. [*Healthy workers can create more productivity*][1] [*They can contribute effectively to the development of the economy*][2]. [*They do not have to spend more time in health checking or treatment*][3]. [*this saves an amount of time as well as cost*][4].

Example 5. People should sometimes do things that they do not enjoy. **[In personal live, we have some responsibilities towards to other people, there is nobody who likes all of these responsibilities]**$_1$. [*Housework is very difficult for me, although my husband helps me some of them, but it is my responsibility*]$_1$. [*I really don't like any of them, however I should do*]$_2$, [*most people's lives are filled with tasks that they don't enjoy doing*]$_3$.

Example 6. Following celebrities can be dangerous for the youth. **[This has an overall effect on personality and future of an individual, following celebrities blindly affects the health of adolescents.]**$_1$ [*Many young people indulge themselves in drugs and start smoking at an early age*]$_1$. [*In a survey carried out in a university, it was asked to students that why did they start smoking, then around forty percent of individuals answered that they wanted to look like their favorite screen actor while smoking cigarettes*]$_2$ [*Imitating celebrities has a negative influence on health of young individuals*]$_3$.

In Example 4 the claim (in bold) appeals to emotions *Pathos* when the author describes how "people are better taken care" in the premises 1 and 3 (in italic). In Example 5 the authors employ *Ethos*, we can notice that the author refers to personal experiences in premises 1 and 2. Example 6 employs *Logos*, the author refers to a formal study, in premise 2, to support its claim.

3.2. Inter-annotator agreement

Before starting the annotation process, three expert annotators carried out a training phase, during which they studied the guidelines and discussed about the ambiguities between the scores for the definitions of Cogency and Reasonableness, amongst others. Then, the annotators were presented with an argument from a persuasive essay and its full argument graph, and they had to annotate the argument quality following the rubric scores.

To ensure the reliability of the annotation task, the inter-annotator agreement (IAA) was calculated on a set of 33 essays, resulting in a Fleiss' kappa of 0.68 for Cogency, 0.78 for Reasonableness Counterargument, 0.84 for Reasonableness Rebuttal, and 0.85 for Argumentation Rhetoric. Despite this substantial agreement, the annotators encountered difficulty in selecting precise scores, such as 25 or 20. To address potential subjectivity issues in the manual annotation, we opted to merge Score 25 with Score 20 and Score 15 with Score 10, resulting in three labels (with Score 0 remaining unchanged).

After recalculating Fleiss' kappa score, we observed an increase only for Cogency (from 0.68 to 0.86). Therefore, we decided to use a three-label score (i.e., 0, 15, 25) for Cogency prediction, while keeping the more fine-grained score (i.e., 0, 10, 15, 20, 25) for Reasonableness. Annotators then engaged in a reconciliation phase, where they resolved disagreements through discussion. One of the expert annotators performed the remaining annotation. Table IV and Table V report on the statistics of the final dataset.

Table IV. Statistics of the dataset, reporting on the percentage and type of Rhetorical arguments.

No Rhetoric	Ethos	Logos	Pathos
76.04%	11.51%	6.79%	5.66%

Table V. Statistics of the dataset, reporting on the percentage of Cogency and Reasonableness for each score.

Dimension	Score 0	Score 10	Score 15	Score 20	Score 25
Cogency	19.70%	9.38%	19.14%	31.71%	20.08%
Reasonableness	27.27%	25.45%	26.36%	16.64%	7.27%
Counterargument Reas. Rebuttal	79.82%	9.65%	4.39%	3.51%	2.63%

4. Automatic assessment of argumentation

An overview of the automatic argument quality assessment framework we propose is visualized in Fig. 2. Starting from the persuasive essays where argument components and their relations are identified, the goal is to assess the quality of each argument (i.e., the quality of each claim). Three scores are computed: a *cogency* score in the range {0, 15, 25}, an *argumentation rhetoric* label among *ethos*, *logos*, and *pathos*, and a *reasonableness* score in the range {0, 10, 15, 20, 25}. Two different methods are combined to assess the quality dimensions of the arguments: *(i)* the cogency score and the argumentation rhetoric labels are predicted using an attention-based neural architecture which employs the argumentation graphs through graph embeddings, and *(ii)* the reasonableness score is computed by means of an algorithm, combining the cogency score predicted at step *(i)* and the graph structure of each persuasive essay.

To feed the argumentative texts into the computational models we need to, first, convert them into vectorial representations called *embeddings*. In the following, we present the textual and graph embeddings we extracted from the persuasive essays to predict the cogency score and argumentation rhetoric labels, the architecture we define to predict these two quality dimensions, and conclude with the reasonableness algorithm used to assess this score.

Table VI summarizes our findings on the technical experiments for the automatic assessment of all quality dimensions, which are further discussed in (Marro et al., 2022)

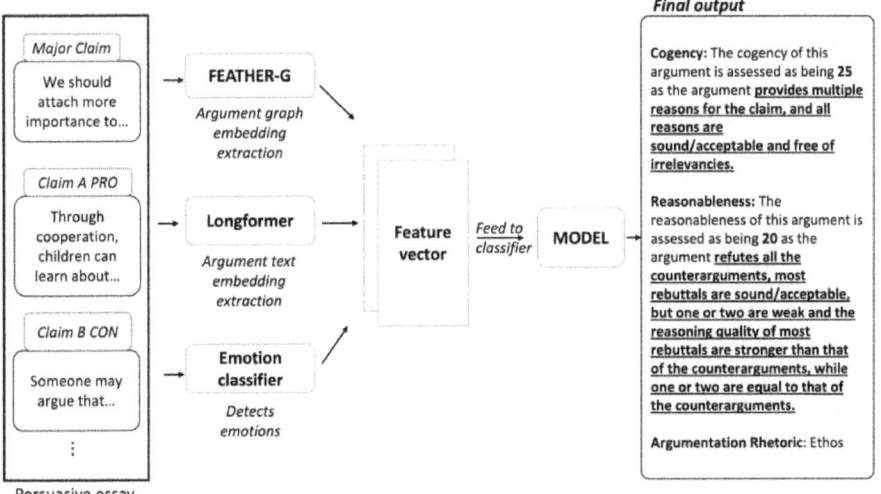

Figure 2. Overview of our natural language argumentation quality prediction model.

4.1. Embeddings

We generate features using *textual and graph embeddings*. Embeddings are low-dimensional, continuous vector representations of real-world data, such as text or graphs, which capture semantics and similarity. For text, the meaning of the words is encoded in such a way that words that are close in the vector space are expected to be similar in meaning. Graph embeddings similarly transform graph properties into vectors to capture topology, vertex-to-vertex relationships, and any other relevant information.

To generate textual features, we employ various embedding approaches, such as GloVe (Pennington et al., 2014) for static methods and Longformer (Beltagy et al., 2020) for contextualized embeddings, among others. To create a textual representation of an argument, we considered not only the sentences in the claim but also those from related components that are linked to the claim by a support or attack relations (e.g., premises, counterarguments, and rebuttals) to reflect human evaluation of argument quality. For graph embeddings, we used the state-of-the-art FEATHER-G (Rozemberczki and Sarkar, 2020) as our primary model.

To enrich our features for the Rhetoric dimension, we explored a way to obtain representations for the emotions present in the arguments. We utilized a state-of-the-art system trained for the Emotion Recognition downstream task. This approach enabled us to obtain an emotion label from a set of six basic emotions (sadness, joy, love, anger, fear, or surprise), which was subsequently used to generate a word embedding via various techniques discussed earlier in this section.

4.2. Cogency and rhetoric scoring assessment

Following feature generation, we proceed to perform an automated assessment of each quality attribute. With regards to Cogency and Reasonableness, we present a range of models in our experimentation, including both standard baselines and advanced methods. Specifically, we evaluate our models using textual embeddings alone, as well as a combination of textual and graph embeddings.

Our findings, presented in Table VI, demonstrate a significant improvement in the performance of our system upon the inclusion of argument graph features. However, in our assessment of rhetorical strategies, we did not observe any such improvement with the incorporation of graph features, whereas the inclusion of emotion embeddings resulted in a positive impact.

4.3. Reasonableness scoring assessment

As counterarguments and rebuttals are scarce in our dataset, our models struggle to properly classify reasonableness. To address this, we propose a new approach that takes into account the structure of the argumentation graph, which plays a significant role in assessing reasonableness.

The reasonableness dimension (Stapleton and Wu, 2015) considers *(i)* the cogency of the counterarguments attacking the argument we want to assess the reasonableness of, *(ii)* the cogency of the rebuttals to these counterarguments, and *(iii)* the relative number of rebuttals and counterarguments. This means that to effectively compute the reasonableness dimension, we need to combine the cogency-based quality of the argument components and the structure of the argumentation graph.

In (Marro et al. 2022) we propose an algorithm to compute the reasonableness score of the arguments in the persuasive essays. In this Rebuttal Reasonableness Score algorithm, we define how each score is evaluated by combining the cogency values of the pertinent arguments and the relevant properties of the argument graph.

Table VI. Results for automatic assessment of Cogency, Reasonableness and Rhetoric given in macro F1 (Marro et al., 2022).

Cogency Assessment		Reasonableness assessment		Rhetorical assessment	
Model	F1 Score	Model	F1 Score	Model	F1 Score
textual features	0.72	majority baseline	0.18	textual features	0.57
textual & graph features	0.77	Reasonableness algorithm	0.54	textual & emotion features	0.63

4.4. Final outcome

After automatically assessing the Cogency, Rhetoric, and Reasonableness dimensions, our system leverages the obtained scores to assist students in improving their essays. The pipeline concludes by automatically generating scores based on the following template:

The [QUALITY DIMENSION] of this argument is assessed as being [PREDICTED SCORE] as the argument [DEFINITION] (see Figure 2).

5. Concluding remarks

We presented a novel approach to the task of automatic quality assessment of natural language argumentation. We built a new resource of 402 students' persuasive essays annotated with 3 different quality dimensions. We show that our neural architecture relying on a transformer-based model and graph embeddings can successfully classify arguments along with these quality dimensions. Our quality assessment method conjugates the empirical evaluation of the cogency dimension with the graph-based computation of the reasonableness one, which encompasses the quality (expressed in terms of cogency) of the counterarguments and the argumentation structure.

In the context of AI in education, we aim to include our automatic argument quality assessment pipeline into a larger framework where the system engages the student into an explanatory rule-based dialogue to assess her essays, explain why they obtained a certain quality score and how to improve them along with the considered quality dimensions.

6. Acknowledgements

This work has been supported by the French government, through the 3IA Côte 'Azur Investments in the Future project managed by the National Research Agency (ANR) with the reference number ANR- 19-P3IA-0002

References

Amgoud, L., Doder, D., & Vesic, S. (2022). Evaluation of argument strength in attack graphs: Foundations and semantics. Artificial Intelligence, 302, 103607.

Aristotle (2004). Rhetoric. Translated by Roberts. Mineola, NY: Dover Publications

Baroni, P., Caminada, M., & Giacomin, M. (2011). An introduction to argumentation semantics. The knowledge engineering review, 26(4), 365-410.

Beltagy, I., Peters, M. E., & Cohan, A. (2020). Longformer: The long-document transformer. arXiv preprint arXiv:2004.05150.

Bench-Capon, T. J. (2003). Persuasion in practical argument using value-based argumentation frameworks. Journal of Logic and Computation, 13(3), 429-448.

Cabrio, E., & Villata, S. (2018, July). Five years of argument mining: A data-driven analysis. In IJCAI (Vol. 18, pp. 5427-5433).

Coertjens, L., Lesterhuis, M., Verhavert, S., & De Maeyer, S. (2017). Judging texts with rubrics and comparative judgement: Taking into account reliability and time investment. Pedagogische Studien, 94(4), 283-303.

da Costa Pereira, C., Tettamanzi, A. G., & Villata, S. (2011, June). Changing one's mind: Erase or rewind? possibilistic belief revision with fuzzy argumentation based on trust. In Twenty-Second International Joint Conference on Artificial Intelligence.

Habernal, I., & Gurevych, I. (2016, August). Which argument is more convincing? analyzing and predicting convincingness of web arguments using bidirectional lstm. In Proceedings of the 54th Annual Meeting of the Association for Computational Linguistics (Volume 1: Long Papers) (pp. 1589-1599).

Lauscher, A., Wachsmuth, H., Gurevych, I., & Glavaš, G. (2022). Scientia Potentia Est—On the Role of Knowledge in Computational Argumentation. Transactions of the Association for Computational Linguistics, 10, 1392-1422.

Lawrence, J., & Reed, C. (2020). Argument mining: A survey. Computational Linguistics, 45(4), 765-818.

Marro, S., Cabrio, E., & Villata, S. (2022). Graph Embeddings for argumentation Quality Assessment. In Findings of the Association for Computational

Linguistics: EMNLP 2022, pages 4154–4164, Abu Dhabi, United Arab Emirates. Association for Computational Linguistics.

Pennington, J., Socher, R., & Manning, C. D. (2014, October). Glove: Global vectors for word representation. In Proceedings of the 2014 conference on empirical methods in natural language processing (EMNLP) (pp. 1532-1543).

Rozemberczki, B., & Sarkar, R. (2020). Characteristic functions on graphs: Birds of a feather, from statistical descriptors to parametric models. In Proceedings of the 29th ACM international conference on information & knowledge management.

Stab, C., & Gurevych, I. (2017). Parsing argumentation structures in persuasive essays. Computational Linguistics, 43(3), 619-659.

Stapleton, P., & Wu, Y. A. (2015). Assessing the quality of arguments in students' persuasive writing: A case study analyzing the relationship between surface structure and substance. Journal of English for Academic Purposes, 17, 12-23.Tindale, C. W. (2007). Fallacies and argument appraisal. Cambridge.

Toledo, A., Gretz, S., Cohen-Karlik, E., Friedman, R., Venezian, E., Lahav, D., ... & Slonim, N. (2019, November). Automatic Argument Quality Assessment-New Datasets and Methods. In Proceedings of the 2019 Conference on Empirical Methods in Natural Language Processing and the 9th International Joint Conference on Natural Language Processing (EMNLP-IJCNLP) (pp. 5625-5635).

Wachsmuth, H., Naderi, N., Habernal, I., Hou, Y., Hirst, G., Gurevych, I., & Stein, B. (2017, July). Argumentation quality assessment: Theory vs. practice. In Proceedings of the 55th Annual Meeting of the Association for Computational Linguistics (Volume 2: Short Papers) (pp. 250-255).

MODULATING EPISTEMIC COMMITMENT TO DISAGREEING AND AGGRESSIVE CONTENTS EXPLOITING THE EVIDENTIAL FUNCTION OF IMPLICIT COMMUNICATION

VIVIANA MASIA
Roma Tre University
viviana.masia@uniroma3.it

Abstract

The idea that manipulation relies more heavily on implicit than on explicit language has been the plank of several earlier and recent debates on argumentation and speaker roles in interactions. The present contribution will inquire the selective nature of the use of implicit communication in political discourse; notably, analyzing the distribution of presuppositions and implicatures in two political debates, it will be argued that the use of these two implicit communicative devices – and, particularly, of presupposition - is likely to correlate with aggressive contents more often than with other content types. This tendency will be accounted for considering the evidential meaning presuppositions and implicatures add to an utterance, which contributes to modulating both speaker's commitment to truth and source identification on the part of the receiver. Data also show that, when face-threatening contents are exchanged, presuppositions epitomize by far the most preferred strategy in both debates.

Keywords: implicit communication, evidentiality, aggression, manipulation, political discourse.

1. Introduction

The advent of social networks has marked a real turning point in the study of political discourse and in its attempts to effectively forge consensus (Garassino et al. 2019; Masia 2021; Garassino et al. in press). Among other reasons, this could be partly put down to the speed of interactions on online platforms, and partly to the informal dimension created by the medium itself, which allows strengthening bonds between the politician and her followers. Nonetheless, the web has also become the seat of unguarded language uses, especially in the field of political propaganda. Indeed, Social Network Sites (SNSs) can nowadays be regarded as "virtual" Parliaments, where politicians do not miss the chance to enhance their own reputation or even smear the reputation of political opponents, with the view to weakening their credibility in the eye of a potential electorate. If this practice has become common in institutional Parliamentary speeches, it has become far more pervasive on SNSs, where the effects of aggressive and conflicting language – combined with the speed and conciseness of the messages - turns out to be even more manipulative (Brocca et al. 2020; Masia 2021; Pietrandrea 2021; Garassino et al. 2022).

Over the last decade, a newly emerged line of research known as experimental pragmatics (Noveck 2018) has produced a body of evidence highlighting the strong correlation that implicit communication bears to utterance understanding. Most of these studies have inquired the behavior of presuppositions and implicatures in directing addressees' attentional processes in sentence comprehension (Cory et al. 2016; Schwarz 2014). Although they are both classified as presumptive meanings, opposed to explicit ones, implicatures and presuppositions display a different discourse behavior, in that they differently interact with the manifestation of speaker commitment and source of information. Based on a comparative analysis of two political debates (one held between Matteo Salvini and Matteo Renzi in 2019, and one between Donald Trump and Hillary Clinton in 2016), it will be argued that the use of presuppositions and implicatures by the four political leaders is often driven by the types of contents being conveyed (attack, stance-taking or praise) and that their distribution between the two strategies is often contingent on how challengeable some content is. The analysis will show that presupposition is by far the most preferred strategy, especially when it comes to conveying attacking contents. The hypothesis is put forward that this trend most probably correlates with the evidential properties of presuppositions and implicatures, with presuppositions epitomizing a strategy to encode *mutual evidentiality* (Hintz & Hintz 2017), because they present some content as to be taken for granted by the addressee, and thus not further addressable by her, while implicature would expose the speaker as *first-*

hand source of some information (Masia 2021), which makes its content slightly more challengeable.

2. Presupposition, implicature and the encoding of evidentiality in discourse

2.1. Presupposition: a working definition

According to Stalnaker (Stalnaker 1973, 1974), presupposition is taken for granted information, which the speaker assumes to already hold in the common ground shared with the receiver. In an utterance, it is typically associated to the use of specific categories of presupposition triggers, the most common of which are reported in (1) (Kiparsky & Kiparsky 1971; Fillmore 1971; Lombardi Vallauri 2009).

DEFINITE DESCRIPTION
1) *The green bottle is over that table*
CHANGE OF STATE VERB
2) *Mary has recently stopped making physical exercise*
FOCUSING ADVERBS
3) *Also Norah has bought a Maserati*
DEFINING RELATIVE CLAUSES
4) *The book my friend wrote for her mother is very touching*
ADVERBIAL SUBORDINATE CLAUSES
5) *When Amber moved to Ireland, she sold her country house*
FACTIVE PREDICATES
6) *It's strange that John has not arrived yet*

In (1), to be presupposed by means of the definite description *The green bottle* is that there exists a green bottle, identifiable by both speaker and receiver, which is over the table. In (2), the change of state verb *stop* presupposes that Mary used to make physical exercise before. In (3), the focusing adverb *also* presupposes that someone else, besides Norah, has bought a Maserati. The defining relative clause in (4) takes for granted that my friend wrote a book for her mother. In (5), the temporal subordinate clause presupposes that Amber moved to Ireland, while, finally, the factive predicate *It's strange* in (6) conveys the presupposition that John has not arrived yet.

One well-known feature of presupposition is their resistance to negation. In this sense, by saying *It is not true that the green bottle is over that table*, I am still presupposing the existence of the green bottle, while the rest of the sentence is being denied. Among other reasons, this property could be seen as hinging on the fact that presuppositions are tacitly accepted as true in an interaction because they are interpreted as being already known (Lombardi Vallauri & Masia 2014). This means that defeating or even denying them would eventuate in an uncooperative communicative move.

2.2. Implicature: a working definition

When some content is conveyed implicitly by exploiting discourse rules of cooperation (Grice 1975), it may often take the form of *implicature*. Grice (1975) defines implicatures as hidden intentional meanings which speakers seek to express by concealing them behind other literal propositions. So, for example, in the short dialogue below,

> (7)
> A: Are you joining us for a beer?
> B: I have to study for the exam

B does not straightforwardly say *Yes, I'm coming* or *No, I'm not coming*; she rather opts for an indirect move which, in any case, allows the interlocutor to reconstruct the refusal to the invitation. Speaker B *exploits* (in Grice's terms) the conversational Maxim of Relation to convey a proposition which is apparently not relevant to the purpose of the interaction, yet the resulting implicit meaning – namely the derived implicature – is interpreted as anyway cooperative in the fulfilment of the communicative task at hand.

2.3. Evidentiality and its pragmatic dimension

In the mainstream literature, evidentiality has been defined as the linguistic manifestation of information source (Willett 1988; Aikhenvald 2004), although, in a broader perspective, it has also been associated to the speaker's degree of commitment to truth (Chafe & Nichols 1986). The most basic evidential systems distinguish at least between first-hand, second-hand (reported) and conjectural (inferential) knowledge (Aikhenvald 2004), but there are also languages with up to six or seven-fold evidential distinctions.

In a recent fieldwork study on Conchucos Quechua (Peru), Hintz & Hintz (2017) have discovered another evidentiality marker signaling that some knowledge is shared by all participants in an interaction prior to utterance time. This evidential category is opposed to direct evidentials whose function is to manifest the speaker as the unique source of some information. Hintz & Hintz called this knowledge state *mutual* to indicate that some information is within all interlocutors' territory of knowledge (Kamio 1997). So, for example, in (8) the fact that ancestors pastured animals is information presented as known by everybody.

(8)
b. *Tsay-pa-**cha**:*   *qati-ya-ra-n*   *mama-yki-kuna*
   That-gen-**mut** follow-pl-pst-3   mother-2-pl
   'By that route, your ancestors pastured animals' (as we all know)

According to Hintz & Hintz (2017), *-cha* indicates that the speaker "shares epistemic authority for the assertion with others with whom common ground has been established" (Hintz & Hintz 2017: 20). From a general perspective, this discourse function of mutual evidentiality could be regarded as paralleling that of presupposition laid out above, in that also presupposition is expected to convey already shared content. Implicatures, on the other hand, seem to perform more similarly to direct evidentials, because – even if computed inferentially – they represent content the speaker conveys as first-hand source. As a matter of fact, once an implicature is computed, the speaker is eventually identified as its unique committed source.

This evidential characterization of implicatures and presuppositions will be zoomed in on later in this work, in order to account for the trends observed in the two political debates and, notably, the interaction between the use of the two implicit communicative devices and different content types. For the purpose of the analysis, the following evidential outlines of presupposition and implicature can be considered:

As a strategy hiding the main content of a message (Lombardi Vallauri & Masia 2014), yet keeping the speaker's communicative intention as equally manifest as it is with direct assertions, an *implicature* can be equated with direct evidential markers, in that they always induce to identify the speaker as the source of the implicit information conveyed.

A *presupposition* is a discourse strategy that signals that some content is already shared by all participants in an interaction. This makes its function akin to that of mutual evidentials in other languages, in that it encodes some information as holding in everybody's territory of knowledge.

Building on these properties, it is reasonable to expect that politicians' recourse to presupposition and implicature strategies may be content-oriented, and that the more critical or tendentious some content is, the more likely it may be encoded by means of presupposition, as opposed to implicature, the former more strongly concealing the speaker's communicative responsibility. And we expect such correlation to appear particularly strong for those contents which are perceived as more face-threatening, such as attacks.

3. The corpus

The analysis described in this section will discuss some trends in the use of presuppositions and implicatures in association to attacking, stance-taking and praising content types in two political debates: an Italian debate held between *Matteo Renzi* (leader of Italia Viva) and *Matteo Salvini* (leader of the League party) in 2019 on Porta Porta, a popular Italian late-evening talk show, and a U.S. debate disputed by *Hillary Clinton* and *Donald Trump* at Hofstra University in Hampstead in 2016.[1]

In order to compare the two speeches on a more reliable basis, an analogous playing time has been extracted for the four politicians (about 10 minutes per each).[2] Furthermore, the four political leaders have been chosen according to their ideological similarity to their foreign counterparts (Matteo Renzi and Hillary Clinton as both belonging to left-wing parties, and Matteo Salvini and Donald Trump as both exponents of right-wing parties). Then, within the two debates general topics related to the politician's agenda, promises or work up to the time of the debate are dealt with, which also helps striking a more desirable balance in the distribution of content types throughout the four speeches.

Comparing attacking/aggressive contents to stance-taking and praising ones allows, on the one hand, to better appreciate how implicit strategies are differently used according to content types and, on the other hand, to unveil whether a different distribution of content types among presuppositions and implicatures could be accounted for on the basis of their different position along a *challengeability scale*. According to this scale, some contents can be expected to be packaged either as overt assertions, implicatures or presuppositions based on how likely they might raise the recipient's critical reaction in a dialogic context.[3]

Below, some examples from the corpus are reported.

(9)
PRESUPPOSITION
Clinton: «*I **know** you live in your own reality*» (factive predicate)

---

[1] The U.S debate is available here: Donald Trump vs Hillary Clinton full Hofstra debate - YouTube, while the Italian debate can be watched here: Il confronto tra Renzi e Salvini - Porta a porta 15/10/2019 - YouTube.
[2] This entailed cutting a relevant part of the U.S. debate which was far longer than the Italian one.
[3] Needless to say, this scale is based on plausible intuitions on what the receiver's attitude towards these content types is expected to be. A further refinement of the proposal put forth in Masia (2020), for example, should involve testing the addressee's natural disposition towards questioning either one or the other content type.

Trump: «*the very nasty commercials **that** you do on me in so many different ways*» (relative clause)

(10)
IMPLICATURE
Salvini: «*Io non mi sognerei mai di creare partiti dalla sera alla mattina*»
   'I would never dream of creating parties overnight'
Renzi: «*Non si va avanti a colpi di gomitate in faccia a Macron e alla Merkel*»
'Getting into conflict with Macron and Merkel repeatedly does not bring us any further'

In (9), through a factive predicate, Clinton conveys as taken for granted information that Trump lives in his own reality. By converse, Trump uses a defining relative clause to present as already known information that Clinton does "very nasty commercials" on him. The Italian occurrences in (10) exemplify cases of attacking implicatures. Notably, by saying "I would never dream of creating parties overnight", Salvini implies that Renzi is used to creating new political parties overnight. On the contrary, Renzi implies that Salvini has been repeatedly getting into conflict with Macron and Merkel.

Below, some examples of the two strategies associated with stance-taking and praising contents are given.

(11)
STANCE-TAKING
Salvini: «*Posso voler cambiare questa Europa?* » 'May I desire to change this Europe?'
Renzi: «*Dev'esserci un'emergenza legalità, un'emergenza sicurezza, non un'emergenza immigrazione*»
'There should be an emergency of legality, not of immigration'

(12)
PRAISE
Clinton: «*I'm **glad** that we're ending private prisons in the federal system*» (factive)
Trump: «*I want to **get on** to creating jobs, because I want to **get on** to having a strong border*» (change-of-state)

The interrogative act used by Salvini in (11) conveys the implicature that he wants a different Europe; and, in so doing, he also provocatively replies to political opponents objecting to him his not accepting European governance as it is. Conversely, Renzi's utterance implies that the focus of emergencies, so far, has mainly been on immigration – what he disapproves – rather than on safety and legality.

4. Method

In the selection of the relevant occurrences, the following criterion has been complied with. First of all, the total number of instances of attacking, stance-taking and praising contents has been computed, irrespective of the presuppositional or implicatural packaging they displayed. Then, the number of presuppositions and implicatures associated to each of the three content types has been counted, as illustrated in the proportion below.

$$\frac{\text{n. of occurrences of PPP/IMPL} \cap \text{content type}}{\text{total n. of occurrences of content type (in any packaging)}}$$

For convenience, the value resulting from the above computation can be seen as epitomizing an index of association frequency (FoA index), indicating how frequently each content type is associated to either a presupposition or an implicature.

5. Results

The association indexes yielded by the performed computations are reported in Table 1. To better highlight differences between the packaging and the content type parameter, the indexes have been reported separately for presupposition and implicature strategies and for each content type.

An ANOVA run on the distribution values of presuppositions and implicatures yielded quite a strong interaction ($F=8.6$, $p=0.01$), with presuppositions appearing much more diffuse than implicatures in the debates. Conversely, the distribution of attacking contents does not significantly differ between the Italian and the US debate. The distribution of the three content types is overall homogeneous between the two debates, with no significant differences between Attack and Praise ($F<1$; $p=0.7$), nor between Attack and Stance-Taking ($F<1$; $p=0.8$), which is also consonant with the assumption that they all are characteristic of the argumentative type instantiated by political speeches in a debate context.

	FoA indexes of Presupposition	FoA indexes of Implicature	Tot. n. occurrences of content type[4]
ITALIAN DEBATE			
Attack	0,72	0,27	45
Praise	1	---	24
Stance-taking	0,44	0,27	36
U.S. DEBATE			
Attack	0,74	0,26	50
Praise	1	-	26
Stance-taking	0,54	0,45	42

Table 1. Frequency of Association indexes of implicatures/presuppositions and content types

As can be noticed in Table 1, more remarkable differences in the association between implicit strategies and content types can be observed for attacking and praising contents, with presuppositions remarkably outnumbering implicatures. The higher indexes shown by presuppositions for praising contents hinges on the fact that no (if very few) instances of praising implicatures have been found. This trend also finds support in findings yielded by other recent corpus-based analyses (Garassino et al. 2019; Masia 2020).

Speakers' preference for presuppositions over implicatures may be grounded in their function of "weakening" the speaker's commitment to their truth. Indeed, since an attack typically zooms in on an issue on which the speaker has more credibility but on which the opponent is weaker (Lee & Xu 2018), conveying it as taken for granted information allows passing it more subtly and possibly below the level of conscious awareness. Among other relevant aims, an attack is expectably meant to induce potential voters to think badly of an opponent (Masia 2020: 139) and persuade them to think of the speaker as someone worth being trusted.

To a certain extent, also praise is likely to be more challengeable than Stance-taking, due to its function of entailing a "raising of the self and a lowering of the other" (Dayter 2014: 92). So, on the whole, the weaker challengeability of stance-taking may be at the basis of its more balanced distribution between presuppositions and implicatures. As a matter of fact, since stance-taking contents can more predictably be associated to a given political leader, the politician may not feel urged to reduce her

---

[4] This results from the sum of all occurrences gathered from each politician's playing time.

commitment to contents which merely reiterate her own perspective on certain critical issues.

6. Persuasiveness of presupposition and implicature: observations on their retractability properties

An interesting aspect associated with the persuasive potential of presuppositions and implicatures in a message correlates with their behavior in retraction contexts. Assessing this behavior involves addressing issues like: (a) to what extent and on what criteria can a speaker be assumed to endorse the truth of a proposition? (b) to what extent and on what criteria can a speaker retract from a belief state? (a) and (b) are closely related in that the more strongly a speaker commits to a proposition, the more easily she can be expected to retract from it. In fact, in negating openly asserted contents or contents which are conveyed as presupposed, the effect would not be the same.

(13)
a. It is not true that *John broke a Chinese vase*
b. It is not true that *when John broke a Chinese vase* his mother was at work

In (13a) the proposition "John broke a Chinese vase" entirely falls within the scope of the negative expression "It is not", while the same proposition is outside of this scope in (13b). In fact, to be denied here is not the presupposition but the asserted unit of the sentence.

As already said, while assertion requires the speaker to commit and endorse its truth (in virtue of the illocutionary function of assertive speech acts, Searle 1969), similarly to what happens with implicatures, with presupposition both commitment and endorsement are shared between all participants in the conversation, which means that the speaker is not free to retract from it since it concerns content already agreed upon by the receiver as well. As a matter of fact, a sentence like (13b) above would only be interpreted as denying what is uttered in the independent proposition and not what is conveyed in the dependent one. From an evidential perspective, it could be argued that while "John broke a Chinese vase" is knowledge which the speaker conveys more subjectively; on the contrary, in (b) the proposition "John went to China" is knowledge conveyed as intersubjective, because it is presented as already established common ground. An analogous difference can also be assumed between presuppositions and implicatures. From an evidential perspective, the discourse behavior of implicatures is quite strikingly similar to that of assertions because also with implied information the speaker's communicative intention is made manifest, although it has to be inferentially reconstructed.

Other differences between presuppositions and implicatures also emerge on the evidential level of utterances and, notably, with respect to their compatibility with expressions of conjectural/inferential evidentiality, such as "It seems", "probably", or the like. Consider (14)a. and b. as an example.

(14)
a. *It seems that / Probably Mary has just arrived home*
IMPL > It seems that/Probably you can ask her out for a date

b. *It's strange that Mary has just arrived home*
PPP > It's strange [*it seems/*probably] that Mary has just arrived home

What can be noticed in the two sentences above is that while evidential expressions can be *inherited* by contents that are implied in a sentence, with presuppositions the same does not obtain. This behavior is on the whole consonant with the property of presupposition to package some content as to be taken for granted by the receiver, which means that its truth need not be further questioned in an interaction. Indeed, any attempt at addressing some presupposed content would turn out to be a conversationally costly operation to handle (Sbisà 2007). On a closer look, presuppositions appear to be somewhat incompatible with certain types of evidential expressions (Anderson 1986; Kiparsky & Kiparsky 1971) and, particularly, with those classifying some content as first-hand knowledge, as exemplified in (15).

(15) *I can *maintain/assure* your believing Mary to be drunk

Both the declarative verbs *maintain* and *assure* present the speaker as committed source of the receiver's belief state; nonetheless, since the state of "believing Mary to be drunk" is encoded as already known information (by means of a genitive definite phrase), it cannot fall within the scope of predicates whose meaning is to commit the speaker to the truth of the presupposition. Basically, what renders such a discourse operation pragmatically infelicitous is the fact that the speaker commits to an event that has already been accepted as true by all interlocutors.

The response of belief retraction to implicatures and presuppositions is also worth considering. In contexts of reported speech, it can be noticed that the speaker can easily retract her belief state from the content implied by an utterance, while the same is not possible for presupposed contents contained in a reported proposition. By way of illustration, let us take the reported assertion in (16) ('Mary has been travelling to Paris a lot lately') to convey the implicature that "Mary might have an affair in Paris", and the temporal adverbial clause in (17) to presuppose that Mary's husband discovered her affair in Paris.

(16) *I've been told that Mary has been travelling to Paris a lot lately, but I don't believe it*
[SCOPE OF DISBELIEF = I don't believe that Mary might have an affair in Paris]

(17) *I've been told that when Mary's husband discovered her affair in Paris, he went ballistic, but I don't believe it.*
[SCOPE OF DISBELIEF = I don't believe that he went ballistic/*that Mary's husband discovered her affair in Paris]

In (16), not only is the speaker withdrawing her belief state from the fact that Mary has been travelling to Paris a lot lately, but also from what this statement implies, namely that she might have an affair in Paris. Conversely, belief retraction in (17) cannot be interpreted as doubting that Mary's husband discovered her affair in Paris – which is the presupposed unit of the sentence – but the independent proposition "he went ballistic", which instead features the assertion of the reported utterance. Therefore, when we say that presupposition is resistant to challenge, not only do we mean that we expect the receiver to not take issue with its content, but also that it is fairly unlikely that any belief-retraction communicative move might address what an utterance encodes as taken-for-granted information, while we expect such a move to have scope on what an utterance conveys as the intentional meaning of the speaker.

Category	Belief retraction	Scope of deniability	Source
IMPLICATURE	**Allowed** since the implicit content is conveyed as only belonging to the speaker's domain of experience.	*It's not true that Mary has been travelling a lot to Paris lately* (> Mary is having an affair)	SPEAKER
PRESUPPOSITION	**Not allowed** due to the fact that the implicit content is conveyed as already holding in the shared common ground.	*It's not true that when Mary's husband discovered her affair he went ballistic* (> Mary's husband went ballistic)	SPEAKER AND RECEIVER

		(> *~~Mary's husband discovered that her wife had an affair~~)	

Table 2. Deniability and retraction properties of implicature and presupposition

Category	Speaker's attitude to knowledge	Evidential meaning	Commitment status	Source
IMPLICATURE	The speaker commits to the truth of the implicit proposition and is identified as the unique source of it.	DIRECT	SPEAKER-CENTERED	SPEAKER
PRESUPPOSITION	The speaker shares her sourceness status with the receiver, and both commit to the truth of the implicit content.	SHARED/MUTUAL	SPEAKER-AND-RECEIVER-CENTERED	SPEAKER AND RECEIVER

Table 3. Evidential properties of implicature and presupposition

Building on the foregoing, we might trace a possible evidential profile of presupposition and implicature, which might account for the different use that speakers (and, in our case study, politicians) make of these two strategies to convey attacking content types and, in general, any type of face-threatening content. In Table 2, I report a summary of the retractability properties associated with presupposition and implicature, while Table 3 clarifies the evidential profile of the two strategies of implicit communication.

7. Final remarks

The present study is only a partial picture of the way politicians communicate when potentially face-threatening contents are negotiated. Within Critical Discourse Analysis frameworks (cf. Van Dijk 1997, 2006)

other studies have highlighted the relevance of other types of implicit meanings, such as those associated to figurative and vague expressions (cf. Danler 2005, among others). For example, Oswald et al. (2016) have noticed that, in certain textual dimensions, passive constructions prove to be particularly effective in concealing the identity of an agent that is responsible for a given fact being perpetrated. Consequently, any reconstruction of that agent on the part of the message receiver is obviously less straightforward. In this work, I have tried to recast current reflection on the impact that implicit communication exerts on message comprehension in an evidential perspective, namely, trying to account for the epistemic implications associated with the use of presuppositions and implicatures. In this light, the present study wishes to be a promising follow up of previous research on the content-based use of presuppositions in political discourse (Masia 2020), in which – based on a small-scale corpus of political speeches – evidence has been gathered that presuppositions are more likely to be used to encode face-threatening, and on the whole more challengeable, content types.

What we noticed in the data discussed in the present analysis is that implicatures behave in a remarkably similar way as assertions, in that their use is much less frequent when it comes to highly challengeable contents such as attacks or praises. It can be deduced from this that the choice of the most effective communicative device does not rely on the explicit/implicit divide (given that implicatures are entirely implicit propositions), but on how manifest the speaker's intention is in resorting to one or the other strategy. The evidentiality account I have proposed may clarify the relevance of the speaker's epistemic role in an interaction. On this account, implicatures and overt assertions share the property of presenting the speaker as subjectively involved in the representation of an event, which candidates her as its unique source. On the contrary, presuppositions convey some knowledge as intersubjective, meaning that both speaker and receiver are depicted as possessors – with equal source status – of that knowledge. For this reason, presupposition can be deemed a *more manipulative strategy*, as it better serves the deceptive purpose of skipping the receiver's critical reaction, thereby being tacitly accepted as true.

8. Conclusion

In this paper, two political debates have been compared with a view to assessing how often attacking contents, compared to praising and stance-taking ones, are encoded as presupposition or as implicature. These two discourse strategies are extensively exploited in persuasive texts, yet their manipulative impact is expected to be different due to the distinct evidential values they add to a sentence. The data gathered from the analysis of the politicians' speeches in their respective playing time showed that, when it comes to highly face-threatening contents, such as attacks (but also praises), presupposition strategies are by far the most recurrent strategy, possibly due to their property to distribute epistemic commitments and information sources among all participants in the interaction, which reduces the speaker's responsibility for the truth of some content.

References

Aikhenvald, A. Y. (2004). *Evidentiality*. Oxford: Oxford University Press.

Anderson, L. B. (1986). Evidentials, paths of change, and mental maps: Typologically regular asymmetries. In W. Chafe & J. Nichols (Eds.) *Evidentiality: The linguistic coding of epistemology* (pp. 273–312). Norwood, NJ: Ablex.

Brocca, N., Masia, V. & Kucelman, E. (2020). Didactics of Pragmatics as a way to improve social media literacy: An experiment proposal with Polish and Italian students in L1. In *Social Media in Education and foreign language teaching*. HeiEducation Journal.

Chafe, W., & J. Nichols (Eds.). (1986). *Evidentiality: The Linguistic Coding of Epistemology*. Norwood, Nj: Ablex.

Cory, B., Romoli, J., Schwarz, F., & Crain, S. (2016). Scalar implicatures vs. Presuppositions: The View from Acquisition. *Topoi, 35*, 57-71.

Danler, P. (2005). Morpho-Syntactic and Textual Realizations as Deliberate Pragmatic Argumentative Linguistic Tools? In L. Saussure de & P. Schultz (Eds.), *Manipulation and Ideologies in the Twentieth Century. Discourse, language, mind* (pp. 45–60), Amsterdam/ Philadelphia: John Benjamins. https://doi.org/10.1075/dapsac.17.04dan.

Dayter, D. (2014). Self-Praise in Micro-blogging. *Journal of Pragmatics, 61*, 91-102.

Fillmore, C. J. (1971). Verbs of Judging. An Exercise in Semantic Description. In, In C. J. Fillmore & T. D. Langendoen (Eds.), *Studies in Linguistic Semantics* (pp. 272-289), New York: Holt-Rinehart and Winston.

Garassino, D., Masia, V. & Brocca, N. (2019). Tweet as you speak. The role of implicit strategies and pragmatic functions in political communications: data from a diamesic comparison. *Rassegna Italiana di Linguistica Applicata, 2-3*, 187-208.

Garassino, D., Brocca, N. & Masia, V. (2022). Is implicit communication quantifiable? A corpus-based analysis of British and Italian political tweets". *Journal of Pragmatics, 194*, 9-22.

Grice, H. P. (1975) [1989]. Logic and conversation. In P. Peter & J. Morgan (Eds.), *Syntax and Semantics*, Vol. 9 (pp. 113 – 127). New York: Academic Press.

Hintz, Daniel J. & Hintz, Diane M. (2017). The Evidential Category of Mutual Knowledge in Quechua. *Lingua (186-187)*: 88-109.

Kamio, A. (1997). *Territory of Information*. Amsterdam/Philadelphia: John Benjamins.

Kiparsky, C. & Kiparsky, P. (1971). Fact. In D. D. Steinberg & L. A. Jakobovitz (Eds.) *Semantics: An Interdisciplinary Reader* (pp. 345–369). Cambridge: Cambridge University Press.

Lee, J., & Xu, W. (2018). The More Attacks, the More Retweets: Trump's and Clinton's Agenda Setting on Twitter. *Public Relations Review, 44*, 201-213.

Lombardi Vallauri, E. (2009). *La struttura informativa. Forma e funzione negli enunciati linguistici*. Roma: Carocci.

Lombardi Vallauri, E. & Masia, V. (2014). Implicitness impact: measuring texts. *Journal of Pragmatics, 61*, 161-184.

Masia, V. (2020). Presupposition, assertion and the encoding of evidentiality in political discourse. *Linguistik Online, 102(2)*, 129-153.

Masia, V. (2021). *The Manipulative Disguise of Truth. Tricks and threats of implicit communication*. Amsterdam/Philadelphia: John Benjamins.

Noveck, I. (2018). *Experimental Pragmatics. The Making of a Cognitive Science*. Cambridge: Cambridge University Press.

Oswald, S., Maillat, D. & Saussure, L. de. (2016). Deceptive and Uncooperative Communication. In L. de Saussure & A. Rocci (Eds.),*Verbal Communication* (*Handbook in Communicative Science* 3) (pp. 509-534). Berlin: Walter de Gruyter.

Pietrandrea, P. (2021). *Comunicazione, dibattito pubblico, social media. Come orientarsi con la linguistica*. Roma: Carocci.

Sbisà, M. (2007). *Detto non detto. Le forme della comunicazione implicita*. Roma-Bari: Laterza.

Searle, J. R. (1969). *Speech Acts. An Essay on the Philosophy of Language*. Cambridge: Cambridge University Press.

Stalnaker, R. C. (1973). Presuppositions. *Journal of Philosophical Logic, 2*, 447-457.

Stalnaker, R. C. (1974). Pragmatic Presuppositions. In M. K. Milton & P. Unger (Eds.), *Semantics and Philosophy* (pp. 471-482). New York: University Press.

Van Dijk, T. A. (1997). What is Political Discourse Analysis? *Belgian Journal of Linguistics, 11(1)*, 11-52.

Van Dijk, T. A. (2006). Discourse and Manipulation. *Discourse & Society 17(2)*, 359-383.

Willett, T. (1988). A cross-linguistic survey of grammaticalization of evidentiality. *Studies in Language, 12(1)*, 57-91.

"WE HAVE WEATHERED THE OMICRON STORM…": A CASE STUDY ON THE ARGUMENTATIVE USE OF METAPHOR IN IRELAND'S POLITICAL DISCOURSE ON COVID-19

DAVIDE MAZZI
University of Modena and Reggio Emilia (Italy)
davide.mazzi@unimore.it

Abstract

This paper is aimed at zooming in on and reconstructing the metaphors that took centre stage in the Republic of Ireland's political discourse on Covid-19. The study is based on a small corpus of addresses to the nation by Leo Varadkar and Micheál Martin, the two *Taoisigh* in office over the past two years. The qualitative, data-driven analysis of the speeches provides evidence of metaphorical arguments, where metaphors such as 'war' and 'journey' can be observed to be part of propositions of policy or propositions of value that make a substantial contribution to underpinning the argumentation.

1. Introduction

In the context of the Covid-19 pandemic, research has extensively focused on the use of discourse to construct and justify the legitimacy of governments' response to the pandemic as a basis for the authority behind public-health measures (Amossy, 2022; Mazzi, 2022; Siess and Amossy, 2022; Wahnich, 2022). From such a perspective, metaphors have been investigated within public discourse as framing tools that carry more conceptual content than literal descriptions and are therefore more likely to influence citizens' attitudes towards public policies (Burgers et al., 2016; Brugman et al., 2019; Brugman et al., 2022). In particular, strong emphasis was laid on Covid-related metaphors across countries (Neshkowska and Trajkova, 2020; Negro Alousque, 2021; Sadoun-Kerber and Wahnich, 2022), with a view to evaluating their overall acceptability (Garzone, 2021).

At a more general level, metaphors have often been interpreted as a subtype of arguments establishing the structure of reality, with metaphor itself conceived of as a condensed analogy arising from the fusion of an element of the *phoros* with an element of the theme (Perelman and Olbrechts-Tyteca, 1969). Furthermore, other approaches exploring the relationship between metaphor and argument along the same lines have been premised on a basic idea. This is that metaphor should be viewed in terms of argument schemes gravitating towards analogy, where the concept of similarity plays a pivotal role (Plantin, 2011; Oswald and Rihs, 2014; Santibañez, 2010; Svačinova, 2014).

In recent years, attempts have been made to bridge the gap between pragma-dialectics and metaphor theory, among other things in order to shed light on the capability of metaphors of activating specific material starting points, i.e. those beliefs and values that are shared by a specific audience (Van Poppel, 2021). Furthermore, rather than conceptualising metaphor as a presentational device only, Jean Wagemans (2016b, p. 79) set out to identify the "building blocks" of a reliable "method for analyzing metaphor in argumentative discourse".

Drawing on present-day debate theory, Wagemans (2016a) thus begins by focusing on metaphors as (part of) a standpoint through combinations of propositions of policy (P), propositions of value (V) and propositions of fact (F). With a view to coming up with a periodic table of arguments, secondly, he discusses metaphors as (part of) an argument, thereby distinguishing between predicate and subject arguments along with the related linguistic structure. Thirdly, he differentiates first- from second-order arguments, which he also sees as instrumental in accounting for the function of metaphors in argumentative discourse (Wagemans, 2016a).

Against this backdrop, the aim of this study is to zoom in on and reconstruct the metaphors that took centre stage in the Republic of Ireland's political discourse on Covid-19 as a case study, along with the related argumentative use. For this purpose, the rest of the paper is organised as follows. In Section 2, corpus design criteria are discussed, and the methodological tools are introduced. Section 3 then presents the findings of the study, which are eventually discussed in the light of the relevant literature in Section 4.

2. Materials and methods

The study was based on the *Cov_Éir Corpus*, which includes 40 addresses to the nation about the health emergency. The speeches were delivered between March 2020 and January 2022 by Leo Varadkar, TD and Micheál

Martin, TD, namely the two *Taoisigh*[1] that coordinated Ireland's public-health response to the pandemic. The addresses to the nation in the corpus were all downloaded from the official website of the Department of the *Taoiseach*,[2] from which they were extracted on the basis of a single criterion: they all broached the inter-related topics of Covid, public-health guidelines and restrictions.

From a methodological viewpoint, a qualitative and data-driven approach was adopted (Tognini Bonelli, 2001, pp. 14-18). As far as this work is concerned, the focus was on the most prominent metaphors identified across corpus texts. It is important to note at the outset that the investigation was descriptive, rather than prescriptive: as such, it was not designed to assess how apt each metaphor could be considered to be. Unlike in Wagemans, moreover, the emphasis was not on single texts or isolated fragments. In fact, the purpose of the investigation was to retrieve salient metaphors as extended metaphors observed to be conceptually reiterated through a wide range of metaphorical occurrences on a wider corpus basis, rather than within single speeches only.

By drawing on Wagemans' approach (2016a and 2016b) to the argumentative use of metaphor, this was instrumental in undertaking a summary reconstruction of the argumentation accredited by metaphors within the relevant contexts. Although different reconstructions may have been possible, the analysis conducted here was intended to capture both the role of metaphor in legitimising argumentation, as is the case with arguments from analogy, and its embeddedness within argumentative discourse. Taken together, the data also allowed to establish any pattern of (dis)continuity between the two *Taoisigh* with respect to their discursive handling of the public-health emergency.

3. Findings: Ireland on the frontline and beyond

As regards the findings of the study, the manual in-depth analysis of the addresses in *Cov_Éir* revealed the presence of three salient metaphors across corpus texts. In chronological order, the first of these (fear as a virus) was identified as specific to Varadkar's speeches; the second (handling the emergency as war) was observed to be shared by the two *Taoisigh* throughout the country's public-health response to Covid; and the third one (going through the pandemic as a journey) was identified as distinctive of Martin's discourse. In this section, we will take a closer look at these metaphors, in order to elucidate their respective role in the argumentation in which they were embedded.

---

[1] This is the plural of Taoiseach, which literally translates as "leader, chief, ruler" (Ó Dónaill, 1977, p. 1203) and is used to refer to the Irish Prime Minister.
[2] https://www.gov.ie/en/organisation/department-of-the-taoiseach/ (last accessed 14 October 2022).

To begin with, the metaphor that was found to be associated with Varadkar's addresses is fear as a virus. As can be seen from example (1) below, the *Taoiseach* was emphatic in drawing the public's attention to fear as the most intense feeling people were expected to share as a result of the grim news from abroad, i.e. China and Italy. The excerpt in (1) is from a well-known speech delivered by Varadkar on St. Patrick's Day (17 March) 2020, when he announced that a national lockdown was effectively going to be imposed on the country:

Example 1. *Tonight I know many of you are feeling scared and overwhelmed. That is a normal reaction, but we will get through this and we will prevail. We need to halt the spread of the virus but we also need to halt the spread of fear.*

So please rely only on information from trusted sources. From Government, from the HSE, from the World Health Organisation and from the national media.

Do not forward or share messages that are from other, unreliable sources. So much harm has already been caused by those messages, and we must insulate our communities and the most vulnerable from the contagion of fear. [...]

Reflecting a steady trend from the early section of the corpus, which covers the first few weeks of the health emergency, not only does Varadkar fully assess the strength of fear (cf. *you are feeling scared and overwhelmed*): he also points out how transmissible, as it were, fear can be said to be. He therefore urges people to rely on information from trusted sources only, whether it be the Government themselves, the Health Service Executive (*HSE*) in charge of the country's health services, the WHO or trustworthy national media. In his view, this provides a solid basis *to halt the spread of fear*, which he considers to be as urgent as halting *the spread of the virus*. From an argumentative perspective, the metaphor of fear as a virus therefore pinpoints Varadkar's argumentative discourse, the substance of which can be reconstructed as follows:

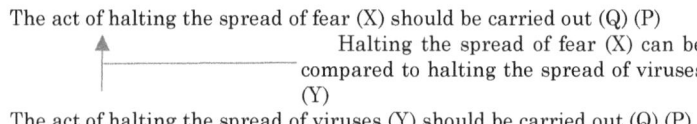

Figure 1. The metaphor of fear as a virus in Varadkar's argumentation

As can be appreciated from the reconstruction in Fig. 1, what is crucial here is a combination of propositions of policy (P), which by definition predicate of a specific act, course of action or policy that they should be carried out (Wagemans, 2016b). At the same time, the shared element is represented by the predicate (Q) – i.e., *should be carried out*. Furthermore, what Fig. 1 shows is an instance of first-order subject argument, where the transfer of acceptability is facilitated by the alleged existence of a specific relation between the subject of the argument (Y) and the subject of the standpoint (X) (Wagemans, 2016b, p. 86). Hence the comparability between halting the spread of the fear and halting the spread of viruses, the necessity of which lies at the heart of the *Taoiseach*'s standpoint.

Moving on to the metaphor shared by Varadkar and Martin, secondly, this can be accounted for as the country's public-health response as war. This comes as no surprise, many a study from the recent literature isolating the war metaphor in the discourse of heads of state or government, particularly at the early stages of the pandemic (Sadoun-Kerber and Wahnich, 2022 on Emmanuel Macron; Wodak, 2022 on a comparison between the French President and Hungary's Prime Minister Viktor Orbán). In Ireland as well as elsewhere, this metaphor had three major implications. First and foremost, it entailed depicting Covid as an enemy (cf. *is cruel and inhumane* in example 2 below; *attacks by stealth* in 3; *a common foe* and *a shared enemy of all of humanity* in 4 and 5, respectively):[3]

Example 2. *The coronavirus is cruel and inhumane. (LV)*

Example 3. *We have more information now than we did at the start of this pandemic. We now know how COVID-19 attacks by stealth and we know how fast it moves. (MM)*

Example 4. *In contrast to what we have seen in some other countries, our political parties have united against a common foe and I want to thank them for their understanding, goodwill and co-operation to date. (LV)*

Example 5. *As we know, the coronavirus is a shared enemy of all of humanity and all governments. I believe the only way we can defeat a global threat is by working together on a multilateral basis. (LV)*

In addition, as is apparent from examples (4) and (5) above as well, the metaphor consistently led the *Taoisigh* to invite the Irish people and indeed their elected representatives to put on a united front against the virus – see *our political parties have united* in (4) and *by working together* in (5). By going through corpus texts in detail, there is ample evidence of an almost systematic correlation between the *Taoiseach* of the day reverting to the metaphor, and the emergence of new Covid variants along with sharp increases in the number of cases, hospitalised patients and eventually Covid-related deaths. Examples (6) and (7) clearly show that, during the acute phases of the pandemic, Ireland's public discourse was geared to meet the need for decisive collective action to minimise the spread of the virus and its circulation at a community level (cf. *working*

---

[3] Wherever appropriate and not otherwise clear from the paper, the source of the numbered examples is revealed in brackets. Thus, (LV) is short for Leo Varadkar, while (MM) stands for Micheál Martin.

together and making the sacrifices...asked of each one of us in 6 and *acting together and holding firm in a spirit of solidarity* in 7):

Example 6. *It is only we as a people, working together and making the sacrifices that are being asked of each one of us that we can slow the new wave of the virus. (MM)*

Example 7. *We will be having those conversations at a national level, and we need these conversations to be taking place in every single home in our nation. Because the Irish Government on its own can't stop the virus spreading. It is only we as a people, acting together and holding firm in a spirit of solidarity that can slow its destructive spread. (MM)*

The third implication of the war metaphor is that it invariably resulted in the *Taoiseach* representing health workers as those *on the frontline*, and the provision of health as well as other essential services as an ongoing *fight* against the virus. This is well illustrated by the passage in (8), where Varadkar stresses that the best way of helping those at work to the general population's benefit is by steadying our nerve and keeping up our guard (*staying the course, and continuing this fight*):

Example 8. *We all know someone who is suffering because of these restrictions, just as we all know someone who is on the frontline or performing an essential service. The best way of helping them is by staying the course, and continuing this fight. (LV)*

In the *Cov_Éir Corpus*, the war metaphor underpins the aggregate standpoint, as it were, that Covid should be fiercely fought. As with all enemies, after all, inaction or ineffective action would lead to the country being overwhelmed by the common foe. Hence the reconstruction of the argumentation in Fig. 2. As we saw from Fig. 1, here too it is on the basis of a comparison that the property *to be carried out* ascribed to the act of fighting a shared and cruel enemy can also be ascribed to the act of fighting Covid:

The act of fighting Covid (X) should be carried out (Q) (P)
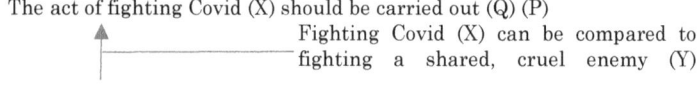
Fighting Covid (X) can be compared to fighting a shared, cruel enemy (Y)

The act of fighting a shared, cruel enemy (Y) should be carried out (Q) (P)

Figure 2. The war metaphor in the *Taoisigh*'s argumentation

As for the metaphor identified as specific to Martin's discourse, in third place, this is the journey metaphor, which was also detected by the literature to be widely spread across the Western world (Neshkowska and Trajkova, 2020). In the corpus under investigation here, the metaphor was closely tied to two main contexts of use. The first is instantiated by the periods where the various Covid storms gradually subsided, with the country ready to return to some form of normality, so to speak. In example (9) below, accordingly, Martin acknowledged that Ireland was not yet out of the woods yet the data provided by the National Public Health Emergency Team (NPHET) enabled it to plan the *route forward*. In (10) and (11), likewise, the *Taoiseach* pointed out that the vaccination rollout would allow the country to gradually leave behind the worst of the emergency and look forward with confidence to all major public-health restrictions being lifted (cf. *will lay* the path *out of this pandemic and put us* on the road *to a safe reopening of our society and our economy* in 10, and *chart* a path *out of restrictions* in 11, [my emphasis]):

Example 9. *We are not at a stage where we can return to normal, but we are at a stage where we can plan our route forward.*

Example 10. *[...] the vaccination programme will lay the path out of this pandemic and put us on the road to a safe reopening of our society and our economy.*

Example 11. *This evening, I want to talk to you about where we are in the management of COVID-19 and how we safely and successfully chart a path out of restrictions in the weeks and months ahead.*

Ireland's response to the pandemic is indeed one Martin often described as a tortuous journey, as highlighted by his recurrent use of the noun 'turns' or the whole phrase 'twists and turns' (examples 12-14):

Example 12. *The road ahead will continue to have many turns. It will challenge us in new ways.*

Example 13. *We know there have been many twists and turns on this journey.*

Example 14. *I know at first hand how difficult the COVID-19 journey has been for the entire higher education sector and I just want to pay tribute this morning to all of you who worked so hard to keep services running through all the twists and turns of this journey.*

The second context in which the journey metaphor frequently appears is the last month covered by the corpus data, i.e. January 2022, when the public-health emergency as such was essentially declared to be over, and even less restrictive regulations such as mandatory mask wearing in indoor settings and public transport began to be relaxed. Nowhere is this more aptly illustrated than in the address to the nation of 21 January 2022. This is one of Martin's best known and most eloquent speeches, where he drew an analogy between Ireland's road to Independence and the country's road through the pandemic. As a result of the stunning success of Irish nationalists in the UK general election of December 1918, the first to be called in the wake of WWI, a Republican parliament (*First Dáil* in example 15) was symbolically convened in Dublin in January 1919 to declare Ireland's independence (Ferriter, 2005). Although Independence proper was only going to come three years later (January 1922), this was a moment of fundamental unity on the nationalist front: as such, it constituted an outstanding achievement that was to galvanise the revolutionary movement and lead to serious engagement in the War of Independence of 1919-1921. On these grounds, a first glimpse of the analogy can be caught from the excerpt in (15):

Example 15. *On this day in 1919, a short distance from here, the representatives of the First Dáil of the Irish Republic met and formally declared our country's independence. [...] The journey to that point was not easy, and our journey as a nation after it was often very difficult. Our journey through the pandemic has brought many twists and turns, and I have stood here and spoken to you on some very dark days.*

The parallel between the *journey to that point* (January 1919) and the *journey as a nation after it* on one side, and the *journey through the pandemic* on the other, may seem somewhat far-fetched. Nonetheless, it bears scrutiny for two reasons. In both scenarios, to begin with, Ireland's path was long, steep and fraught with difficulties: the country regained control over its own destiny after more than seven hundred years of English presence on its soil, while the fight against Covid was not won without devastating losses and moments of untold heartbreak. In both scenarios, moreover, the aim having been accomplished (Independence or quasi-Independence, to be more precise, on one side; reaching a post-emergency stage on the other), the path ahead was not necessarily a straight one. Thus, only a few months into its journey as an independent State would Ireland experience the terrible grief and bitterness of Civil War (1922-1923). Likewise, the Irish people believed on more than one occasion that the worst of the pandemic was over, only to find out that new Covid variants would emerge and more difficulties would have to be surmounted, often at the cost of harsh restrictions including extensive lockdown periods between the autumn of 2020 and the spring of 2021.

Martin's analogy extends into a subsequent fragment of the speech, where he celebrates the milestone of Ireland sitting as an equal among all nations. In particular, the *Taoiseach* argues that the Irish have come *a long way* since January 1919 by being themselves. Similarly, he exhorts people to be themselves again as a precondition for the country to *navigate this new phase of Covid* (example 16):

Example 16. *For all our faults as a country, we have come a long way since this day in 1919. Ireland is now firmly established as an equal among all the nations, and we've been a positive force in the world through our arts, our culture, our peace keeping and our commerce. We've done this by having the confidence to be ourselves. As we face into our second century as a free democracy, and as we navigate this new phase of COVID, it is time to be ourselves again.*

Taken together, passages such as those in (15) and (16) earlier on seem to suggest that journey acts as a powerful metaphor bolstering Martin's argumentation from analogy. Based on examples (9)-(16), more generally, it is possible to reconstruct the *Taoiseach*'s argumentation as shown in Fig. 3:

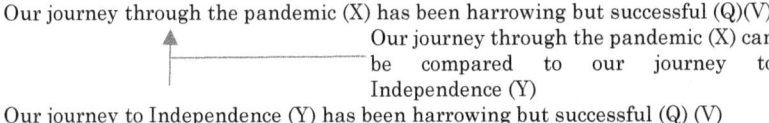

Figure 3. The journey metaphor in Martin's argumentation

Once more, the metaphor is not (part of) the standpoint. Rather, it is deeply rooted in the argument and ultimately supports the standpoint that the country's journey through the pandemic, however harrowing, has been enormously successful. Furthermore, the shared element is again represented by the predicate (Q) but this time, unlike in the reconstructions in Fig. 1 and Fig. 2, within a combination of propositions of value (V). By definition, these "predicate of an entity (thing, person, event or act) that it has a specific value or that it can be judged or evaluated in a specific way" (Wagemans, 2016b, p. 82). In the concrete case the argumentation, corroborated as it is by the journey metaphor, is that Ireland's voyage has been as arduous and successful during Covid, as it was while on its course to Independence.

4.   Discussion and conclusions

This was a case study on the argumentative use of metaphor in the Republic of Ireland's public discourse on Covid-19. The research was less aimed at an extensive coverage of the Irish context, which would have amounted to designing comparable corpora of full parliamentary debates and/or news texts, than at a sample survey of public discourse from above, i.e. at a Government level. By way of a qualitative investigation of addresses to the nation during the health-emergency period proper, first of all, the analysis provided evidence of a circumscribed set of metaphors as argumentative devices enabling the transfer of what was ascribed to the entity belonging to the source domain (e.g., the property *to be carried out* in relation to the act of halting viruses in Fig. 1) to the entity belonging to the target domain (e.g., the act of halting the spread of fear).

In the context of the *Cov_Éir Corpus*, more specifically, analogy was observed to play a key role as "the metaphorical argument supports the justificatory force of an argument characterised as a subject argument" (Wagemans, 2016b, p. 91). This is actually identified by Wagemans as the only situation where the concept of analogy – or, broadly speaking, similarity and comparison – is relevant to appreciating the contribution of metaphors to argumentation. Consistent with that, this is what we saw across the reconstructions of the metaphorical arguments in Section 3.

Secondly, the findings presented earlier on appear to demonstrate the high relevance of first-order metaphorical arguments resting on combinations of propositions of policy and propositions of value. In the context of Ireland's public-health response to the pandemic covered by the corpus, in more detail, the data shed light on a discursive management of Covid opening with the handling of fear, so to speak, focusing on war as a common thread and eventually, as a kind of exit strategy from the emergency, capitalising on the rhetorical strength of journey.

On the whole, the findings reported here uphold the validity of Semino's (2021, p. 52) view that "the more complex and long-term a phenomenon, the more we need different metaphors to capture different facets and phases, and to communicate with different audiences". Accordingly, the data from *Cov_Éir* confirm that, with the sobering realisation that the country would be in it for the long haul, the Irish Government consciously modelled its discourse on different metaphors, the rhetorical effectiveness of which was assessed and shown in Section 3.

In particular, our findings not only indicate that the war metaphor, meticulously examined in the extant literature, was reiterated across speeches. The journey metaphor was also detected and revealed to be firmly rooted in argumentation from analogy where Irish history was drawn on as a highly valuable resource. Micheál Martin was not unique in this choice. Siess and Amossy (2022) note that former German Chancellor Angela Merkel's analogy argumentation also included historical examples. Hence, Merkel's narrative harking back to both the days of national reconstruction in the wake of the Second World War, and the reunification of Germany, was instrumental in drawing people's attention to the gravity of the Covid crisis and urging people to redouble their efforts so that the same kind of collective behaviour would end in the same phenomenal success.

In Ireland as well as in Germany, the analogy between the Covid emergency and watershed moments from the country's history appears to be part of the rhetorical devices tapping into a particular audience's shared cultural heritage, whereby allusion "to an important historical event" – in our case, Ireland's Independence – "creates emotions aroused by memories and pride" (Long, 1983, p. 114). It is all the more exciting a development that such a compelling line of argument may ultimately have been uncovered through a study of the role of metaphor in argumentation such as that undertaken here.

5. References

Amossy, R. (2022). Construire le légitimité et l'autorité politiques en discours. *Argumentation et Analyse du Discours, 28*. https://doi.org/10.4000/aad.5984 (last accessed 13 July 2022).

Brugman, B. C., Burgers, C., & Vis, B. (2019). Metaphorical framing in political discourse. *Language and Cognition, 11*, 41–65.

Brugman, B. C., Droog, E., Reijnierse, W. G., Leymann, S., Frezza, G., & Renardel de Lavalette, K. Y. (2022). Audience perceptions of COVID-19 metaphors: The role of source domain and country context. *Metaphor and Symbol, 37(2)*, 101–113.

Burgers, C., Konijn, E., & Steen, G. J. (2016). Figurative framing: Shaping public discourse through metaphor, hyperbole, and irony. *Communication Theory, 26*, 410–430.

Ferriter, D. (2005). *The transformation of Ireland 1900-2000*. London: Profile Books.

Garzone, G. (2021). Re-thinking metaphors in Covid-19 communication. *Lingue e Linguaggi, 44*, 159–181.

Long, R. (1983). The role of audience in Chaim Perelman's New Rhetoric. *Journal of Advanced Composition, 4*, 107–117.

Mazzi, D. (2022). "…not a substitute for facts": The Irish public discourse on Covid-19 at the intersection of legislation, fake news and judicial argumentation. *International Journal for the Semiotics of Law, 35*, 1233–1252.

Negro Alousque, I. (2021). Les metaphors du virus COVID-19 dans les discours d'Emmanuel Macron et de Pedro Sanchez. *Cédille, 19*, 595–613.

Neshkowska, S., & Trajkova, Z. (2020). Coronavirus-inspired metaphors in political discourse. *THESIS, 9(2)*, 99–132.

Ó Dónaill, N. (1977). Foclóir Gaeilge-Béarla. Dublin: An Gúm.

Oswald, S., & Rihs, A. (2014). Metaphor as argument: Rhetorical and epistemic advantages of extended metaphors. *Argumentation, 28(2)*, 133–159.

Perelman, C., & Olbrechts-Tyteca, L. (1969). *The new rhetoric. A treatise on argumentation*. Notre Dame, IND: University of Notre Dame Press.

Plantin, C. (2011). Analogies et métaphores argumentatives. *A contrario, 16(2)*, 110-130.

Sadoun-Kerber, K., & Wahnich, S. (2022). Emmanuel Macron face à la Covid-19: Un Président en quête de réparation d'image. *Argumentation et Analyse du Discours, 28*. https://doi.org/10.4000/aad.6113 (last accessed 13 July 2022).

Santibáñez, C. (2010). Metaphors and argumentation: The case of the Chilean parliamentarian media participation. *Journal of Pragmatics, 42(4)*, 973–989.

Semino, E. (2021). "Not soldiers but fire-fighters" – Metaphors and Covid-19. *Health Communication, 36(1)*, 50–58.

Siess, J., & Amossy, R. (2022). Légitimité démocratique et autorité en temps de Corona: L'allocution à la nation d'Angela Merkel (18 mars 2020). *Argumentation et Analyse du Discours, 28*. https://doi.org/10.4000/aad.6053 (last accessed 13 July 2022).

Svačinova, I. (2014). Reconstruction of metaphors in argumentation: A case study of Lincoln's metaphor of "swapping horses while crossing a stream". *Cogency, 6(2)*, 67–89.

Tognini Bonelli, E. (2001). *Corpus linguistics at work*. Amsterdam: Benjamins.

Van Poppel, L. (2021). The study of metaphor in argumentation theory. *Argumentation, 35*, 177–208.

Wagemans, J. H. M. (2016a). Constructing a periodic table of arguments. *OSSA Conference Archive, 106*. http://scholar.uwindsor.ca/ossaarchive/OSSA11/papersandcommentaries/106 (last accessed 13 July 2022).

Wagemans, J. H. M. (2016b). Analyzing metaphor in argumentative discourse. *Rivista Italiana di Filosofia del Linguaggio, 2*, 79–94.

Wahnich, S. (2022). Introduction. La construction de la légitimité et de l'autorité: Les démocracies face à la Covid-19. *Argumentation et Analyse du Discours, 28*. https://doi.org/10.4000/aad.6390 (last accessed 13 July 2022).

Wodak, R. (2022). Légitimer la géstion de crise pendant la Covid-19. *Argumentation et Analyse du Discours, 28*. https://doi.org/10.4000/aad.5999 (last accessed 13 July 2022).

FROM DEEP DISAGREEMENT TO RATIONALLY IRRESOLVABLE DISAGREEMENT

GUIDO MELCHIOR
University of Graz
guido.melchior@uni-graz.at

Abstract
In this paper, I will critically review the existing discussion on deep disagreement. I will first reconstruct Fogelin's presentation of deep disagreement and reveal a tension between his systematic analysis and his examples. Second, I will present the two main schools concerning the nature of deep disagreement, hinge theories and fundamental epistemic principles theories, and discuss some major problems for these views. Third, I reflect on different views about the rational resolvability of deep disagreement and the ambiguities of this discussion. Finally, I sketch an alternative methodological proposal of first providing a theory of rationally irresolvable disagreement and, based on this theory, clarifying issues concerning deep disagreement.

1. Fogelin on deep disagreement

The notion of deep disagreement, hereinafter DD, traces back to Fogelin (1985/2005), who introduced the phenomenon in his pioneering work entitled *"The Logic of Deep Disagreements"*.[1] In this section, I reconstruct Fogelin's line of argumentation in detail. Fogelin (1985, 5) stresses the importance of contexts for argumentation and argues that arguments are possible because "the parties to the conversation share a great many beliefs and (if this is different) a great many preferences." According to Fogelin, shared assumptions are a fundamental feature of ordinary, normal argumentation. In his words:

> I shall say that an argument, or better, an argumentative exchange is normal when it takes place within a context of broadly shared beliefs and preferences. I shall further insist that for an argumentative exchange to be normal, there must exist shared

---

[1] Page numbers refer to reprinted version in Informal Logic 25(1), (2005), 3-11.

> procedures for resolving disagreements. People often disagree over simple questions of fact, but, in general, they agree on the method for resolving their disagreement. (Fogelin 1985, 6)

By referring to Wittgenstein's (1969) *On Certainty*, Fogelin (1985, 6) notes that "arguing, i.e., engaging in an argumentative exchange, presupposes a background of shared commitments" and depends "on the fact that together *we* accept many things."[2] Based on the claim that normal argumentation is characterized by a class of assumptions (and preferences) that both parties share, Fogelin (1985, 7) then raises the central question about what "happens to arguments when the context is neither normal nor nearly normal?" In a first attempt, Fogelin answers this question as follows:

> [T]o the extent that the argumentative context becomes less normal, argument, to that extent, become impossible. This is not the weak claim that in such contexts arguments cannot be settled. It is the stronger claim that the conditions for argument do not exist. The language of argument may persist, but it becomes pointless since it makes an appeal to something that does not exist: a shared background of beliefs and preferences. Here I wish to speak about *deep disagreements*. My thesis, or rather Wittgenstein's thesis, is that deep disagreements cannot be resolved through the use of argument, for they undercut the conditions essential to arguing. Fogelin (1985, 7f)

Fogelin characterizes DD by distinguishing it from other forms of disagreement which are irresolvable because of obvious vices of one of the involved parties, such as stubbornness. DD, in contrast, occurs due to differences in underlying principles:

> What is a deep disagreement? First let me say what I don't mean by this notion. A disagreement can be intense without being deep. A disagreement can also be unresolvable without being deep. I can argue myself blue in the face trying to convince you of something without succeeding. The explanation might be that one of us is dense or pig-headed. And this is a matter that could be established beyond doubt to, say, an impartial spectator. But we get a very different sort of disagreement when it proceeds from a clash in underlying principles. Under these circumstances, the parties may be unbiased, free of prejudice, consistent, coherent, precise and rigorous, yet still disagree.

---

[2] Against Fogelin, Phillips (2008) argues that the background required for successful argumentation is not constituted by shared beliefs and commitments but by shared procedural commitments and competencies.

And disagree profoundly, not just marginally. Now when I speak about underlying principles, I am thinking about what others (Putnam) have called framework propositions or what Wittgenstein was inclined to call rules. We get a deep disagreement when the argument is generated by a clash of framework propositions. (Fogelin 1985, 8)

At this point, Fogelin characterizes DD as disagreement about framework propositions. However, he also argues that DD is holistic in that it does not concern single propositions in isolation.

When we inquire into the source of a deep disagreement, we do not simply find isolated propositions [...] but instead a whole system of mutually supporting propositions (and paradigms, models, styles of acting and thinking) that constitute, if I may use the phrase, a form of life. (Fogelin (1985, 9)

After describing DD *in abstracto*, Fogelin presents two examples to illustrate his point. The role of examples will be crucial in further reflection, so let me present his examples in more detail. The first examples are discussions about the morality of abortion. Fogelin notes that these discussions can proceed in the way of normal argumentation, but Fogelin (1985, 8) also notes that discussions on this topic can be deep in the sense that "they persist even when normal criticisms have been answered" and "are immune to appeals to facts." In Fogelin's own words:

Parties on opposite sides of the abortion debate can agree on a wide range of biological facts-when the heartbeat begins in the fetus, when brain waves first appear, when viability occurs, etc., yet continue to disagree on the moral issue. Their disagreement can even survive a general agreement on moral issues: for example, on the sanctity of human life, for the central issue of the abortion debate is the moral status of the fetus and that cannot be settled by an appeal to biological facts or by citing moral principles already limited to moral agents or patients. (Fogelin 1985, 8)

Fogelin argues that this is so because opponents of abortion believe that at some very early point some kind of immortal soul enters into the fertilized egg, a belief that is part of a large set of mutually supporting religious beliefs.

The second example for DD that Fogelin presents concerns affirmative action. Fogelin suggests that opponents and adherents of affirmative action believe in very distinct concepts of fairness, leading to DD about affirmative action that cannot be resolved. In Fogelin's own words:

I want to say that we are here dealing with a deep disagreement because the parties on both sides might agree on all historical and statistical matters, but still disagree. The dispute is, in fact, one concerning moral standing. (In this way it is like disputes concerning abortion, and this may explain, at least in part, its intractability.) The anti-quota argument rests on the assumption that only individuals have moral claims. The pro-quota argument rests upon the assumption that social groups can have moral claims against other social groups. But the word 'assumption' is too weak. The arguments on each side are carried on within the framework of such commitments. Is there any way of adjudicating a clash of this kind? I confess that I do not see how. (Fogelin 1985, 10)

To sum up Fogelin's pioneering take on DD: Fogelin first points out that normal argumentation requires a basis of shared assumptions (and preferences) between the conflicting parties. Based on these shared assumptions the disagreement can either be resolved or at least rationally addressed via argumentation. In a second step, Fogelin presents DD as a kind of disagreement which lacks shared assumptions between the conflicting parties such that no tools are available for resolving disagreement. Third, he presents two examples for DD. Both examples are moral disputes, one about the morality of abortion, and the other about affirmative action. Finally, Fogelin addresses the resolvability of DD, making explicit his negative assessment when concluding the paper with the following quote:

In the end, however, we should tell the truth: there are disagreements, sometimes on important issues, which by their nature, are not subject to rational resolution. (Fogelin 1985, 11)

2. A tension between analysis and examples

There is a tension in Fogelin's paper between his theoretical approach and the examples that he provides, which becomes important for further discussions about DD. On the one hand, he characterizes the phenomenon in abstract terms, arguing that there is a kind of disagreement, which he calls DD, that is holistic, concerns fundamental propositions or even systems of propositions and that cannot be resolved in a rational way. On the other hand, he presents two examples of DD, both forms of moral (and philosophical) disputes about the morality of abortion and about the fairness of affirmative action. These examples are problematic because they plausibly lack some of those features which are, according to Fogelin, necessary for disagreements to be deep. For instance, it is hard to see why two parties who disagree about affirmative action should be unable to agree about some more fundamental principles about fairness that can lead to resolution of the disagreement. Analogously, it is hard to see why there should not be more fundamental moral principles available that can be used to resolve the disagreement about abortion. Both cases seem to be philosophical discussions about deep moral questions rather than instances of DD, and we assume that there are rational resolutions to philosophical discussions, at least in principle, even though they are fundamental and deeply rooted.

Given this tension, one can proceed in two ways. On the one hand, we can stick with Fogelin's systematic analysis. In this case, one accepts Fogelin's analysis of DD as basically correct and then seeks more appropriate examples. This path is chosen by many philosophers who share Fogelin's basic intuition and consequently aim at formulating more plausible examples. A current, widely shared example is a disagreement between a scientist and a creationist about the age of the earth.[3] The scientist believes in scientific evidence whereas the creationist believes that the bible overrides any potential conflicting information. The two parties fundamentally disagree about the reliability of sources, resulting in DD. On the other hand, one can accept that the examples proposed by Fogelin are genuine cases of DD. In this case, one can look for further similar examples, but one has to provide an alternative analysis of these cases than the one proposed by Fogelin.

Thus, Fogelin's paper contains a tension between the systematic analysis provided and concrete examples given. One can resolve this tension either by going with the analysis or by adopting the examples. However, these options do not necessarily result from Fogelin's incoherent account. More generally, we can approach the topic of DD by providing a specific analysis and then searching for examples or by picking out concrete instances of disagreement, which are in some intuitive sense deep, and then investigate their features.

3. Two questions about deep disagreement

Fogelin's pioneering paper has all the ingredients for kicking of a lively discussion. It addresses a phenomenon that we can intuitively access and that has been rather neglected in the literature. Furthermore, it is sufficiently ambiguous to allow various interpretations and analyses of DD. For example, different interpretations of the way DD "undercuts the conditions essential to arguing" are possible. Moreover, Fogelin remains rather unclear about the exact subject of DD. At one place, he talks about underlying principles, Putnamian framework propositions, or Wittgensteinian rules, while in other places about "mutually supporting

---

3 See Lynch (2010), Matheson (2021), and Ranalli (2021).

propositions (and paradigms, models, styles of acting and thinking) that constitute a form of life." Finally, Fogelin clearly denies that DD is rationally resolvable, but this view has not remained uncontested. Accordingly, the current discussion about DD focuses on two central questions:

> Q1: What is deep disagreement?
> Q2: Can deep disagreement be resolved or not, and if so, how?

The first question addresses the metaphysical nature of DD, and the second question addresses one of its central (alleged) features, namely whether it is resolvable or not.

Let us first have a look at Fogelin's own take on these two questions. We can find his answer to Q1 in his characterization of DD. According to Fogelin, DD results from "a clash in underlying principles" which are Putnamian framework propositions or Wittgensteinian rules. Moreover, DD is holistic since it does not concern isolated propositions but systems of mutually supporting propositions. Here we face, again, a certain tension between the claims that DD concerns very fundamental or general propositions and that it concerns whole systems of propositions. A plausible and charitable interpretation of Fogelin is that DD is systematic because disagreement about framework propositions or rules leads to disagreement about systems of propositions. Nevertheless, it is worth noting that Fogelin's characterization of DD is somewhat metaphorical, as he talks about "propositions (paradigms, models, styles of acting and thinking) that constitute form of life." Concerning Q2, Fogelin is a pessimist. As Fogelin (1985, 9) notes: "But if deep disagreements can arise, what rational procedures can be used for their resolution? The drift of this discussion leads to the answer NONE."

3.1. Theories about the nature of deep disagreement

Let us now come to the current debate about Q1 and Q2. Concerning Q1, about the metaphysical nature of DD, a lively philosophical discussion has emerged in recent years. In this debate, we can identify two major schools of thought, the first one defending a Wittgensteinian picture of DD, as proposed by Fogelin (1985), Kusch (2021) and Pritchard (2021), discussed by Ranalli (2020) and criticized by Siegel (2021), and the second one arguing that DD concerns fundamental epistemic principles. Fogelin himself referred to Wittgenstein (1969), extensively citing him, when explaining his understanding of DD. This Wittgensteinian take has been further developed by various authors in terms of hinge propositions or hinge commitments which are not explicitly mentioned by Fogelin. These notions are based on Wittgenstein's (1969) approach that there is a type of fundamental propositions, which he compares to hinges, that we cannot rationally access and which cannot be justifiedly believed.

Hinge epistemology is a growing branch of contemporary epistemology and different views on hinge epistemology are currently defended. First, we can distinguish between epistemic and non-epistemic hinge theories. Epistemic hinge theories hold that hinges or hinge propositions can be somehow epistemically assessed. Wright (2004 and 2014), for example defends an epistemic hinge theory by arguing that we are entitled to accept or trust hinge propositions and, in this sense, they can be rationally assessed. Non-epistemic hinge theories, in contrast, deny the idea of rational assessability of hinge propositions. There are non-epistemic theories that claim that hinge propositions cannot be rationally assessed because they are not truth apt and cannot be true or false. This view is defended by Wright (1985) and Moyal-Sharrock (2004 and 2016). These views reject that hinge propositions are really propositions and can, accordingly be labeled non-propositional views. The alternative non-belief view, as defended by Pritchard (2016 and forthcoming) holds that hinge propositions are real propositions but they cannot be rationally assessed because they cannot be believed.

Hinge theories of DD face various challenges. The most obvious one is that one has to buy into some form of hinge epistemology in order to find hinge theories of DD persuasive, a controversial step in its own right. However, even if one finds the overall project of hinge epistemology persuasive, further obstacles remain. Defenders of hinge epistemological accounts of DD face, for example, the challenge of choosing and defending the correct account of hinge propositions that reveals which propositions are actually hinge propositions. Pritchard (2021) lists the following examples from Wittgenstein (1969): That one has hands, that one is speaking one's native language, that one's name is such-and-such, that one has never been to the moon. Further examples from Wittgenstein (1969) are that there is a brain inside one's skull, that the table is still there when no one sees it, and that the Earth is round. Brueckner (2007, 285) cites as examples of hinge propositions that there is an external world, that sense perception is reliable, that I am not a brain in a vat, that my faculty of reasoning is reliable, that the Earth is more than three minutes old, that testimony is reliable, and that memory is reliable. Wittgenstein's classifies very fundamental propositions as hinges while Brueckner's list concerns negations of skeptical hypotheses. Pritchard (2021, 1120) argues that all of our hinge commitments are manifestations of an overarching "über hinge commitment" that we are not radically and fundamentally in error in our beliefs. Thus, even if we buy into hinge epistemology, we face the challenge of deciding what types of propositions actually are hinge propositions.

Moreover, there is a more specific problem for hinge theories of DD. Fogelin, also a Wittgensteinian concerning DD, presents moral discussions about abortion or about affirmative action as examples of DD. If DD is about hinge propositions, then these propositions should be hinge propositions. However, these propositions concern specific moral principles and are very different from propositions such as that there is an external world. Furthermore, paradigmatic cases of DD such as disagreement between a scientist and a creationist about the age of the earth are not obviously disputes about hinge propositions. Thus, hinge theories of DD are in danger of being too narrow by precluding such disagreements from being deep or having to accept a rather loose understanding of hinge propositions.

Moreover, we have seen that hinge theories defend different views about whether hinge propositions are rationally assessable. Negative views differ concerning the explanation of why they are not rationally assessable and positive theories differ on how they are rationally accessible. These disputes directly affect hinge theories of DD. Ranalli (2021) points out that if DD is about hinge propositions and hinge propositions are not rationally assessable, then one has to explain why proponents nevertheless mistakenly engage in argumentation in cases of DD. One might argue that these problems for hinge theories of DD are solvable but nevertheless one has to admit that the challenges are severe.

The second branch of theories about the metaphysical nature of DD are fundamental epistemic principles theories. Proponents of this view are Lynch (2010 and 2016) and Kappel (2012 and 2021), among others. The basic idea of this approach is that two parties deeply disagree when they have a disagreement about fundamental epistemic principles. Here, the pressing question to be answered is what fundamental epistemic principles are. The paradigmatic case of a fundamental epistemic principle concerns the reliability of our sense apparatus. Thus, fundamental epistemic principles can concern the reliability of sources, but it is an open question which sources they can concern. Lynch (2010) argues that fundamental epistemic principles are principles that can only be justified via epistemically circular reasoning, which is reasoning about the reliability of a source via information delivered from that very source. Accordingly, a fundamental epistemic principle concerns the reliability of one's own sense-apparatus, which can only be justified via empirical information from one's own sense apparatus. According to Lynch (2010, 264) disagreement is deep when it meets the following conditions:

1. Commonality: The parties to the disagreement share common epistemic goal(s).

2. Competition: If the parties affirm distinct principles with regard to a given domain, those principles (a) pronounce different methods to be the most reliable in a given domain; and (b) these methods are capable of producing incompatible beliefs about that domain.

3. Non-arbitration: There is no further epistemic principle, accepted by both parties, which would settle the disagreement.

4. Mutual Circularity: The epistemic principle(s) in question can be justified only by means of an epistemically circular argument.

One problem for Lynch's view is that the meaning of non-arbitration remains unclear. Suppose Crusoe and Freitag are stranded on a lonely island and have two hygrometers with them that deliver different but coherent outcomes. Crusoe trusts H1, while Freitag trusts H2. There is no further hygrometer available to settle the question about the reliability of H1 and H2. Suppose that Crusoe and Freitag disagree about humidity and engage in an exchange. Their disagreement fulfills all four of Lynch's criteria for DD. (1) The two parties share common epistemic goals; (2) they refer to two different principles concerning the humidity and the two principles can produce conflicting beliefs; (3) there is no further principle or source available for settling the disagreement; and (4) the epistemic principle(s) in question can be justified only by means of an epistemically circular argument. Nevertheless, that a particular hygrometer is reliable is intuitively not a fundamental epistemic principle, since in this particular case the conditions for DD are fulfilled for purely contextual reasons. Thus, Lynch's definition of disagreement does not succeed in capturing only fundamental epistemic principles.

Kappel (2021, 1039), another protagonist of the epistemic principles approach, rejects the idea that fundamental epistemic principles are a single phenomenon and claims more generally that in "deep disagreements local disagreements are intertwined with more general basic disagreements about the relevant evidence, standards of argument or proper methods of inquiry in that domain." We see that defenders of the fundamental epistemic principles approach face the challenge of providing a precise specification of fundamental epistemic principles and of spelling out an account of DD that fits with this specification. This endeavor is problematic. Lagewaard (forthcoming) criticizes Lynch's approach by arguing that even paradigmatic instances of DD do not rely on fundamental but only on derived epistemic principles. Moreover, Ranalli (2021) argues that fundamental epistemic principles theories are in danger of being too narrow and of not doing justice to the plurality of DD since there are normative and non-normative instances of DD that do not concern fundamental epistemic principles.

3.2. Theories about the resolvability of deep disagreement

There is wide philosophical disagreement about how to answer the first question about the nature of DD. The second question concerning the resolvability of DD is slightly less controversial, since the majority of theorists answer it in the negative. Fogelin (2005) is very explicit that DD cannot be rationally resolved and most authors agree with him on that point. Lynch (2010, 273), uses Wittgenstein's metaphor, noting that where "there is deep epistemic disagreement over some fundamental principle, the disagreement has hit bedrock, the spade has turned."[4] However, it must also be noted that some authors argue that DD can be rationally resolved. Resolving disagreement means that two parties which initially hold different doxastic states (or credences) concerning a proposition end up, after an exchange, holding the same doxastic attitude. Thus, two parties can also resolve disagreement both by one party convincing the other or both parties suspending judgment about the target proposition(s). Following this line of thinking, Feldman (2005) argues that DD is rationally resolved if the two parties suspend judgment about the target proposition.[5] Matheson (2021), another optimist about the resolvability of DD, argues that DDs are strongly rationally resolvable because epistemic principles are self-supporting in that they are justified when they meet their own standards for justification. However, it has to be noted that Feldman and Matheson defend rational resolvability in an objective sense, namely that there is an objective rule stating what subjects in case of DD should rationally do. This approach does not cover the fact that the deeply disagreeing parties, because of their own convictions about epistemic rules, might not follow this objective rule on a subjective level.[6]

Finally, let me note a connection between DD and skepticism: DD points towards fundamental epistemic problems and limits, and, consequently, many authors observe a strong connection between DD and skeptical problems. Aikin (2021) interprets DDs as relying on traditional epistemic problems, namely the regress problem and the problem of the criterion. I argue in Melchior (2023a) that traditional skeptical problems concerning regresses about premises, epistemic circularity, and meta-regresses about the rationality of arguments be reinterpreted as instances of DD with a skeptic and I conclude that different types of DD must be distinguished.

To sum up, the existing discussion on DD faces various problems. First, there is a methodological tension between going by examples and going by general theories. Moreover, the two major schools concerning the nature of DD, hinge theories and fundamental epistemic principles theories, each face severe challenges in providing a cogent and explanatorily fruitful

---

[4] Ranalli (2021) criticizes Lynch's incorporation of rational irresolvability into his account of deep disagreement, a question that Ranalli holds should be open to further investigation.

[5] For a similar proposal, see Lugg (1986). For a criticism of Feldman's solution, see Henderson (2020).

[6] Pritchard (2021) suggests a resolution to deep disagreement in "an indirect, and side-on fashion" by looking to common ground (common beliefs, common hinges) in order to change an interlocutor's wider set of beliefs, thereby changing over time the other person's hinge commitment which manifest the unchangeable über hinge commitment concerning one's own reliability.

theory of DD. Finally, there is an ongoing debate about whether DD is resolvable and on how to interpret rational irresolvability.

4. Rationally irresolvable disagreement

There is a further deficit of the current discussion on DD, which points towards an alternative methodological approach. As Fogelin claims, not any kind of disagreement that is irresolvable or hard to resolve is deep. Disagreement can be intense and irresolvable without being deep, for example if one of the disputing parties (or both) is stubborn. What we are interested in are disagreements that are deep in some systematic way and that "are not subject to rational resolution." (Fogelin 2005, 11) Thus, Fogelin already mentions the idea of rationally irresolvable disagreement, and many papers explicitly or implicitly defend the idea that rational irresolvability is a crucial mark of DD.[7] There is a lively philosophical debate about the metaphysics of DDs. In contrast, there is surprisingly no theory on the market that analyzes rationally irresolvable disagreement and its central features, despite the acknowledged connection between DD and rational irresolvability.[8]

The shortcomings of the current debate about DD and the lacuna of developing a theory of rationally irresolvable disagreement motivate an alternative methodological approach towards DD. According to this approach, one can develop a theory of rationally irresolvable disagreement and then investigate its connection to DD. This is an open project and various outcomes are possible. First, it could turn out that DD simply *is* rationally irresolvable disagreement. This project would be reductionist in spirit since it reduces DD to rationally irresolvable disagreement.[9] In this case, further investigation into the nature of DD is superfluous. Second, it could be shown that rationally irresolvable disagreement should be carefully distinguished from DD. In this case, we can also acquire new insights about the specific sense in which DD is rationally irresolvable. Moreover, if DD is rationally *irresolvable* then we can understand in what respects DD is more than mere rationally irresolvable disagreement. If it is rationally *resolvable,* we can then see how DD substantially differs from rationally irresolvable disagreement. This approach starts from a different angle than existing theories about DD and can shed new light on the

---

7 The rational irresolvability of deep disagreement is assumed by Adams (2005), Lynch (2010), and Aikin (2021), discussed by Ranalli (2021) and Ranalli and Lagewaard (2022), and discussed but rejected by Lugg (1986), Feldman (2005), and Matheson (2021).
8 For some suggestions of how to interpret rational resolvability, see Ranalli (2021) and Ranalli and Lagewaard (2022).
9 Ranalli (2021) expresses some skepticism concerning this project.

ongoing discussion about DD. In this paper, such an account has been motivated by stressing problems and challenges for existing theories about DD.[10]

Acknowledgments I am thankful to the audience of the ECA 2022 for their helpful feedback and to Wes Siscoe for his comments on this paper. The research was funded by the Austrian Science Fund (FWF): P 33710.

References

Adams, D. M. (2005). Knowing when disagreements are deep. *Informal Logic, 25(1)*, 65–77.
Aikin, S. (2021). Deep disagreement and the problem of the criterion. *Topoi, 40(5)*, 1017–1024.
Feldman, R. (2005). Deep disagreement, rational resolutions, and critical thinking. *Informal Logic, 25(1)*, 12–23.
Fogelin R. (1985). The logic of deep disagreements. *Informal Logic, 7*, 1–8.
Godden, D., & Brenner W. (2010). Wittgenstein and the logic of deep disagreement. *Cogency, 2*, 41–80.
Henderson, L. (2020). Resolution of deep disagreement: Not simply consensus. *Informal Logic, 40(3)*, 359–382.
Kappel K. (2012). The problem of deep disagreement. *Discipline Filosofiche, 22(2)*, 7–25.
Kappel, K. (2021). Higher order evidence and deep disagreement. *Topoi, 40(5)*, 1039–1050.
Lagewaard, T. J. (2021). Epistemic injustice and deepened disagreement. *Philosophical Studies, 178(5)*, 1571–1592.
Kusch, M. (2021). Disagreement, certainties, relativism. *Topoi, 40(5)*, 1097–1105.
Lugg, A. (1986). Deep disagreement and informal logic: no cause for alarm. *Informal Logic, 8*, 47–51.
Lynch, M. (2010). Epistemic circularity and epistemic disagreement. In A. Haddock, A. Millar, & D. Pritchard (Eds.) *Social Epistemology* (pp. 262–277). Oxford: Oxford University Press.
Lynch, M. (2016). After the spade turns: disagreement, first principles and epistemic contractarianism. *International Journal for the Study of Skepticism, 6*, 248–259.
Matheson, J. (2021). Deep disagreements and rational resolution. *Topoi, 40(5)*, 1025–1037.
Melchior, G. (2023a). Skeptical arguments and deep disagreement. *Erkenntnis, 88(5)*, 1869-1893.
Melchior, G. (2023b). Rationally irresolvable disagreement. *Philosophical Studies, 180(4)*, 1277–1304.

---

10 Developing a fleshed out theory of rationally irresolvable disagreement is beyond the scope of this paper. For such a theory of rationally irresolvable disagreement, see Melchior (2023b).

Moyal-Sharrock, D. (2004). *Understanding Wittgenstein's on certainty*. Cham: Palgrave Macmillan.
Moyal-Sharrock, D. (2016). The animal in epistemology. *International Journal for the Study of Skepticism, 6,* 97–119.
Phillips, D. (2008). Investigating the shared background required for argument: a critique of Fogelin's thesis on deep disagreement. *Informal Logic, 28(2),* 86–101.
Pritchard, D. (2016). *Epistemic angst: radical skepticism and the groundlessness of our believing*. Princeton: Princeton University Press.
Pritchard, D. (2021). Wittgensteinian hinge epistemology and deep disagreement. *Topoi, 40(5),* 1117–1125.
Pritchard, D. (forthcoming). Disagreement, of belief and otherwise. In C. Johnson (Ed.) *Voicing dissent*. New York: Routledge.
Ranalli C. (2020). Deep disagreement and hinge epistemology. *Synthese* 197, 4975–5007.
Ranalli C. (2021). What is deep disagreement? *Topoi,* 40, 983–998.
Ranalli C. & T. Lagewaard (2022). Deep disagreement (part 2): epistemology of deep disagreement. *Philosophy Compass, 17(12),* 1–14.
Siegel, H. (2021). Hinges, disagreements, and arguments: believing hinge propositions and arguing across deep disagreements. *Topoi, 40(5),* 1107–1116.
Wittgenstein, L. (1969). *On Certainty*. Oxford: Basil Blackwell.
Wright, C. (2004). Warrant for nothing (and foundations for free)? *Aristotelian Society Supplementary, 78(1),* 167–212.
Wright, C. (2014). On epistemic entitlement II: Welfare state epistemology. In D. Dodd & E. Zardini (Eds.), *Scepticism and perceptual justification* (pp. 213–247). Oxford: Oxford University Press.

CONFLICTING CHARACTERIZATION FRAMES AND ARGUMENTATION IN THE PUBLIC CONTROVERSY SURROUNDING FASHION SUSTAINABILITY

CHIARA MERCURI
Università della Svizzera Italiana
chiara.mercuri@usi.ch

Abstract

This paper investigates the relationship between frames and argumentation in the public controversy surrounding fashion sustainability. In this context, which constitutes an instance of argumentative polylogue, the players involved tend to hold conflicting characterization frames about themselves and others, which lead their positions to become polarized and in turn exacerbate the conflict.

While frame analysis is considered crucial to understand the interests at issue within controversial situations, how to reconstruct the reasons underlying conflicting characterization frames by means of linguistic and discursive tools is yet to be investigated. This paper claims that an argumentation-based approach can help to make a step forward, that is, to reconstruct the partially implicit arguments laying behind conflicting frames.

In order to investigate the argumentation related to conflicting frames, this paper is based on an empirical corpus composed of texts issued by different players across various places, including social media posts published by activists, sustainability reports released by global fashion brands and official documents produced by public institutions. After retrieving the conflicting characterization frames present in the corpus and identifying their underlying implicit premises, from the findings it emerges that there are three different levels at which these frames are related to argumentation. Overall, this paper brings forward the theoretical reflection on the relationship between frames, material starting points in argumentation and their potential to analyze public controversies.

1. Introduction: the public controversy surrounding fashion sustainability

In recent years, as sustainability is becoming of growing interest in the public sphere, the fashion industry has been put under increasing scrutiny for its practices. This is due mainly to the environmental impact linked to the production and the consumption of clothing items (Niinimäki et al., 2020) as well as to concerns related to garment workers' treatment, especially in developing countries (Hibbert, 2018).

Following Dascal (2003), the debate surrounding fashion sustainability can be defined as a public controversy, as it is protracted over time and contains an element of polarization. Moreover, according to Greco and De Cock (2021), this public controversy represents an instance of *argumentative polylogue*, that is, a discussion which involves multiple players, positions and places (Aakhus & Lewiński, 2017). In fact, this complex polylogue involves various players including global fashion brands, citizens, small businesses, NGOs and public institutions; each group of players communicate through different places, e.g. social media, sustainability reports and institutional documents (Greco & De Cock, 2021).

According to Greco and De Cock (2021), this controversy is characterized by the presence of *argumentative misalignments* among the different players, that is, inconsistencies in the material starting points advanced in the opening stage of a critical discussion (van Eemeren, 2010). In this paper, I argue that one of the argumentative misalignments in this controversy consists in the existence of *conflicting characterization frames* among the different players (see also Mercuri, 2023), a notion that comes originally from conflict resolution studies and that refers to how conflicting groups interpret conflictual situations related to themselves and others (see Lewicki & Gray, 2003; Shmueli, et al., 2006; Shmueli, 2008). According to these studies, the analysis of conflicting characterization frames represents a crucial instrument for understanding where the deep clashing points between the parties are situated, that is, what leads the controversy to be protracted over time. However, this stream of research lacks the linguistic-discursive tools to justify how frames can be reconstructed, starting from discourse.

On the other hand, studies in argumentation (Greco Morasso, 2012; Bigi & Greco Morasso, 2012) have shown that frames play an argumentative role, specifically as they are linked to the selection of material starting points through certain frame-activating words (Fillmore, 1976).

Drawing on studies proposing an argumentative approach to frames, in this paper I consider that argumentation can help to reconstruct the reasons underlying conflicting characterization frames, thus allowing to expose the deep clashing points among players in the public controversy surrounding fashion sustainability.

The research questions I address in this paper are the following:

RQ1. *From an argumentative perspective, at what level can characterization frames be conflicting?*

RQ2. *How are these conflicting characterization frames related to argumentation and specifically to material starting points?*

The paper develops as follows. In section 2, I present my theoretical framework; in section 3, I explain the methodology, while in section 4 I discuss my findings and in section 5 I expose conclusions and perspectives for future research.

2. Theoretical framework

2.1. Conflicting characterization frames

In conflict resolution studies, frames are seen as "lenses through which disputants interpret conflict dynamics" (Lewicki & Gray, 2003, p. 5). Broadly speaking, frames act as *filters*, as they entail a simplification of reality while overlooking details about a given situation (Kaufman & Smith, 1999). According to Shmueli et al. (2006), frames are particularly useful in situations of complexity, since they help to eliminate details deemed unnecessary.

However, the simplification of reality entailed by frames can become problematic. In situations of conflict, according to Shmueli et al. (2006), the parties involved are likely to hold strongly divergent frames of a given event, i.e. *conflicting frames*; as these frames are present at a deep level and are often unconscious, they tend to lead conflicts to become polarized and exacerbated (Shmueli et al., 2006). According to Shmueli (2008), for example, in the context of the Israeli-Palestinian conflict, Israeli Jews identify Jewish small towns as the results of the fight for independence, while Israeli Arabs see them as usurpation of their lands. Thus, the conflicting frames held by the conflicting parties are 'Independence' and 'Catastrophe' respectively (see Shmueli, 2008).

Scholars have proposed various empirically-based taxonomies of conflicting frames; according to Lewicki and Gray (2003), for example, participants in a conflict hold frames about what the conflict is about, how it originated, its foreseen developments and the different parties involved. In this paper, I focus on a specific category of conflicting frames, that is, *characterization frames* (see Kaufman & Smith, 1999; Shmueli et al., 2006; Shmueli, 2008).

Following the definition proposed by Shmueli (2008), characterization frames refer to how players involved in the controversy view their own behaviour, that of others, as well as the relationships among the parties; these characterizations can be either positive or negative (Shmueli et al., 2006). Moreover, characterization frames also concern the evaluation of the players' own attributes, i.e. *self-characterization frames* (Kaufman & Smith, 1999). Since I consider characterization frames as closely connected to another category of conflicting frames identified by Shmueli (2008), that is, *values and identity frames,* in this research I adopt an understanding of characterization frames which encompasses both categories (see also Mercuri, 2023).

According to Kaufman and Smith (1999) and Shmueli (et al., 2006; 2008), identifying the conflicting frames and how they were constructed is crucial to understand the controversial situation at issue, in order to then decide possible interventions. However, while scholars in conflict resolution studies agree that frame analysis represents an important tool to gain a deep understanding of the controversy at issue, they lack the discursive and linguistic fine-grained tools to justify how these conflicting frames can be reconstructed, starting from discourse: for example, Lewicki and Gray (2003, p. 7) state that their unit of analysis "was the 'thought unit'- that is, the words, sentences, or paragraphs used to express an identifiable thought". As a consequence, from these contributions it is not clear what it means exactly to say that frames are divergent between conflicting parties, that is, if frames are conflicting at the semantic, at the discursive or at the argumentative level. To address this gap, I propose to adopt an argumentative approach to the analysis of conflicting frames.

2.2. Argumentative perspectives on framing and frames

Previous studies in argumentation have investigated the relationship between frames, or framing, and argumentation (see Bigi & Greco Morasso, 2012; van Eemeren, 2010; Fairclough & Mădroane, 2020; Greco Morasso, 2012). A common aspect between all these contributions is the recognition that frames and framing play an important role in the argumentative process, as the presentation of reality according to a certain frame or framing tends to orient the audience towards the acceptance of a given standpoint.

Beyond this common aspect, these studies can be differentiated on the basis of whether they focus on the process of creating frames, i.e. *framing*, or on the results of it, i.e. *frames*. Argumentation scholars who concentrate on *framing* adopt the definitions of the concept proposed by Goffman (1974) and Entman (1993). They include van Eemeren (2010), who views framing as related to strategic manoeuvring, and Fairclough and

Mădroane (2020), who examine the role of framing in the selection of reasons throughout the deliberation process.

On the other hand, contributions focusing on *frames* include Bigi and Greco Morasso (2012) and Greco Morasso (2012), who follow the linguistic definition of the concept developed by Fillmore (1976). In Fillmore's perspective (1976), which I also adopt in this paper, frames are cognitive structures that are triggered through certain lexical features, i.e. specific frame-activating words.

Within a methodological framework that combines Pragma-dialectics (van Eemeren, 2010) with the Argumentum Model of Topics (Rigotti & Greco, 2019), two studies by Bigi and Greco Morasso (2012) and Greco Morasso (2012) determine that frames are related to material starting points in argumentation. More specifically, they claim that frames contribute to selecting *endoxa*, i.e. implicit cultural premises (Rigotti & Greco, 2019), that are then employed in the construction of an argument.

While the work by Bigi and Greco Morasso (2012) and Greco Morasso (2012) represents an important step for clarifying the role played by frames in respect to argumentative premises, research about the specific relationship occurring between conflicting frames and argumentation remains yet to be done. Thus, the potential of argumentation for investigating the reasons underlying conflicting frames still needs to be explored. In this paper, drawing on the contributions combining Pragma-dialectics with the Argumentum Model of Topics (henceforward: AMT), I propose an argumentative approach to the analysis of conflicting frames, in order to gain a better understanding of the material starting points at work in the public controversy surrounding fashion sustainability.

3. Methodology

3.1. Corpus

To empirically address my research questions, I collected a composite corpus composed of texts belonging to different genres, following Greco and De Cock (2021). In the corpus, I inserted documents published across multiple places by three different groups of players participating to the polylogue (Table I): social media posts by concerned citizens, small businesses and NGOs; sustainability reports by major fashion brands; official documents by a specific public institution, i.e. the EU. All documents were issued between 2020 and 2022.

For the first group of players, which encompasses citizens, small businesses and NGOs, I considered Instagram posts and Tweets containing #fashionrevolution that were published during Fashion Revolution Week, an annual campaign that promotes awareness about the current issues related to fashion and that encourages all members of society to take part in the radical transformation of the industry[1]. In this paper, I consider all the authors of posts containing #fashionrevolution under the encompassing label 'activists', as they all participated to the campaign (see Mercuri, 2023).

For major fashion brands, I examined the first sections of five sustainability reports issued by high street brands that were included among the largest public fashion companies by the Business of Fashion (2021).

For the EU, I selected two communications produced by the European Commission that concerned sustainability in the fashion industry.

Table I. Types of documents included in the corpus

Players	Places
Activists Citizens, small businesses, NGOs	*Social media posts* 200 Instagram posts 200 Tweets #fashionrevolution
Major fashion brands H&M Inditex Levi Strauss & Co Fast Retailing Gap	*Sustainability reports* H&M Sustainability Report Inditex Annual Report Levi Strauss & Co Sustainability Report Fast Retailing Sustainability Report Gap Global Sustainability report
Institutional bodies European Commission	*Official documents* "A New Circular Economy Action Plan for a Cleaner and more Competitive Europe" "EU Strategy for Sustainable and Circular Textiles"

3.2. Method of analysis

The method of analysis is comprised of three steps. First, I annotated the characterization frames present across the corpus, following the refined definition of the concept proposed by Mercuri (2023), according to which characterization frames can be discursively activated through the use of frame-activating expressions that refer to a *persona*, a *stable behaviour* or a *state/condition*. For the aim of this paper, I selected characterization frames related to the groups of players participating to the polylogue, i.e. garment workers, small businesses, activists and NGOs, public institutions, big brands and consumers, in line with Mercuri (2023). After identifying the frame-activating expressions, I aggregated them in

---

[1] https://www.fashionrevolution.org/frw-2022/

categories of characterization frames according to their overarching predicate (see Mercuri, 2023).

Then, I compared the documents released by conflicting players, to identify where the identified characterization frames appeared divergent: for example, I examined whether the characterization frames associated to a specific player, e.g. garment workers, differed between activists' social media posts and sustainability reports by big brands. As the exact meaning of the term 'conflicting frames' is per se not definite, but on the contrary is one of the points I aim to clarify, in this phase I considered *conflictingness* between frames in the broad sense of 'being divergent', before moving to the argumentative analysis.

In the last step of analysis, I investigated the cases in which characterization frames appeared divergent from an argumentative perspective, in order to better understand at what level they were conflicting (RQ1). To this end, for all characterization frames that appeared conflicting I analyzed the portion of discourse containing the relevant frame-activating expressions, by performing an analytical overview (van Eemeren & Gootendorst, 2004) that allowed to see if the characterization frame was located within an argumentation structure. Afterwards, for those cases in which the characterization frame was indeed situated within an argumentative structure, either in the standpoint or in the argument, I reconstructed the identified arguments by applying the AMT (Rigotti & Greco, 2019), which allows for a fine-grained level of analysis. With this model, I was able to reconstruct the *inferential configuration* of argumentative moves containing characterization frames, by making explicit both the procedural and the material starting points. Thus, I could determine how conflicting characterization frames in my corpus were related to argumentative material starting points (RQ2).

4. Preliminary findings

In response to RQ1, from the analysis of the corpus it emerges that there are three different levels at which characterization frames can be conflicting.

The first level of *conflictingness* that I identified is located at the level of the argumentative issues that arise when comparing documents by conflicting players, while the second is related to the contradictory factual premises that are selected and the third to the clashing cultural premises that are constructed concerning the definition of 'being sustainable'. These three levels are discussed in the following sections, together with relevant examples.

4.1. Misalignments between argumentative issues raised by conflicting players

When looking at the different *places*, i.e. types of documents produced by conflicting players, it can be noticed that the *frequency* of characterization frames related to a specific player changes. In other words, the participants to the polylogue that are deemed worthy of characterization, and are thus put at the centre of discussion, vary in the discourse of conflicting players.

For example, the number of occurrences of characterization frames related to the group of players 'garment workers' varies greatly depending on the type of document that is considered. While characterization frames related to garment workers are very frequent in social media posts published by activists during Fashion Revolution Week, they are much less frequent in big brands' reports and even less in official documents by the European Commission (Table II).

Table II. Frequency of characterization frames related to 'garment workers'

Occurrences of characterization frames for 'garment workers'	Social media posts by activists	Sustainability reports by big brands	Communications by the EU
291	192	82	17

From an argumentative perspective, this suggests that the discourse of each player is centred around a different *issue,* i.e. the "question around which an argumentative discussion revolves" (Schär, 2021, p. 10): for activists, the characterization frames associated to garment workers constitute a crucial point in the controversy, while this is less the case for the other players. This finding indicates that activists participating to Fashion Revolution are the most concerned about the current treatment of workers, i.e. the social dimension of sustainability, while this is less the case for other players, who tend to focus more on other aspects of sustainability, e.g. the environmental implications.

Thus, for this first level of *conflictingness,* characterization frames employed by different players are conflicting at the level of the argumentative issue addressed in their discourse.

4.2. Contradictory *data* between conflicting players

The second level of *conflictingness* concerns the cases in which the characterization frames associated to the same player are contradictory across documents by conflicting players, especially between activists' social media posts and big brands sustainability reports. In fact, each player chooses to highlight specific attributes of the other players, thus generating contradictory interpretations of the same piece of reality (see section 2).

For example, activists represent big brands through characterization frames such as "Fast fashion is a key driver of climate breakdown", while big brands employ self-characterization frames as in "We promote circularity". In the analytical overview reconstructed for the portions of discourse containing these characterization frames, it can be noticed that they are employed as arguments in support of two opposite standpoints which intend to lead the audience towards opposite courses of action, respectively "You should stop buying from fast fashion" in social media posts and "You should continue supporting big brands" in sustainability reports. Upon closer examination of these examples, the inferential configurations reveal that in both cases the argument is linked to the standpoint by means of a locus from termination and setting up, in which characterization frames work as *data*, i.e. factual premises (Rigotti & Greco, 2019), in support of the two standpoints (Fig. 1 and Fig. 2).

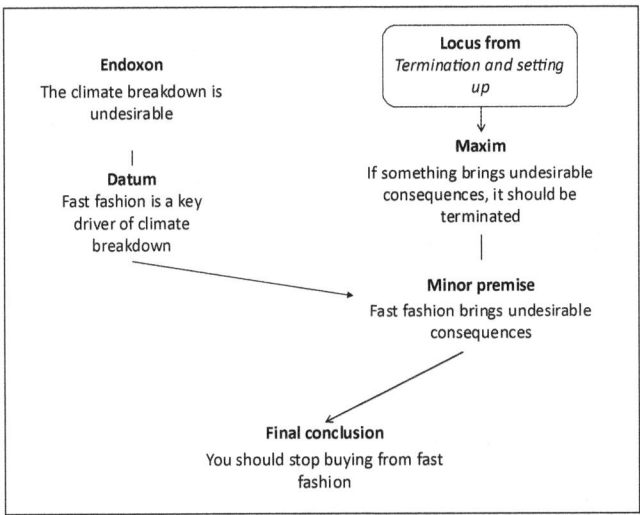

Figure 1. Inferential configuration of locus from termination and setting up: activists

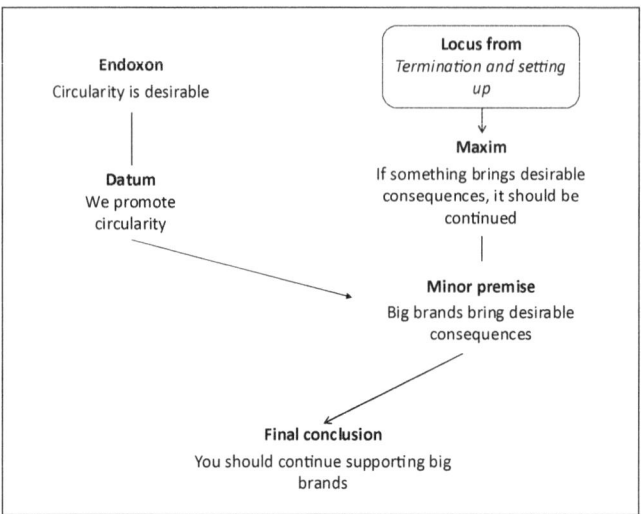

Figure 2. Inferential configuration of locus from termination and setting up: big brands

From the reconstructed inferential configurations, it emerges that the whole argumentation that is advanced is conflicting between the two groups of players, starting from the *datum*, in which the characterization frame is found. By highlighting a certain aspect of reality related to a specific player, which contradicts the characteristics of that same player highlighted by others, characterization frames select argumentative moves to guide the audience towards certain course of actions. The courses of action that are proposed are in line with each player' overarching interests, which are incompatible between various players: in Fig. 1, activists are interested in changing the current fashion system, while big brands want to continue growing their businesses (Fig. 2).

Thus, for this second case it can be said that the *conflictingness* arises from the *data* that are selected, which then construct the whole argumentative moves in the discourse of conflicting players.

4.3. Clashing *endoxa* between conflicting players

The third level at which characterization frames are found to be conflicting concerns the definition that each player attributes to the term 'being sustainable', which is expressed through divergent categories of self-characterization frames, that is, the ways in which players characterize their own *identity, stable behaviour* and *state/condition* (see section 3.2)

For example, while both small businesses participating to the Fashion Revolution campaign and big brands claim to be sustainable, small businesses characterize themselves through the category "being revolutionary", whereas big brands adopt the self-characterization frames included in the category "continuing to work towards sustainability". As in the previous section (4.2), both characterization frames work as arguments, more precisely as *data*, and are connected to the standpoint by means of the same locus, this time a locus from definition. However, in this case both characterization frames justify the same standpoint, which is "This business can legitimately claim to be sustainable", as reported in the two inferential configurations reconstructed in Fig. 3 and Fig. 4.

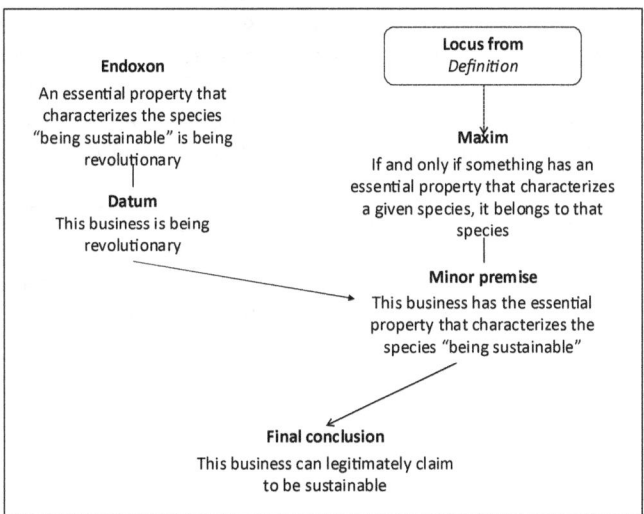

Figure 3. Inferential configuration of locus from definition: small businesses

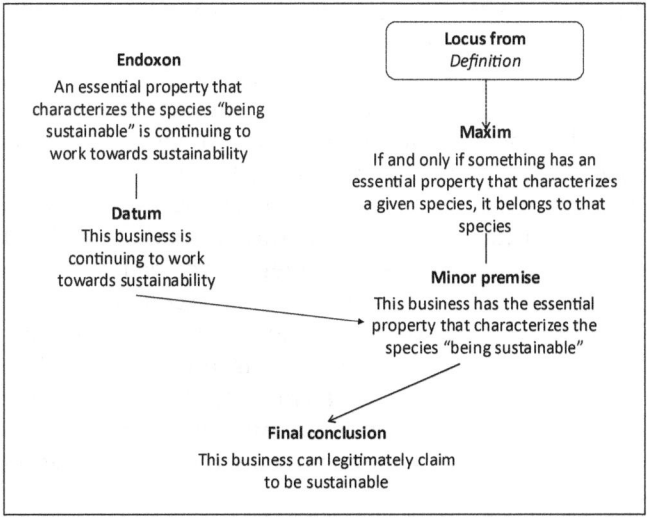

Figure 4. Inferential configuration of locus from definition: big brands

From the reconstructed inferential configurations, it appears that, according to activists participating to Fashion Revolution, the current fashion system can only be changed through a revolution, which aims to disrupt all current practices. In this perspective, a sustainable business is a brand that breaks with the existing state of affairs, as stated in the *endoxon*, i.e. the shared cultural premise (Rigotti & Greco, 2019) in Fig. 3. On the other hand, big brands are opposed to radical change, because the current system is the one where they own the majority of the market. Therefore, they present themselves as agents that have undertaken and are presently undertaking steps towards sustainability within the existing system. Accordingly, big brands propose an *endoxon* in which a sustainable brand is a business that is already making advancements on the sustainability path and that plans to continue doing so, as shown in Fig. 4.

Thus, in this third case characterization frames by different players contribute to construct clashing *endoxa* (see also Mercuri, 2023) concerning the definition of 'being sustainable', which means that they are conflicting at the level of both material premises, i.e. the *datum* and the *endoxon*.

5. Conclusions

This paper has proposed an argumentative approach to the analysis of conflicting characterization frames in the context of the public controversy surrounding fashion sustainability, in order to offer linguistic-discursive tools for better identifying the deep clashing points present in this controversy and the reasons underlying them.

The preliminary findings discussed in this paper have shown that there are three levels at which characterization frames appear to be conflicting between the different players involved. These three levels are directly related to the type of relationship occurring between conflicting characterization frames and argumentation. At the first level, conflicting frames are connected to the argumentative issues that arise in the discourse of conflicting players, while at the second level characterization frames appear to be conflicting depending on the specific aspects of reality that are highlighted, which select certain *data* that then construct the whole argumentation that is advanced. Last, at the third level characterization frames are conflicting in terms of both material starting points, as each group of players propose their own *endoxon* about the definition of 'being sustainable'.

At the theoretical level, this paper contributes to argumentative studies by advancing the reflection on the relationship between the discursive interpretation of frames and their connection to argumentation, especially to material starting points. Moreover, it begins to explore the potential of argumentation combined with the study of conflicting frames to analyze public controversies.

At the current stage of analysis, most conflicting frames are found in comparing characterization frames employed by activists and by global fashion brands. In the future, I plan to expand the corpus; with the inclusion of further documents, more conflicting frames in respect to the discourse by public institutions could also be identified. Moreover, completing a taxonomy of the conflicting characterization frames present in the controversy could offer an overview of the most important issues that need to be addressed by actors willing to intervene.

References

Aakhus, M., & Lewiński, M. (2017). Advancing polylogical analysis of large-scale argumentation: disagreement management in the fracking controversy. *Argumentation, 31*, 179-207.

Bigi, S., & Greco Morasso, S. (2012). Keywords, frames and the reconstruction of material starting points in argumentation. *Journal of Pragmatics, 44*(10), 1135-1149.

Business of Fashion (2021). *Measuring fashion's sustainability gap.* https://www.businessoffashion.com/reports/sustainability/measuring-fashions-sustainability-gap-download-the-report-now/

Dascal, M. (1989/2003). Understanding controversies. Reprinted in M. Dascal (2003), *Interpretation and understanding* (pp. 280-292). Amsterdam/Philadelphia: John Benjamins.

Eemeren, F.H., van. (2010). *Strategic manoeuvring in argumentative discourse: Extending the pragma-dialectical theory of argumentation.* Amsterdam/Philadelphia: John Benjamins.

Eemeren, F.H, van, & Grootendorst, R. (2004). *A Systematic Theory of Argumentation.* Cambridge: Cambridge University Press.

Entman, R. M. (1993). Framing; toward clarification of a fractured paradigm. *Journal of Communication, 43*(4), 51-58.

Fairclough, I. & Madroăne, I. (2020). An Argumentative Approach to "Framing". Framing, Deliberation and Action in an Environmental Conflict. *Co-herencia, 17*(32), 119-158.

Fillmore, J. (1976). Frame semantics and the nature of language. *Annals of the New York Academy of Science, 280,* 20-32.

Goffman, E. (1974). Frame analysis: an essay on the organization of experience. New York: Harper & Row.

Greco Morasso, S. (2012). Contextual frames and their argumentative implications: A case study in media argumentation. *Discourse studies, 14*(2), 197-216.

Greco, S., & De Cock, B. (2021). Argumentative misalignments in the controversy surrounding fashion sustainability. *Journal of Pragmatics, 174,* 55-67.

Hibbert, M. (2018). Key challenges for the fashion industry in tackling climate change. *Studies in Communication Sciences, 18*(2), 383-397.

Kaufman, S. & Smith, J. (1999). Framing and reframing in land use change conflicts. *Journal of Architectural and Planning Research, 16(2)*, 164-180.

Lewicki, R. J., & Gray, B. (2003). Introduction. In R. J. Lewicki, B. Gray, & M. Elliott (Eds), *Making sense of intractable environmental conflicts. Concepts and cases* (pp. 1-9). Washington/Covelo/London: Island Press.

Mercuri, C. (2023). Characterization frames constructing *endoxa* in activists' discourse about the public controversy surrounding fashion sustainability. [Manuscript submitted for publication].

Niinimäki, K., Peters, G., Dahlbo, H., et al. (2020). The environmental price of fast fashion. *Nature Reviews: Earth and Environment, 1,* 189-200.

Rigotti, E., & Greco, S. (2019). *Inference in argumentation: A topics-based approach to argument schemes.* Cham: Springer.

Schär, R. G. (2021). *An argumentative analysis of the emergence of "issues" in adult-children discussions.* Amsterdam/Philadelphia: John Benjamins.

Shmueli, D. (2008). Framing in geographical analysis of environmental conflicts: Theory, methodology and three case studies. *Geoforum, 39(6)*, 2048-2061.

Shmueli, D., Elliott, M., & Kaufman, S. (2006). Frames changes and the management of intractable conflicts. *Conflict Resolution Quarterly, 24,* 207-2018.

PERSUASION WITHIN DELIBERATION: SYMBIOSIS OR PARASITISM?

HENRI MÜTSCHELE
Department for Communication and Media Science,
Heinrich- Heinrich Heine University Düsseldorf, Germany
henri.muetschele@hhu.de

Abstract

While deliberative and persuasive modes have long been regarded as mutually incompatible, the role of persuasion within deliberation is now evaluated in a more differentiated way. This paper claims that persuasive speech modes can be a permissible form of deliberative talk. By contrasting theoretical reflections on deliberative reasoning with empirical results on persuasive effects in deliberative settings, it is argued that contextual conditions of deliberation are decisive for the normative evaluation of persuasive modes within deliberation. A typology is introduced distinguishing between decision- and epistemic-oriented deliberation and homogenous and heterogeneous group structures.

1. Introduction

Deliberation has become one central theoretical framework for dispute resolution and collective decision-making. The question which forms of speech are normatively permissible within deliberation has raised considerable scientific interest within the last years. Persuasion is arguably the most common talk form within deliberation. As deliberation is based on collective reasoning, interlocutors ought to convince each other through arguments. Participants are asked to shift their preferences towards the position which is best argued for. However, persuasion does not necessarily fulfill a deliberative function, as persuasive messages are often discussed under the suspicion of manipulation.

While most studies in this research field focus on the assessment of various types of deliberative speech and the generation of criteria to distinguish between permissible and harmful ways of deliberative talk, the contextual conditions of deliberation have hardly been taken into consideration. This perspective is especially important because artificial deliberative settings are frequently created, which cannot meet all deliberative standards. Additionally, communication processes are analyzed regarding their deliberative quality in a variety of contexts, even in the unstructured online sphere. For a normative evaluation of persuasion within deliberation the following research question needs to be addressed: Under which conditions is persuasion permissible within deliberation?

This paper makes two claims. First, I argue that persuasion can be a permissible form of a deliberative talk, but that there are deliberative and non-deliberative forms of persuasion. Second, I argue that for the normative evaluation of persuasion within deliberation the group structure (homogenous vs. heterogenous) and deliberative function (epistemic vs. decision-oriented) need to be taken into consideration. Based on the theoretical understanding of reasoning in deliberation and influencing factors of persuasive speech from communication science, I refer to empirical observations in deliberation research to support this change in perspective.

This approach is relevant for the evaluation of empirical results on deliberative talk, which has moved into the spotlight of research within the social sciences. The exact reasons for post-deliberative opinion formation, or possibly opinion change, still represent a black box in the research landscape on deliberation.

Addressing the question what exactly drives deliberative opinion change is also important for practical implementations of deliberative processes. When persuasive modes are used, which do not fulfill deliberative standards, the deliberative procedure may be counterproductive for the solution of the discussed issue, or the legitimation of decisions made.

2. Deliberation and Reasoning

Deliberation generally entails the notion of *collective reasoning* enabling rational decision-making and highly justified actions (Chambers, 2003). The question what makes humans and their actions rational, is a central one within philosophy. The term rationality is fundamentally vague. While Habermas and other representatives of the Frankfurt school criticize the understanding of rationality as a pure rationality of purpose, that is a focus on the means rather than questioning the ends, it becomes evident that from a communicative point of view rational actions require the provision and explication of reasons through speech. Argumentation is broadly understood the field, in which reasons for certain positions are given. However, not every form of argumentation is rational, as reasons might not support a position or conditions for reason-giving are not provided.

The question what the criterion of rationality is within argumentation, can be answered regarding deliberation by agreement-building and the solution of a dissensus on a conflicting issue. "A language user taking part in an argumentative discussion is a rational language user if in the course of the discussion he performs only speech acts which accord with a system of rules acceptable to all discussants which furthers the creation of a dialectic which can lead to a resolution of the dispute at the centre of discussion" (van Eemeren/Grootendorst, 1984, 48).

This perspective is compatible with Habermas' notion of *communicative rationality*, in which utterances need to be critically reflected "by the unforced force of the better argument" (Habermas, 1996, 305) and transformed to a consensual standpoint. Following speech act theory, Habermas argues that every utterance enfolds several claims of validity. If individuals want to find acceptance for their claims, they need to justify these. Habermas distinguishes between three claims of validity: *truth*, which refers to the factual accuracy of a statement, *correctness*, meaning the conformity of statements made with socially accepted moral norms and values, and *sincerity*, which refers to the agreement with the actual underlying intention of the speaker. In his later works, Habermas focuses on the claims of validity of truth and correctness, as these are discursive and can be, in contrast to sincerity, guaranteed through reasons. This illustrates that individuals have a demanding role within deliberation.

For Habermas, argumentation is the thematization of conflicting validity claims, which need to be based on reasons, evaluated based on their cogency (1987, 37f). This leads to the question, in which situations and for which utterances the provision of reasons is required in deliberation and how these reasons ought to be processed by the participants. The understanding of reasoning as a cognitive mechanism has induced, among others, two opposing theoretical perspectives, which are relevant for deliberation.

2.1. The Argumentative Theory of Reasoning

The *argumentative theory of reasoning* (Mercier/Sperber, 2011) is based on the notion of a systematic processing of a message content by the receiver through the identification and evaluation of reasons. This perspective was a paradigm shift for the understanding of reasoning, as its function is no longer understood as serving better decision-making and improving the receivers' own epistemic status. Thus, social and interpersonal conditions for reasoning move into the spotlight, compatible with the idea of deliberation, as it requires the fulfilment of several normative socio-political conditions.

Through the exchange of reasons for and against positions, deliberators are able to weigh up the provided arguments and evidence in a conscious manner (ibid.). Following this notion, two important conditions for successful reasoning within deliberation can be derived. First, reasons need to be *exchanged*. Arguments and counterarguments for at least two diverging opinions need to be exchanged in a clear argumentation chain allowing for feedback on each of the provided utterances. Deliberation is a fundamentally reciprocal procedure. Second, arguments need to be *generated* in deliberation. Reasoning first and foremost requires the identification of arguments in a message content by the receiver (Mercier/Landemore, 2012, 246). I would add that the latter condition is based on a strong precondition. Arguments need to be made *transparent*. This means that the premise, the warrant, and the conclusion of a given claim (following Toulmin, 2003) ought to be explicit to allow for an evaluation of a raised position. All steps of a single argument must potentially be challenged in deliberation.

Lupia (2002) defines further conditions for successful cognitive deliberative reasoning. He notes that the better argument not only needs to win the audience's attention, but also needs to be processed through storage in memory. Only new and distinct memories can potentially lead to an opinion change (144f). Moreover, this argument needs to be shifted into a new belief, by rejecting an old one and actively transferring it into a corresponding action.

2.2. The Threat of Motivated Reasoning

In contrast to argumentative reasoning, the phenomenon when receivers evaluate arguments not in an open manner is called *motivated reasoning*. This means that reasoning is based on the clear intention of maintaining pre-existing preferences and values (Kahan, 2015). This aspect is obviously a challenge for persuasion, as provided arguments are not reflected upon their objective cogency. Receivers will disregard the quality of any argument, which contradicts their own belief system. Although motivated reasoning will arguably happen unconsciously to a certain extent, it follows a different mechanism than other forms of cognitive biases, as individuals are not even willing to consider opposing viewpoints.

The effective strength and the normative evaluation of motivated reasoning within deliberation is controversially discussed, as motivated reasoning might still produce belief change within groups, although some pre-existing beliefs remain resistant (Richey, 2012).

Following these understandings of reasoning, the question needs to be answered to what extent and under which conditions persuasive speech induces deliberative reasoning. The definition and effects of persuasion are briefly described first.

3. Influencing Factors of Persuasive Messages

The general element of persuasion is an intentional change of the mental status of another. "We use the term persuasion to refer to any instance in which an active attempt is made to change a person's mind" (Petty/Cacioppo, 1996, 4). These changes can take many different forms. Despite opinion change, persuadees can also be influenced regarding their beliefs, attitudes, or values. Persuasion might even lead to a behavioral change, or at least a shift in the behavioral intentions. Another important aspect is the intensity of persuasive effects. A change implies a successful form of persuasion, while persuasive speech might at first result in shaping beliefs. It is also conceivable that pre-existing positions are reinforced (Perloff, 2010).

Another principle of persuasion is that a receiver of a persuasive message has a free choice to accept or reject it (O'Keefe, 2002). This is the main difference between persuasion and coercion. Interlocutors in persuasive discourses generally have adversarial individual goals, whereby the persuader is required to treat the persuadee as an autonomous individual. Poggi argues that this requires a persuader to take the opponent's viewpoint into account to convince the persuadee that the intended goal is adequate for the pre-existing belief system (2005, 310f). In this sense, persuasion merely means to raise the likelihood for a persuadee to pursue some goal.

The aspect of the intentional use of persuasive speech remains problematic. This raises the question whether such intentions might exist without being perceived by the communicator, as non-intentional speech might nonetheless have a persuasive effect. Examples for non-intentional persuasion can be found in narration research (Green et al., 2002).

Research on persuasion has received a renaissance in communication science. For a long time, the research focus was lying on linear and direct cognitive effects based on social psychological concepts but within the last years more complex, mediatized and potentially weaker persuasive effects have been identified (Holbert et al., 2010). There are various influencing factors in the persuasion process that can explain limitations of persuasive messages. These are relevant for a normative evaluation of persuasive speech within deliberation.

One group of factors takes characteristics of the persuader into account. Decisive are mainly the perceived competence and trustworthiness. But aspects like physical attractivity or likability play a significant role as well. It is conceivable that persuaders use or even change their prior external perception in a strategic manner through ethos-construction (Amossy, 2001).

Another group of factors centers around the receivers. Singled out here is the aspect of consonance. This refers to the content level, for which the consonance and cumulation of persuasive messages determines the intensity of their processing. But consonance is also relevant for the individual level because the existence of strong pre-existing attitudes and beliefs as well as the knowledge of counterarguments hinders the unfolding of an argument's quality. This is dependent on a persuadee's individual characteristics, i.e., the consistence of affective and cognitive attitude components (Eagly/Chaicken, 1993).

As deliberation aims at resolving dispute through collective reasoning, persuasive discourse can lead to a common ground. This means that at least one side needs to concede to a certain extent by accepting reasons from others. Deliberators are not only allowed but even asked to raise and argue in favor of their preferences they genuinely believe in and they assume their audience finds compelling, if there are willing to revise it considering the discussion (Chambers, 2003). As Manin (1987, 351) puts it, "the process of deliberation, the confrontation of various points of view, helps to clarify information and to sharpen their [participants'] own preferences. They may even modify their initial objectives, should that prove necessary."

In sum, in deliberation pro- and contra arguments for certain proposals need to be exchanged which generally allows for persuasive speech. Accordingly, a persuader in deliberative settings needs to provide compelling reasons for claims and positions, formulated in a way that it can be in principle accepted by others. Persuadees are asked to shift their opinion towards the most convincing position. This requires an active engagement in deliberation and not only passive listening. However, opinion shifts must be grounded in objective argumentative superiority, impartial from individual cognitive and affective constraints.

4. Two Approaches of Implementing Persuasive Modes within Deliberation

Although opinion shifts are in order within successful deliberation, it remains to be seen how these changes of pre-existing positions ought to occur through persuasion. Walton and Krabbe (1995) distinguish between six formal-descriptive dialogue types based on three stages: the *initial situation, individual goals,* and the *collective goal.* For deliberation dialogues, the initial situation is characterized by a practical choice, a choice of at least two diverging courses. The aim of deliberation dialogue is to find agreement which is best for the group. Persuasion dialogues are merely based on disagreement, a conflict of opinion. The dialogue partners prove their own or attack the opposing standpoint through the exchange of arguments. Decisive is the use of premises the other party finds acceptable to persuade it of a conclusion (ibid., 79f). Therefore, persuaders have different individual goals than deliberators. While the latter wants to determine the best result for the group as a whole (which can diverge from the individual preference), persuaders wish to solely convince the persuadee by his/her preferred option or an option that is as close as possible to it and that the persuadee finds acceptable.

While Walton and Krabbe argue that deliberation dialogues will often entail shifts to persuasion dialogues, they do not offer a clear concept what forms of persuasion are permissible or harmful within deliberation. The exchange of arguments is more complex within deliberation. Walton (2006) argues that individual goals need to be clarified regarding the circumstances of the participants, as proposals need to be put forward before the explication of reasons. Some proposals might be modified or discarded through rational criticism. It is evident that the exact speech acts will be quite similar between persuasion and deliberation dialogue, which makes it very difficult for observers (Kirchev et al., 2019). There is a wide variety of persuasive modes which are not solely based on the use of arguments but can include different rhetorical elements. All entail forms of an exertion of influence, which can have positive or negative effects for the principle of collective deliberative reasoning and the establishment of a common course of action within deliberation.

In my view, there are two main avenues in the research landscape on the disentanglement of speech modes within deliberation. The first approach focuses on *relaxing the requirements of the deliberative ideal*. With the development of type-II-deliberation (Bächtiger et al., 2010) a more inclusive notion of deliberation permits the use of non-argumentative utterances. These include speech forms like personal experiences, narratives, or emotions, which are arguably used more frequently in everyday talk. As these might entail persuasive messages which are relevant for the issue at state, it can favor deliberative exchange with fewer hurdles for participants with a lower argumentative repertoire and less rhetorical capacities regarding the explication of their preferences and reasons.

The second approach on the investigation of deliberative talk emphasizes the *distinction between deliberative and non-deliberative forms of speech*. I refer to one example here. Close to Habermas' distinction between strategic and communicative action, Klemp (2012) develops criteria for different versions of rhetorical speech modes: *manipulation*, *strategic persuasion*, and *deliberative persuasion*. For him, rhetoric is a general type of speech with a persuasive notion.

Klemp proposes that "in deliberative persuasion, the speaker seeks to induce agreement with an orientation toward mutual understanding" (ibid., 55) and defines three criteria: *Openness to revision*, which can be described as a general willingness to revise the pre-existing preferences. The second criterion is *sincerity*, whereby he refers to one of the validity claims by Habermas. By bringing forward reasons one genuinely believes in, a commitment of understanding is achieved. The last criterion is *focus on the merits*, meaning the serious and reflective consideration of all given arguments (ibid., 56f). On the contrary, strategic persuasion refers to the notion of persuasion for intentionally winning the debate to change the viewpoint of others. While Klemp is aware that even within deliberative persuasion successful convincing is possible through agreement-building, he claims that the intention of winning can be more dominant than the goal of achieving mutual understanding (ibid, 57). Finally, he defines manipulation as the intentional bypassing of the listener's rational capacities for a certain end (ibid., 54). In this sense, manipulation exerts a form of hidden power by the exclusion of other judgements.

By formulating more detailed criteria allowing for a further distinction within strategic action, Klemp, like Habermas, highlights the role of intentional language use for its normative force regarding the establishment of agreement. However, the focus on the role of intentions remains problematic, as it can hardly be assessed regarding the normative evaluation of deliberative potentials and limitations of specific utterances.

5. Ideal-Typical Model: Considering Group Structure and Deliberative Function

The presentation of the complexity of deliberative reasoning and the variety of persuasive modes including its influencing factors shows the need for a distinction between deliberative and non-deliberative persuasive speech. Following this avenue of research, language-based criteria are significant, but for the normative evaluation of their permissibility the contextual conditions of deliberation need to be taken into consideration as well. Therefore, I propose an ideal-typical model in which two dimensions, the *group structure* and *deliberative function* are introduced. I argue that both have a major impact on the structure of the deliberative process and for the desired deliberative outcome. To illustrate the significance of these factors on deliberative reasoning I also refer to exemplary empirical results on the effects of the deliberative throughput on the deliberative outcome. Due to space constraints, I can only shed some light on the benefits of this new perspective for the evaluation of persuasive speech modes within deliberation.

The theoretical model consists of the two dimensions – group structure and deliberative function – with two ideal-typical specifications each (Table I). For group structure, I propose a distinction between *homogenous* and *heterogenous groups*. While the latter are characterized by conflicting interests with a higher level of dissonance among the deliberators, the former entails more forms of common ground, as the pre-existing positions of the participants are closer in terms of beliefs and values. The probability of the explication of opinion-confirming arguments and the level of consonance are correspondingly higher.[1]

Additionally, the intended deliberative function is decisive for the evaluation of the used persuasive modes. There is a difference between *decision-* and *epistemic-oriented deliberation*. Both functions are grounded in different perspectives on the legitimation of deliberative processes within society. Decision-oriented deliberation aims at producing highly justified decisions. These entail quite immediate real-world effects, as participants are required to choose a certain course of action. On the contrary, deliberation can fulfill an epistemic function. Here, I refer to the notion of epistemic proceduralism (Estlund, 1997). Rejecting the pure intrinsic value of the democratic procedure, which focuses on general criteria like fairness and equality, Estlund argues that democratic decision-making is legitimate, because its procedure tends to produce correct decisions. Such an understanding of epistemic deliberation can lead to a greater acceptance of plurality, as agreement on belief or value systems could be established more easily (see Dryzek/Niemeyer, 2006 for a discussion on meta-consensus). Here, the preferred deliberative outcome does not necessarily need to be a consensus, as even forms of disagreement can have an epistemic value. This perspective is important, because participants arguably tend to talk in a different manner, in cases in which no ultimate course of action is determined.

Table I. Ideal Types of Deliberative Settings

	Homogenous	Heterogenous
Group Structure	Disagreement is nearly absent (forms of common ground exist) More opinion-conforming arguments → higher probability of consonance	Forms of disagreement dominate More opinion-diverging arguments → higher probability of dissonance
Deliberative Function	**Decision-Oriented** Deciding course of action with immediate real-world effects → potential change of individual preferences	**Epistemic-Oriented** Epistemic Proceduralism → potential change of individual belief- and value-systems

As pointed out above, within more homogenous groups more likeminded people are reasoning together. Nonetheless, minority positions still exist even within more homogenous groups and participants holding

---

1 I am aware that deliberation in a normative sense requires a certain level of heterogeneity to allow for the inclusion of diverging positions and the existence of a conflict to be resolved. As the presented categories represent ideal types, in the real-world there will be gradations of these deliberative settings. Nonetheless, conflicts and deliberative discussion can generally arise in more homogenous groups with one dominant position.

such a position might less often raise it due to an aversion to disagree. Through social influence mechanisms, they might feel an immanent pressure to comply with the perceived dominant position (Sunstein, 2000). Here, a distinction between the actual and perceived majority position is important. It is conceivable that representatives of certain positions give the impression of being in the majority. The silencing of minority positions can imply the tendency for groups to make more extreme decisions than can be expected based on the participants' initial positions, which is described by the phenomenon of group polarization (ibid.). On an individual level, pre-existing positions are then strengthened or intensified.

Gerber et al. (2014, 416) point out that the number of individuals offering a certain argument ought not to play a role in deliberation, but the dominance of one position can be harmful for deliberative talk. The simple repetition of arguments can lead to a validity effect of supporting this position, mainly for rather uninformed participants (Moons et al., 2009). Simultaneously, minority positions might be more unexpected and therefore require more explanation, making it even more difficult to receive proper attention. It can be argued that individuals are allowed to use non-argumentative persuasive force to ensure the consideration of all proposals in this situation. This is an essential requirement for decision-oriented deliberation.

Moreover, homogenous groups tend to have a strong group identification which might imply a stronger adherence to held beliefs. The danger of motivated reasoning is increased through a strong social identity, as the acceptance of opposing opinions, beliefs or norms is limited (Strickler, 2019). This aspect also applies to epistemic-oriented deliberation and the establishment of meta-consensus. Persuasive modes in homogenous groups can imply the tendency of negatively evaluating outgroups, which is often discussed under the term of affective polarization. However, new ideas and beliefs can emerge within deliberation, which do not require abandoning pre-existing beliefs. Despite argumentative persuasion, rhetorical modes can contribute to this purpose and lead to an exchange of individual beliefs.

Within more heterogenous groups, conflicts are more common and empirical findings indicate that deliberation can result in depolarization or even homogenization of opinions (Mercier/Landemore, 2012, 252). Existing disagreements need to be made transparent and require reason-giving in an adversarial manner. For decision-making, individuals might be less willing to concede and discard their initial position. Again, for the epistemic deliberative function of knowledge gain, rhetorical modes can have a positive impact (Bächtiger et al., 2010). Nonetheless, individual reasoning remains important for epistemic deliberation. This means that deliberators must be able to identify and actively process the given claims and conclusions, regardless of whether these are based on arguments or rhetorical modes.

6. Implications for Empirical Deliberation Research

The presented empirical observations on the structures and effects of deliberative reasoning demonstrate that the question of this paper, whether persuasion can be regarded as a symbiosis or parasitism within deliberation, must be answered based on the contextual conditions of deliberation. I hope to have illustrated the need for a more differentiated evaluation of persuasion within deliberation by considering both the group structure and the deliberative function. This perspective entails some implications for further avenues in empirical deliberation research.

The structures of used interactions and their effects on opinion formation have received more attention in deliberation research (Rosenberg, 2007), because the deliberative quality is not solely dependent on the deliberative outcome like the frequency and intensity of individual opinion shifts. Neblo (2007) has already pointed out that the empirical assessment and evaluation of the deliberative output need to be based on normative standards. As the dynamics of the deliberative throughput are very complex, the exact processes of both interpersonal and personal reasoning need be investigated more closely.

The role of persuasion and its normative evaluation regarding deliberation are promising research avenues in this field. Only very few studies have empirically tried to do so (e.g. Westwood, 2015). The distinction between an epistemic- and decision-oriented deliberative function and a homogenous or heterogenous group structure can contribute to an improvement of empirical operationalization schemes to measure the quality of the deliberative throughput.

Furthermore, most studies on the role of persuasion as a driving force for opinion change focus on artificial deliberative settings. Other contexts including more forms of everyday talk have hardly been considered for empirical analysis so far. Although persuasion might be less likely in these settings, these forms of discourses happen more regularly and are therefore more relevant for deliberation research. This is especially true for deliberation in the algorithmically curated online sphere.

References

Amossy, R. (2001). Ethos at the Crossroads of Disciplines: Rhetoric, Pragmatics and Sociology. *Poetics Today, 22(1),* 1–23.

Bächtiger, A., Niemeyer, S., Neblo, M., Steenbergen, M.R., & Steiner, J. (2010). Disentangling Diversity in Deliberative Democracy: Competing Theories, Their Blind Spots and Complementarities. *The Journal of Political Philosophy, 18(1),* 32–63.

Chambers, S. (2003). Deliberative Democratic Theory. *Annual Review of Political Science, 6,* 307–326.

Dryzek, J.S., & Niemeyer, S. (2006). Reconciling Pluralism and Consensus. *American Journal of Political Science, 50(3),* 634–649.

Eagly, A.H., & Chaiken, S. (1993). *The psychology of attitudes.* Fort Worth: Harcourt Brace Jovanovich.

Estlund, D. (1997). Beyond Fairness and Deliberation: The Epistemic Dimension of Democratic Authority. In J. Bohman, & W. Rehg (Eds.), *Deliberative Democracy: Essays on Reason and Politics* (pp. 173–204). Cambridge: MIT Press.

Gerber, M., Bächtiger, A., Fiket, I., Steenbergen, M.R., & Steiner, J. (2014). Deliberative and non-deliberative persuasion: Mechanisms of opinion formation in EuroPolis. *European Union Politics, 15(3),* 410–429.

Green, M.C., Strange, J.J., & Brock, T.C. (Eds.). (2002). *Narrative impact: Social and cognitive foundations.* Mahwah: Erlbaum.

Habermas, J. (1996). *Between Facts and Norms. Contributions to a Discourse Theory of Law and Democracy.* Cambridge: MIT Press.

Habermas, J. (1987). *The Theory of Communicative Action, Volume I.* Cambridge: MIT Press.

Holbert, R.L., Garrett, R.K., & Gleason, L.S. (2010). A new era of minimal effects? A response to Bennett and Iyengar. *Journal of Communication, 60(1),* 15–34.

Kahan, D.M. (2015). The Politically Motivated Reasoning Paradigm, Part 1: What Politically Motivated Reasoning Is and How to Measure It. In R.A. Scott, S.M. Kosslyn, & M.C. Buchmann (Eds.), *Emerging Trends in the Social and Behavioral Sciences.* Hoboken: John Wiley & Sons.

Kirchev, Y., Atkinson, K., & Bench-Capon, T. (2019). Demonstrating the Distinctions Between Persuasion and Deliberation Dialogues. In M. Bramer, & M. Petridis (Eds.), *Artificial Intelligence XXXVI. 39th SGAI International Conference on Artificial Intelligence* (pp. 93–106). Cham: Springer.

Klemp, N. (2012). *The Morality of Spin: Virtue and Vice in Political Rhetoric and the Christian Right.* Lanham: Rowan & Littlefield.

Lupia, A. (2002). Deliberation disconnected: What it takes to improve civic competence. *Law and Contemporary Problems, 65(3),* 133–150.

Manin, B. (1987). On Legitimacy and Political Deliberation. *Political Theory, 15(3),* 338–368.

Mercier, H., & Landemore, H. (2012). Reasoning Is for Arguing: Understanding the Successes and Failures of Deliberation. *Political Psychology, 33(2),* 243–258.

Mercier, H., & Sperber, D. (2011). Why do humans reason? Arguments for an argumentative theory. *Behavioral and Brain Sciences, 34,* 57–111.

Moons, W.G., Mackie, D.M., & Garcia-Marques, T. (2009). The impact of repetition-induced familiarity on agreement with weak and strong arguments. *Journal of Personality and Social Psychology, 96(1)*, 32–44.

Neblo, M.A. (2007). Change for the better? Linking the mechanisms of deliberative opinion change to normative theory. Available at: https://polisci.osu.edu/sites/polisci.osu.edu/files/NebloChange4B063014.pdf

O'Keefe, D.J. (2002). *Persuasion: Theory and research* (2nd edition). Thousand Oaks: Sage Publications.

Perloff, R.M. (2010). *The dynamics of persuasion: Communication and attitudes in the 21st century* (4th edition). New York: Routledge.

Petty, R.E., & Cacioppo, J.T. (1996). *Attitudes and persuasion: Classic and contemporary approaches.* Boulder: Westview.

Poggi, I. (2005). The goals of persuasion. *Pragmatics & Cognition, 13(2)*, 297–336.

Richey, M. (2012). Motivated Reasoning in Political Information Processing: The Death Knell of Deliberative Democracy? *Philosophy of the Social Sciences, 42(4)*, 511–542.

Rosenberg, S.W. (2007). Types of Discourse and the Democracy of Deliberation. In Id. (Ed.), *Deliberation, participation and democracy: Can the people govern?* (pp. 130–158). London: Palgrave Macmillan.

Strickler, R. (2018). Deliberate with the Enemy? Polarization, Social Identity, and Attitudes toward Disagreement. *Political Research Quarterly, 71(1)*, 3–18.

Sunstein, C.R. (2000). Deliberative Trouble? Why Groups Go to Extremes. *The Yale Law Journal, 110(1)*, 71–119.

Toulmin, S.E. (2003). *The Uses of Argument* (updated edition). Cambridge: Cambridge University Press.

van Eemeren, F.H., & Grootendorst, R. (1984). *Speech Acts in Argumentative Discussions. A Theoretical Model for the Analysis of Discussions Directed towards Solving Conflicts of Opinion.* Dordrecht: Foris Publications.

Walton, D. (2006). How to make and defend a proposal in a deliberation dialogue. *Artificial Intelligence and Law, 14*, 177–239.

Walton, D., & Krabbe, E. (1995). *Commitment in Dialogue. Basic Concepts of Interpersonal Reasoning.* Albany: State University of New York Press.

Westwood, S.J. (2015). The Role of Persuasion in Deliberative Opinion Change. *Political Communication, 32(4)*, 509–528.

AN ARGUMENTATIVE RECONSTRUCTION OF MENCIUS'S VIRTUE THEORY

ZI-HAN NIU
Department of Philosophy, Sun Yat-sen University, China
niuzh@mail2.sysu.edu.cn

MING-HUI XIONG
Guanghua Law School, Zhejiang University, China
xiongminghui@zju.edu.cn

Abstract

Virtue has always been the core concept in Chinese philosophy. Mencius (372 BC - 289 BC) constructed the Confucian system of moral theory based on his list of four innate ethical dispositions: benevolence (*rén*), righteousness (*yì*), propriety (*lǐ*), and wisdom (*zhì*). In Mencius's ethical system, moral norms are not only external ethical means and moral rules but also the embodiment of individual moral character and moral subjectivity. In this paper, we will show that Mencius's perspective on the virtues and their extension into argumentation echoes the virtue argumentation theory (VAT) that emerged in the west. Some scholars like Cohen and Aberdein (2010) have tried to compile the typology of argumentational virtues, including intellectual empathy, intellectual humility, intellectual integrity, etc., which can be associated with Mencius's ethical dispositions. We will reveal the interaction between VAT and Mencius's virtue theory from the following two perspectives. On the one hand, VAT brings a new perspective for us to understand Mencius's virtue theory in an argumentative approach, and in reverse, virtues advocated by Mencius can contribute to the understanding of argumentational virtues raised by VAT. On the other hand, we hope to discover the role of virtue in Mencius's argumentation.

1. Introduction

Mencius (372 BC - 289 BC) was a representative Chinese Confucian philosopher whose development of Confucianism earned him the title "Second Sage", meaning second in importance only to Confucius himself. One of Mencius's most influential views was his list of four innate ethical dispositions. Mencius extended Confucius's *"ren-yi-li"* ethical system into "the four sprouts or virtues", namely, the *"ren-yi-li-zhi"* ethical system, among which *ren* refers to benevolence or humanness, *yi* to righteousness, *li* to propriety, and *zhi* to wisdom or knowledge. Besides, Mencius was also known as "being fond of argumentation". Mencius's thinking was recorded in the eponymous collection of his dialogues, debates, and sayings, the *Mengzi* (*Mencius*)[1]. Mencius linked moral value judgment with moral practice, which provided inherent rationality for moral thoughts such as benevolence, righteousness, and goodness of nature. In this paper, we will show that Mencius's perspectives on the virtues and their extension into argumentation echo the virtue argumentation theory that emerged in the west.

Virtue Argumentation Theory is a branch of argumentation theory that aims to take the character of the arguer into consideration, which infuses the perspective of contemporary virtue ethics and epistemology into argumentation analysis and evaluation. In the past decades, it has been developed by some scholars like Cohen and Aberdein (Cohen, 2007; Aberdein, 2010). In the virtue argumentation theory, some scholars like Cohen and Aberdein have tried to compile the typology of argumentational virtues including intellectual empathy, intellectual humility, intellectual integrity, etc. (Aberdein, 2010), which can be associated with Mencius's ethical dispositions.

We will reveal the interaction between virtue argumentation theory and Mencius's virtue theory from the following two perspectives: (a) virtue argumentation theory brings a new perspective for us to understand Mencius's virtue theory in an argumentative approach, and in reverse, virtues advocated by Mencius can bring new connotations to argumentational virtues raised by virtue argumentation theory; (b) in terms of argumentation evaluation, many scholars have explored Mencius's argumentative discourses from different perspectives like analogical reasoning (Lucas, 2019) and the pragma-dialectical model (Yan & Xiong, 2020), but few have done it by the virtue approach; we hope to discover the role of virtue in his argumentation.

In the next section, we will try to construct a Chinese philosophical version of virtue argumentation theory — *"ren-yi-li-zhi"* of argumentation — based on Mencius's ethical system; following the elaboration of this construction, section 3 will turn to a comparative study of Aberdein's classification and Mencius's classification of argumentational virtue to highlight the characteristics of each classification and show how each classification can work together by conducting a dialogue between them. Section 4 will conclude the paper.

---

[1] For the original text of Mencius, we refer to Yang Bojun's Translation and Annotation of Mencius (2018).

2. A Chinese Philosophical Version of Virtue Argumentation Theory — "ren-yi-li-zhi" of Argumentation

The insight of "aretaic turn" in virtue argumentation theory derives from ancient western classic thoughts. "An emphasis on virtue, or *aretê*, was characteristic of ancient Greek thought from the time of Homer, if not earlier." (Aberdein, 2010). Both Socrates and Plato have virtue theories, while the virtue theory in argumentation takes more Aristotelian ideas. Aristotle's focus on virtue pervaded much of his work, contributing a lot to virtue epistemology. Indeed, in the same era as Aristotle, the Confucian school in the pre-Qin period of China also paid great attention to virtue and built a complete ethical system. Mencius was one of the representative figures. Therefore, we are considering whether there is a possibility of applying Mencius's virtue theory to the theory of argumentation and developing a Chinese philosophical version of virtue argumentation theory.

By developing a Chinese philosophical version of virtue argumentation theory with Mencius's, we hope to find theoretical resources that can contribute to the development of virtue argumentation theory. Besides, by constructing a Chinese philosophical version of virtue argumentation theory, Mencius's argumentative virtues will be more familiar and applicable to arguers who are deeply influenced by Confucian culture.

To construct such a Chinese philosophical version of virtue argumentation theory, first, we need to be clear with the connotations of "*ren-yi-li-zhi*" and their extension into argumentational virtues. In Mencius's ethical system, each of the four virtues is associated with a characteristic emotion or motivational attitude: "The feeling of compassion is benevolence. The feeling of disdain is righteousness. The feeling of respect is propriety. The feeling of approval and disapproval is wisdom" (Mengzi 6A6; van Norden, 2019)[2]. Everyone has four ethical attributes, among which the feeling of approving and disapproving is the beginning of wisdom or knowledge, the feeling of shame and dislike is the sprout of propriety, the feeling of modesty and complaisance is the start of observance of rites, and the feeling of commiseration is the origin of benevolence (Xiong, 2013). In the following subsections, we will give a separately detailed explanation of the extension of "*ren-yi-li-zhi*" into argumentational virtues.

---

[2] For the English version of *Mencius*, we refer to Irene Bloom's translation of *Mencius* (2009).

2.1. "*ren*" of Argumentation

"*ren*" of argumentation can be regarded as the argumentative benevolence. According to Mencius's interpretation, benevolence is not simply a matter of feeling a certain way: it also has cognitive and behavioral aspects. A fully benevolent person will be disposed to recognize the suffering of others and to act appropriately (van Norden, 2019).

Applying it to argumentation, a benevolent arguer should be able to fully empathize with other people's situations and have excellent insight and psychological analysis ability; she has the sensitivity to communicate in the way that the other person is comfortable with. In Book 6A, Mencius and Gaozi had several arguments over human nature. Mencius is well known for his claim that "human nature is good", while this claim was challenged by another theorist Gaozi, who declared that human nature was neither good nor not-good. In Book 6A6, there is a dialogue about human nature between Mencius and his disciple Gong Duzi. Gong Duzi listed four viewpoints on human nature that included Mencius's and Gaozi's claims and asked Mencius which one was correct. Mencius answered: "One's natural tendencies enable one to do good; this is what I mean by human nature being good. When one does what is not good, it is not the fault of one's native capacities. The mind of pity and commiseration is possessed by all human beings…Humaneness, rightness, propriety, and wisdom are not infused into us from without. We definitely possess them. It is just that we do not think about it, that is all. Therefore it is said, 'Seek and you will get it; let go and you will lose it.'…" (Mengzi 6A6). In his answer, Mencius did not deny that immoral people were born with good human nature. They behaved immorally because they did not seek humaneness, rightness, propriety, and wisdom. We think that Mencius's debate about human nature rests on a critical assumption—people are willing to fully empathize with others only when they believe that human nature is good. Mencius's defence of people who did bad things is motivated by his belief in the innate goodness of nature and his empathy for these people. From the perspective of virtue argumentation theory, Mencius himself is a virtuous arguer for believing human nature is good and seeking benevolence.

2.2. "*yi*" of Argumentation

"*yi*" of argumentation can be regarded as argumentative righteousness. Righteousness is a disposition to disdain or regard as shameful dishonorable behavior or demeaning treatment (van Norden 2019). As is the case with benevolence, righteousness has cognitive and behavioral aspects. Thus, a righteous person would object to being addressed disrespectfully (7B31), and would not engage in an illicit sexual relationship (3B3). Argumentative righteousness is the idea of arguing for the sake of what one believes in. It is not a matter of trying to make a point but defending one's values or beliefs.

Mencius was representative of a group that believed that one should participate in politics with adherence to the principle of righteousness. Indeed, throughout his life, Mencius was trying his best to promote the humanistic spirit of Confucian tradition (Yang, 2017). In Book 3B9, there is a discussion between Mencius and his disciple Gong Duzi where Mencius explains why he was said to be fond of argumentation. Gong Duzi asked Mencius if he was really fond of argumentation. In his answer to this question, Mencius answered: "How should I be fond of argumentation? I am compelled to do it...As the Duke of Zhou would have chastised those who denied fathers and rulers, I, too, want to correct people's minds, to stop deviant speech, to resist perverse actions, to banish licentious words, and so to carry on the work of the three Sages. In what way am I fond of argumentation? I am compelled to do it. Whoever can resist Yang and Mo with words is a follower of the Sage." (3B9) Mencius said he was not fond of argumentation and stressed twice that he was compelled to do it. "Clearly, this was not a choice, arising from the bottom of his heart, but simply a reluctant response to the challenge posed by the heretical doctrines of Yang and Mo. In other words, he was doing not what he loved to do, but what needed to be done, and he did so in a helpless and reluctant manner" (Niu & Zheng, 2020). Mencius's explanation of his "being fond of argumentation" happens to be the best embodiment of the virtue of argumentative righteousness. Some people are fond of argumentation because they are competitive and argumentative by nature. However, Mencius's debate was not motivated by his own selfish desires but by something he had to do in order to defend his beliefs.

2.3. "*li*" of Argumentation

"*li*" of argumentation can be regarded as argumentative propriety. Mencius suggests that wisdom and propriety are secondary in importance to benevolence and righteousness: "The core of benevolence is serving one's parents. The core of righteousness is obeying one's elder brother. The core of wisdom is knowing these two and not abandoning them. The core of ritual propriety is the adornment of these two" (4A27; van Norden, 2019). Therefore, we can regard propriety as a means of practicing benevolence and righteousness. In other words, benevolence and righteousness are the kernels, and ritual is the externalization of this kernel.

Analogously to argumentation, the rite of argument is the means, scheme, and strategy of argumentation. The arguer should be adept at using different argumentation skills. Mencius's classification of argumentational virtue is able to introduce the rhetorical dimension — that is, to take audience reception into consideration.

Xiong & Yan (2018) have summarized three main types of argument schemes used in Mencius's political argumentation — argument from consequences, argument by analogy and argument from authority. Here is an example of argument by analogy in *Mengzi*.

In Book 1B1, Mencius heard that King Xuan of Qi enjoyed listening to music; therefore, he asked the king: "Which is more pleasurable — enjoying music by oneself or enjoying it in the company of others?" the king answered: "Enjoying it with others." he continued asking: "which is more pleasurable—enjoying it in the company of a few or enjoying it in the company of many?" The king responded: "Enjoying it with many." then he said to the king: "Now suppose the king is making music here. Hearing the echoes of the king's bells and drums and the sounds of his pipes and flutes, the people, joyfully and with delighted countenances, all tell one another, 'Our king must be quite free of illness, for if he were ill, how would he be able to make music?' ...This is solely because he shares his enjoyment with the people. Now, if Your Majesty simply will share your enjoyment with the people, you shall be a true king." (1B1) In fact, Mencius chose an analogy to illustrate the principle of governing the country. The king should not be separated from his people but should let everyone lead a happy and prosperous life together. In this passage, concerning this particular audience, King Xuan of Qi, Mencius succeeded in making a persuasive argument based on the king's love of music. Sometimes, in order to achieve the purpose of persuading the target audience, the arguer could take the current situation or needs of the target audience as the starting point of the argument, which is often conducive to the realization of the purpose of persuasion.

2.4. *"zhi"* of Argumentation

"zhi" of argumentation can be regarded as argumentative wisdom. As Mencius said in Book 5A9, wisdom involves an understanding of and commitment to the other virtues, especially benevolence and righteousness. Other passages indicate that a wise person has the ability to properly assess individuals and has skill at means-end deliberation (5A9; van Norden, 2019).

Accordingly, a wise arguer should have the ability to distinguish right from wrong; she should have faith in reason and fairmindness; meanwhile, she has the ability to recognize reliable authority but also has her own value judgement.

In Book 4A17, Mencius and Chunyu Kun had an argument about a matter of ritual propriety. Back in their time, unmarried men and women should not touch one another in giving and receiving things. However, Chunyu Kun put forward an implied argumentation by asking Mencius if one's sister-in-law is drowning, may one save her with his hand? Mencius answered: "If one's sister-in-law were drowning and one did not save her, one would be a wolf. For men and women, in giving and receiving, not touching one another is according to ritual. To save a sister-in-law from drowning by using one's hand is a matter of expedience." Although Mencius realized it was a violation of the ritual prohibition, he chose to defend for the more important value. Knowing when to violate ritual is a matter of wisdom (van Norden, 2019). For the arguer, proper respect for authority and public opinion is necessary, but they should not be blindly followed, because authority and public opinion are not universal. In this sense, Mencius is a wise arguer by justifying his own beliefs without blindly following the authority of rites.

3. A Comparative Study: Aberdein's Classification and Mencius's Classification of Argumentational Virtue

In this section, we will do a comparative study of Aberdein's classification and Mencius's classification of argumentational virtue. By comparing the two types of classification, we are going to highlight the characteristics of each classification; and show how each classification can work together by conducting a dialogue between them.

Daniel Cohen (2005, 64) identifies four principal virtues in the ideal arguer: (1) *willingness to engage in argumentation*; (2) *willingness to listen to others*; (3) *willingness to modify one's own position*; and (4) *willingness to question the obvious*. Aberdein (2010) made a first attempt at compiling a typology of argumentational virtues (see Table 1 below) by employing Cohen's framework.

Table 1 A tentative typology of argumentational virtue
Willingness to engage in argumentation
  Being communicative
  Faith in reason
  Intellectual courage
    Sense of duty
Willingness to listen to others
  Intellectual empathy
    Insight into persons
    Insight into problems
    Insight into theories
  Fairmindedness
    Justice
    Fairness in evaluating the arguments of others
    Open-mindedness in collecting and appraising evidence
  Recognition of reliable authority

    Recognition of salient facts
        Sensitivity to detail
Willingness to modify one's own position
    Common sense
    Intellectual candour
    Intellectual humility
    Intellectual integrity
        Honour
        Responsibility
        Sincerity
Willingness to question the obvious
    Appropriate respect for public opinion
    Autonomy
    Intellectual perseverance
        Diligence
        Care
        Thoroughness

Although this is a tentative typology of argumentational virtue, it gives us a starting point to make argumentational virtues more detailed and intuitive. As Aberdein (2010) said: "...None of this work can be completed within the scope of a single paper. But Table 1 goes far enough to indicate how the project might be tackled, and to suggest that this is well worth doing."

3.1. A Re-classification of Aberdein's Typology into the Framework of "*ren-yi-li-zhi*"

Aberdein (2010) also pointed out that: "It makes no claim to be exhaustive or definitive: doubtless further virtues may be added, and some of the classificatory assumptions may be challenged." One way we think we can refine the understanding of Table 1 is to reclassify it with Mencius's virtue theory, that is, under the framework of "*ren-yi-li-zhi*". We believe that some subclasses in his typology are able to have better correspondence in Mencius's framework. Besides, it will be helpful to make Mencius's framework more specific and understandable and also make a contrast between the two. Based on the connotation of four principal virtues and each subcategories, we are able to find corresponding positions for Aberdein's subclasses in Mencius's classification system. Here we establish Table 2:

Table 2 A Re-classification of Table 1

Benevolence (ren) of argumentation
  Intellectual empathy
  Intellectual candour
  Being communicative

Righteousness (yi) of argumentation
  Intellectual courage
    Sense of duty
  Intellectual integrity
    Honour
    Responsibility
    Sincerity
  Intellectual perseverance
    Diligence
    Care
    Thoroughness
  Intellectual humility

Propriety (li) of argumentation
  Proper use of rhetoric
    Sense of rhetorical devices
    Sensitivity to context
    Consideration of the audience
      Respect for audience
    Analysis of audience
    Recoginition of specific target

Wisdom (zhi) of argumentation
  Faith in reason
  Fairmindedness
    Justice
    Fairness in evaluating the arguments of others
    Open-mindedness in collecting and appraising evidence
  Recognition of reliable authority
  Recognition of salient facts
    Sensitivity to detail
  Common sense
  Autonomy
  Appropriate respect for public opinion

Mencius's framework can cover the subcategories in Aberdein's typology, and after reclassification, the original subclasses have better correspondence in Mencius's framework. In Table 1, Aberdein adds fairmindedness as the subclasses of willingness to listen to others; however, we think fairmindedness has more to do with the arguer's sanity and objectivity; thus, it would be proper to consider it as one of the virtues of wisdom (*zhi*) of argumentation. Similarly, the virtue of appropriate respect for public opinion seems to have a slight contradiction with the willingness to question the obvious. It would be more agreeable with the propriety (*li*) of argumentation. Moreover, we find a lack of attention to the audience in Aberdein's typology, while the propriety (*li*) of argumentation can fill this gap. Therefore, we add new virtues under the propriety (*li*) of argumentation — proper use of rhetoric, which includes sense of rhetorical devices, sensitivity to context and consideration of the audience. We will discuss this in more detail in the following subsection.

3.2. Mencius's Virtue Theory Brings New Insights into Argumentational Virtues

As we interpreted before, the rite of argumentation is the means, scheme, and strategy of argumentation. The virtue of propriety (*li*) of argumentation emphasizes the audience. A good arguer should consider her audience during the argumentation. She should keep the sense of rhetorical devices and sensitivity to context. For example, the arguer needs to know the characteristic of her target audience by asking herself questions like: does the audience already have an advance understanding of my purpose? what does the audience know about the scope of the topic? what is the audience's experience of proposals similar to my proposal? what impact will my proposal have on the audience? what are the audience's beliefs, prejudices, or tendencies related to my purpose? etc.

The virtue of propriety (*li*) in Mencius's classification of argumentational virtue is an essential addition to Aberdein's typology and brings new insights into argumentational virtues by introducing the rhetorical dimension — to consider the audience, especially from the following two dimensions.

One is from the rhetorical dimension. The arguer needs to consider how to make the argument more acceptable to the audience. In *Mengzi*, we can find out that Mencius uses many persuasion skills facing different audiences. Mencius is particularly adept at strategic maneuvering.

The other is from the sociocultural dimension. Since different cultures have different languages, knowledge, value systems, social norms, and customs, the arguments of different sociocultural groups follow different norms, have different forms, and obey different standards of rationality. In *Mengzi*, we can find various discourses where Mencius took different strategies for facing audiences with different social backgrounds. When arguing with the rulers, Mencius skillfully used metaphors and twists in his argument. He talked about the problem first, hiding the main idea, arguing from the side, the back, and the periphery, gradually leading to the topic. If he intended to talk about the practice of benevolent government, he started with a campaign metaphor and argued from the opposite side so that King Hui of Liang unconsciously followed him. Such arguments are fascinating and winding; When arguing with his students, Mencius would adapt an educational way to help students shape their cultivation and character; When arguing with his opponents, Mencius was good at making his opponents unable to avoid his questions by pressing them with continuing questioning and doubting, and he also enlightened them with affection and motivated them with reason. This also made Mencius's argumentation rigid and flexible and able to advance and retreat freely, appearing mature and steady.

To sum up, there are two implications of this comparative study. On the one hand, Aberdein's typology and Mencius's virtue framework are compatible; they can work together to provide multidimensional argumentational virtues. On the other hand, the two types of classification are reciprocal; they can help to improve each other. Aberdein's typology promotes a better understanding of the meaning of *"ren-yi-li-zhi"* of argumentation in the sense of giving further clarity and interpretation by adding the subclasses into each principal virtue, while Mencius's virtue theory brings new insights into the argumentational virtue by introducing the rhetorical dimension.

4. Conclusion

To conclude, virtue argumentation theory brings a fresh perspective to understand Mencius's virtue theory in an argumentative approach, while virtues advocated by Mencius bring new connotations to argumentational virtues raised by virtue argumentation theory. The construction of the Chinese philosophical version of virtue argumentation theory and the comparative analysis between two kinds of classification indicate that ancient Chinese philosophy has theoretical resources that can contribute to the development of virtue argumentation theory and two frameworks can improve each other. What we have done in section 2 and 3 is a preliminary try; we believe there is more to be explored. We will do further detailed research.

As we can see, virtue argumentation theory can serve an essential role in the meeting between Chinese and Western argumentation theories and also help to start the virtue turn in the study of ancient Chinese argumentation studies.

The work we have done so far is only a preliminary exploration, and there are some issues that deserve more in-depth study but which we have not yet covered in this paper. For example, we put forward the framework of *"ren-yi-li-zhi"* based on Mencius's virtue theory; however, we find some inner tension between different virtues like *"yi"* and *"li"*. Argumentative righteousness (*yi*) is the idea of arguing for the sake of what one believes in, while argumentative propriety (*li*) emphasizes taking audience reception into consideration. So how to balance these two virtues during the arguing is a question worth thinking about. Besides, The most crucial theory of Mencius is applied in political philosophy, and its transplantation to the theory of argumentation is an initial attempt. We are wondering whether we can find more evidence in the literature to support this attempt or whether there is some misinterpretation of his theory. We hope to be able to continue to address these issues in future research.[3]

References

Aberdein, A. (2010). "Virtue in Argument," *Argumentation*, 24, p. 171.
Cohen, D. (2007). Virtue epistemology and critical inquiry: Open-mindedness and a sense of proportion as critical virtues. *Presented at the Association for Informal Logic and Critical Thinking, APA Central Division Meetings*, Chicago, IL.
Lucas, T. (2019). Analogies and Analogical Reasonings in the Mencius. *Studies in Logic, Vol. 12*, no.6, p. 46-48.
Mencius. (2009). *Mencius*. Trans. by Irene Bloom. Edited and with an introduction by Philip J. Ivanhoe. New York: Columbia University Press.
Niu, Z., Zheng, S. (2020). Argumentation in Mencius: A Philosophical Commentary on Haiwen Yang's The World of Mencius. *Argumentation 34,* 275–284. https://doi.org/10.1007/s10503-018-9452-3
Xiong, M. (2013). Confucian philosophical argumentation skills. In Mohammed, D. & Lewiński, M. (Eds.). *Virtues of Argumentation. Proceedings of the 10th International Conference of the Ontario Society for the Study of Argumentation (OSSA)*. ON: OSSA, pp. 1-11.
Xiong, M., Yan, L. (2019). Mencius's Strategies of Political Argumentation. *Argumentation 33*, 365–389. https://doi.org/10.1007/s10503-018-9463-0
Van Norden, B. (2019). "Mencius", The Stanford Encyclopedia of Philosophy, Edward N. Zalta (ed.),
URL =<https://plato.stanford.edu/archives/fall2019/entries/mencius/>.
Yan, L., Xiong, M. (2020). Dissociation and Its Philosophical Foundation in Mencius's Argumentation. In Catarina Dutilh Novaes, Henrike Jansen, Jan Albert van Laar and Bart Verheij, eds. *Proceedings of the 3rd European Conference on Argumentation., Studies in Logic and Argumentation vol. 87.* (pp.115-126) . Lodon: College Publications .
Yang, B. (2018). *Translation and Annotation of Mencius.* (《孟子译注》). Beijing, Zhonghua Shuju.
Yang, H. (2017). *I am good at norishing my vast flowing: the world of Mencius.* (《我善養吾浩然之氣: 孟子的世界》). Jinan, Qilu Shushe.

---

3 We appreciate the ECA 2022 Prof. Fabio Paglieri for providing us with an opportunity to present this research work at the conference online. This work was supported by the Major Program of National Social Science Foundation of China (Grant No. 19ZDA042).

WITTGENSTEIN'S HINGES AND THE LIMITS OF ARGUMENT

PAULA OLMOS
Universidad Autónoma de Madrid
paula.olmos@uam.es

Abstract

In this paper I sketch an answer to H. Siegel's recent (2013, 2021) criticism of the Wittgenstenian notion of unchallengeable *hinge propositions* and the use R. Fogelin (1985) made of it. Siegel adopts a pancritical fallibilism and argues for the challegeability and open scrutiny of all propositions and commitments, considering undesirable and deleterious for argumentation the idea that some privileged beliefs would, on the one hand, be exempt from argumentative support and preclude it (when we agree on them) and, on the other, impede or end our argumentative exchanges (when we disagree on them). My response does not oppose Siegel's main point about general content-challengeability but does first emphasize the positive aspects (rather than the negative ones) that functionally hinge propositions have on our argumentative life. For that end, I propose to apply many of Wittgenstein's ideas on hinge commitments and certainties to certain argumentative relations between contents rather than to specific (atomic) contents, approaching Wittgenstein's hinges under the model of Toulmin's warrants and accounting for their role in the recognition of reasons and kinds of reasons. Such hinges would be the basis of our *normal* argumentative exchanges. Secondly, about the non-permanent status of basic assumptions and their challengeability, I claim that Wittgenstein already assumes them as possibilities. But Wittgenstein's main point is that that we are able to distinguish practices in which we challenge basic assumptions from practices in which we assume and rely on them. A sensibly stratified proposal of argumentative practices may pay attention to the kinds of normative responses that the different levels of argumentative activity call for and account for the variable operation, from intuitive to discussable, of our standards of rationality.

1. Introduction

As is well known, R. Fogelin's 1985 paper on "The Logic of Deep Disagreements" makes ample use of Wittgensteinian ideas, quoting *On Certainty* profusely and relying on the notion of framework (or background, or underlying) propositions (or principles, or commitments)[1] precisely to define what a *deep disagreement* is.

These propositions, the most basic general assumptions on which we base our world views, are functionally characterized by Wittgenstein as *hinges*:

> The questions that we raise and our doubts depend on the fact that some propositions are exempt from doubt, are as it were like hinges on which those turn (*OC* §341, quoted by Fogelin, 1984: 4).

They might be either shared or not by different people, groups, cultures or *forms of life*, but they cannot be (or at least are usually not) justified or debated. According to Fogelin, shared framework propositions are central for the practices of argumentation and argument evaluation and define the workings of a *normal* argumentative exchange:

> An argument, or better, an argumentative exchange is *normal* when it takes place within a context of broadly shared beliefs and preferences (Fogelin, 1985: 3).

But when there is a clash between background principles, then we face a *deep disagreement* that *cannot be argumentatively solved*:

> [i]f deep disagreements can arise, what rational procedures can be used for their resolution? The drift of this discussion leads to the answer NONE (Fogelin, 1985: 6).

Therefore, according to Fogelin, hinge propositions (or principles, or commitments) both make possible and set limits to argumentative exchanges.

Now H. Siegel has recently (2013, 2021) published some critical papers against this whole idea of propositions that cannot be "rationally challenged, defended, and evaluated" (Siegel, 2021: 1107), of commitments that "lie apart from the route traveled by enquiry" (Wittgenstein, *OC* §88). Siegel adopts instead a *pancritical fallibilism* and argues for the challegeability and open scrutiny of all propositions and commitments, including those that could be seen as framework/hinge propositions.

For Siegel, the idea that some *privileged beliefs* would, on the one hand, be exempt from argumentative support and preclude it (when we agree on them) and, on the other, impede or end our argumentative exchanges (when we disagree on them) seems undesirable and deleterious for argumentation.

---

[1] Fogelin is not very consistent with the use of all these expressions that could obviously lead to differently nuanced hinge theories.

Siegel's arguments in these papers try to support his general thesis/conclusion about challengeability [A] and also oppose two possible sources of qualification, regarding the permanent/non-permanent status of hinge propositions [B] and their possibly relative status regarding specific practices or (in Wittgenstein's terms) *language games* [C]. Siegel's main tenets are:

[A] [f]ramework/hinge propositions are just as open to critical scrutiny as everything else. When we argue, there is no free lunch; no starting point immune from critical scrutiny. We should opt for fallibilism, not Wittgensteinian 'forms of life' or unchallengeable 'hinge propositions.' (Siegel, 2013: 166).

[B] [e]ven if Wittgenstein is right that hinge propositions are unchallengeable, there is nothing necessary or permanent about that status (Siegel, 2013: 165).

[C] One plausible reading of Wittgenstein is that hinge propositions are groundless within a given *language game* or system of presuppositions/ judgment, but are evaluable/criticizable/supportable from some other language game or perspective. I am myself in some sympathy with this way of thinking about hinges [...] But this sort of relativity to a language game is innocuous, epistemologically (Siegel, 2021: 1111).

My idea in this paper is not really challenging Siegel's main thesis [A] which, in very broad, general terms, as a universal, abstract statement, could be accepted, but trying to establish, instead, the not so *innocuous* character of the qualifications brought up by reflecting on [B] and most specially [C].

2. A Sketch Response to Siegel

The first thing I would like to point out is that Siegel's remark on the "non-permanent status" of basic commitments and of hinge propositions can already be found in Wittgenstein. As is well known, the *certainties* on which we base our reasoning and justifying habits allegedly work as the *unjustifiable limit of justification*, the *bedrock* on which the spade turns (Wittgenstein, *PI* §217). But *On certainty* also uses the not-so-permanent image of the *river-bed*:

It might be imagined that some propositions [...] were hardened and functioned as channels for such empirical propositions as were not hardened but fluid; and that this relation *altered with time*, in that fluid propositions hardened, and hard ones became fluid. [...] The mythology *may change* [...] But I distinguish between the movement of the waters on the river-bed and the shift of the bed itself; though *there is not a sharp division* of the one from the other. [...] Yet this is right: the same proposition may get treated *at one time* as something to test by experience, *at another* as a rule of testing [...] And the bank of that river consists partly of hard rock, subject to no alteration *or only to an imperceptible one*, partly of sand, which now in one place now in another gets washed away, or deposited. (Wittgenstein, *OC* § 96-99, emphases added).

So it seems that the non-permanent (and therefore not necessary) status of what Wittgenstein still considers basic certainties (at some point) does to affect his theory about their mode of functioning or his conviction about their epistemological import (i.e. their core influence on how our epistemic life really operates).

It is noticeable that one of Wittgenstein's choice examples ("No one has ever been on the Moon", *OC* § 106, 108, etc.)[2] cannot be taken anymore as a *certainty* on which to base other beliefs and practical decisions. And this has not deter philosophers from trying to make sense of his hinge theory.

Something similar happens with [C]. Siegel says he is sympathetic to the possibility of conceiving of hinge propositions as relative to concrete practices or language games. But he does not consider this feature something *epistemologically significant*. For Siegel, so it seems, epistemology regards the ultimate philosophical establishment of certain deep/universal principles. And thus, if what in some particular context works as a hinge proposition can still be discussed and challenged in an epistemological inquiry (within a philosophy seminar), then its basic or *serious* status is what is revealed of it in such an inquiry:

> What should be said of the fact that the examples of 'hinge' propositions offered are in fact regularly discussed and challenged/defended in many Introduction to Philosophy courses, thus seemingly undermining the Wittgensteinian claims that in these cases the question of justification is inappropriate and that calling such propositions into question is 'unintelligible'? (Siegel, 2013: 165).

But, of course, Wittgenstein was aware of that, taking in account that in *OC* he was responding/reacting to G.E. Moore's ostensive argument about the existence of the external world (1993 [1939]), and yet, he seemed to think that what can be challenged and established in an epistemological discussion has not *so much* bearing on our current epistemic life, or on the way we usually understand terms as *knowing* or *knowledge*:

> For when Moore says 'I know that that's...' I want to reply 'you don't know anything!'-and yet I would not say that to anyone who was speaking without philosophical intention. That is, I feel (rightly?) that these two mean to say something different (Wittgenstein, *OC* §407).

Wittgenstein's main point here (as in *OC* § 97) is that we are able to distinguish between epistemic practices in which we challenge certain

---

[2] *On Certainty* was written between 1949 and 1951.

hinge propositions from the ones in which our cognition *hinges* on them. And what remains to be seen is first what is it to *work as a hinge proposition within* a particular practice and whether there are certain of these propositions (the *hardest*), for which even if the possibility of challenge is available, it is in *very special practices* that somehow change the usual grammar of certain terms, bearing just slightly on our epistemic life.

Within this approach, it seems better to characterize hinges more functionally than ontologically. The question would be, then, what is that role played by certain principles or propositions, within certain practices, that has consequences for their (*at least local*) challengeability? Wittgenstein suggest they get hardened to become "rules of testing" (*OC* § 98), that is, rules that define how we test, prove, argue for, or establish the correctness or acceptability of other contents.

In this sense, I want to explore the idea that a suitable (and argumentatively relevant) way to understand hinges would make of them not exactly profound convictions that we are not prepared to renounce but rather relational principles that play a more distinct role in the *recognition of something as a reason for something else,* as Marraud's definition of arguing states (2020: 11), shaping thus the way we argue and understand argument. Their positive role for (Fogelin's *normal*) argument would not just be that one needs some starting point or some basic beliefs in order to justify others as, in the view I will defend here, hinges would not directly work as reasons (or evidence) but as *rules of what can be taken as a reason* for some particular content, within a particular practice. What I clearly want to suggest is that a suitable approach to Wittgenstein's hinge theory could be based on S. Toulmin's theory of warrants (1958).

3. Coliva's framework reading of hinges

In her paper "Which Hinge Epistemology?" (2016), A. Coliva construes her understanding of hinges in a way that can be very useful for our purposes, even though in her approach hinges still have a propositional nature.

Coliva sensibly leaves aside propositions as "No one has ever been to the Moon". She does not feel so confident, either, in treating "Here is my hand" as a good example of a hinge proposition, and chooses instead, more general ("There are physical objects," or "My sense organs are generally reliable,") or, at least, more generally applicable ("The Earth has existed for a very long time,") certainties. Coliva offers a relevant list of Wittgenstein's claims about hinge propositions being:

(i) neither true nor false (*OC* 196–206);(ii) neither justified nor unjustified (*OC* 110, 130, 166, 359); (iii) neither reasonable nor unreasonable (*OC* 559); (iv) therefore, they are neither known nor unknown (*OC* 4); (v) they cannot sensibly be called into doubt (*OC*

123, 231); (vi) thus, finally, for these very reasons, they aren't empirical propositions but rules (*OC* 95, 98, 494).(Coliva, 2016: p. 82)

She also offers (Coliva 2016: 84) a very clear metaphilosophical classification of different readings or interpretations of both Wittgenstein's hinges and of the epistemological anti-skeptical import of *On Certainty:* a) the *therapeutic interpretation* (Conant, 1998) for which skeptical doubts about such certainties are nonsensical because meaningless; b)the *framework reading* (Moyal-Sharrock, 2004; Coliva, 2010) that takes hinges to be rules, so that skeptical doubts about them are not meaningless but misplaced; c) the *naturalist reading* (Strawson 1985), claiming that we naturally take hinges for granted, so skeptical doubts are unnatural and, finally, d) the *epistemic reading* (Wright 2004; Williams 2004), basically opposed to the framework reading and the rule (vs. propositional) interpretation of hinges, and encouraging a broader idea of justification that would made hinges justifiable (even if not *evidentially justifiable*).

In this paper, Coliva defends what she calls "a qualified framework reading". She thinks hinges are, or better *function basically as rules* that, nevertheless, also have a propositional character (on a fairly relaxed notion of proposition). The first aspect (their functioning as rules) is what governs their normative role and exemption from doubt. The second aspect (their retaining some propositional content or determination) is what makes sense of their being the object of the propositional attitude of certainty. In Coliva's terms they are "propositions that *play a rule-like role"* (Coliva, 2016: 86).

Following Wittgenstein's idea of the "rules of testing", Coliva suggest that they are "rules of evidential significance" (what makes them *evidentially* unjustifiable):

[t]hey allow us to acquire and assess evidence [...] Yet that evidence, in its turn, being dependent on holding them fast, can't give them any epistemic support. [...] Therefore, hinges can't be justified (Coliva, 2016: 86).

Skeptical doubts about them are not meaningless but really ill-posed, as such decontextualizing doubts do not take into account their rule-like functioning nor the specific role they play in "acquiring and assessing evidence" (Coliva, 2016: 88). Coliva also claims that hinges are not metaphysically necessary and that they are "entrenched in a practice that, quite clearly, could have been different from what it in fact is" (89). Nevertheless, she tries to identify especially entrenched hinges that would sustain an anti-skeptical epistemology. Such as those that would allow us "to take mind-dependent evidence to bear onto beliefs about mind-independent objects" (Coliva, 2016: 92).

4. Toulmin on natural laws and Toulmin's warrants as hinges

Coliva's subtle qualifications about hinges match Toulmin's (1953) *clearly Wittgensteinian* ideas on "scientific laws". Against the idea of the laws of nature being "restricted empirical generalizations" (Toulmin 1953: 88ss), Toulmin qualifiedly seems to embrace M. Schlick's and F. P. Ramsey's suggestions that they are not "propositions" but "instructions for the formation of propositions", "maxims", "directions for the investigator to find his way about in reality" (1953: 90). If this is so, words like "true", "false" and "probable" are not applicable to laws but only to the statements that constitute the application of the law.

However, just as Coliva with hinges, Toulmin wants "laws of nature" to *retain* some empirical (propositional) content or at least dependence. So against Schlick's insistence on the "instruction/directive" character of laws (based on the assumed dichotomy between theoretical and practical contents), Toulmin claims that this view excessively "severs laws of nature from the world" (1953: 91), as if they were only speaking about the physicist's (conventional) conduct.

Thus, he prefers to conceive of laws as "principles", "inference-licenses" or "inference-tickets" (instead of mere instructions) and claims that these constitute a third category between propositions and directions:

[t]he fact that some inferences rather than others come to be licensed usually tells us much more about the world than about the physicist and his methods (Toulmin, 1953: 91).

It is not too far-fetched to see in Toulmin's reflections on *laws of nature* the seed of what just five years later (Toulmin, 2003 [1958]) would be the cornerstone of his model of argument, namely those very particular components of argument called *warrants*.

Toulmin's warrants seem to nicely fit some of Wittgenstein's (and Wittgensteinian scholars's) ideas on hinges and to play the same kind of role in being "rules of testing". They also present the double aspect of positively making possible (Fogelin's *normal*) argument and of negatively limiting justifying practices, because of their (at least *local* or *practice-dependent*) unchallengeability:

[u]nless, in any particular field of argument, we are prepared to work with warrants of some kind, it will become impossible in that field to subject arguments to rational assessment (Toulmin, 2003 [1958]: 93).

Indeed, if we demanded the credentials of all warrants at sight and never let one pass unchallenged, argument could scarcely begin (Toulmin, 2003 [1958]: 98).

Toulmin is also aware that such (local) unchallengeability is usually a contested feature, precisely, by philosophers and by specific philosophical stances:

> But supposing a man rejects all warrants whatever authorising (say) steps from data about the present and past to conclusions about the future, then for him rational prediction will become impossible; and many philosophers have in fact denied the possibility of rational prediction just because they thought they could discredit equally the claims of all past-to-future warrants (Toulmin, 2003 [1958]: 93).

This skeptical position is not incomprehensible for Toulmin but the price to be paid for it is obviously too high.

The limitations of warrant justification and challengeability also have to do with their specific "logical role" and their rule-like functioning. When Toulmin traces the distinction between the Data and the Warrant in his model of argument (a strategy that sensibly breaks with the traditional *functionally undifferenciated* premises-plus-conclusion model), the distinction is precisely based on the (typically) propositional character of the data *vs. the rule-like nature* of the warrant:

> [t]he nature of this distinction is hinted at if one contrasts the two sentences, 'Whenever A, one *has found* that B' and 'Whenever A, one *may take it* that B.' (Toulmin, 2003 [1958]: 91).

Warrants, not being propositions, are not to be evaluated/justified in terms of their truth/falsity. Still their applicability and acceptability might be established/defended as *depending* on some propositional content, a usual feature (at least within some fields/practices) of what Toulmin calls Backings:

> Certain types of warrants thus rely for their soundness and relevance (as in statute law) on deliberate collective human decisions. Others (as in natural science) rely on *our recognition of general patterns in the world of nature*. Others (as in much of our everyday understanding of human actions and motives) rely on familiar, recognized regularities in human affairs (Toulmin *et al.* 1984 [1978]: 68, emphasis added).

Now, the relation between Backing and Warrant is an argumentative not a restrictedly inferential one and, thus, the non-propositional nature (even though tied to and based on certain facts) of the warrants can still be maintained:

> Though the facts about the statute may provide all the backing required by this warrant, the explicit statement of the warrant itself is more than a repetition of these facts: it is a general moral of a practical character, about the ways in which we can safely argue in view of these facts (Toulmin 2003 [1958]: 98).

Toulmin's warrants perform the role of ruling what kind of data/evidence allow us *to take it that* our claims are justified. They, therefore, delimit the recognition of some content as a reason for something else.

Warrants can typically be challenged and justified. That's what Backings are for and the reason why Toulmin talks about *warrant-establishing arguments* (2003 [1958]: 111). But their specific logical role in ruling what kind of evidence allow us *to take it that* our claims are justified make of them foundation stones of evidential practices that are usually not challenged *within* those practices.

Warrant's contextual or local unchallengeability defines, thus, recognizable "argumentative practices". Changes in legal laws or in laws of nature (i.e. paradigm changes) change the way we assess evidence and establish claims within those same practices.

5. Are there absolutely unchallengeable warrants?

However, there might still possibly be some *very general (very basic) warrants* (we could call them *deep warrants*), whose rejection amounts to the kind of skepticism that can factually be advocated in a philosophical discussion and yet not bear onto the way skeptics argue everywhere else (or in most of their epistemic life).

This does not mean that they are *absolutely* unchallengeable, but that challenging them implies a frame of mind that requires changes in the way we currently operate and in the way we understand certain terms. In Wittgenstein's terms requires speaking *with* philosophical intention (*OC* §407).

A symptom of our having reached this kind of epistemically fundamental principles occurs when we challenge and discuss basic, abstracted, modes of arguing in philosophical exchanges, approaching problems and questions as the long-standing "justification of induction" or the "justification of abduction", a kind of philosophical discussion which I have recently explored (Olmos 2021). I will use my findings in this particular topic to try to establish what a *deep warrant* might be.

One thing is to account for the way we model and discuss a *regular* (first-level) abductive argument, in which the following components are identifiable:

Warrant: Hypothesis could explain data

Data: Shared (usually empirical) data

So

Claim: Hypothesis

And a different thing is to make sense of the philosophical questioning of the sheer possibility of using abductive reasons. I have proposed modelling this second-level kind of exchange as the discussion of the following scheme:

Warrant: That a hypothesis explains some data is a reason to consider those data a justifying reason for that hypothesis:

Reason: Hypothesis explains Data

So

Claim: Data justify Hypothesis

Following Toulmin's suggestion, this candidate *deep warrant* would better be expressed as: "Whenever a hypothesis explains some data, *one may take it* that those data justify the hypothesis". In my own work there is a noticeable struggle in the determination of this supposedly challengeable and justifiable but clearly engrained principle. First of all, it does not seem to respond to some of the usual characteristics of Toulminian warrants:

> [i]f you ask, what kind of reason could that one be for such a claim? the answer should be expressed by the *warrant* I have provided which is, I must agree, *frustratingly* redundant, and so kind of useless. It might be considered [...] a "more general" statement than the argument it covers, but it surely does not add any new substance to it, beyond formal or informal subsumption. This is a problem according to my own reading of Toulmin's *warrants* (Olmos, 2021, 143-144).

One of my excuses for it is that with such explorations: "we might be reaching a really deep (cognitive, logical) level of what counts as a reason, not really based on more reasons-for-reason-being than sheer "intuition"." (Olmos, 2021, 144). And yet, philosophers keep trying to look for backings for such a warranting principle even if sometimes they have to settle for the mere realization that:

> The familiarity and ubiquity of this pattern of argumentation shows that speakers and hearers share some common understanding of it, and also share a basic agreement that it has persuasive force (Josephson, 2001: 1623).

What Fisher (2014, 1070) has called "a basic source of justification in our theorizing about the world" is not easy to discuss, although we might philosophically conceive of its challengeability.

Moreover, this kind of *deep principles*, inference-licenses, etc. are generally not explicitly learned but acquired in practice and even conceiving of challenging them is difficult for us as they are usually and for most of our practices the object of our *certainty*. As Coliva (2016, p. 86) has put it:

> That is to say, we do accept them and hold them fast, we behave in accord with them, or else accept criticism for not doing so and, finally, pass them on to our children through education and training.

6. How does practice-dependent challengeability work?

And yet we are philosophical creatures that actually (but only sometimes) challenge our most profound and ingrained convictions.

A possible way to understand how we do that and still act upon them can come from our awareness of the *stratified* character of our constellation of reason-giving-and-receiving practices. When we focus on these, the normativity of the rational relations that are presented through them surfaces in three ways or levels of argument reception and reason assessment.[3]

1. First, we may rely on (hinge on) the kind of *tacit or implicit normativity* shown in our argumentative normative conduct. That's how our usual recognition of reasons work, without necessarily resorting to underlying general principles. Following tacit (usually *deep*) warrants is something we learn to do together with the common uses of language and discourse. And this allows us to recognize basic likenesses between modes of arguing, to present analogies and counter-analogies, and to apply basic normative standards as Perelman's "rule of justice" (Tindale, 2015: 73) or Woods and Hudak (1989) criterion of a "parity of reasoning".

2. Argumentative normativity may also be *made explicit, expressed and used* in argumentative practice (without, in principle challenging it). We explain and justify our concrete reason-giving acts through the expression of recognizable warrants that define the *standards of reasonability* that are presented as currently valid within a community or a particular argumentative practice. We learn to recognize and use them in higher-level, more reflective, typically specialized and professionalized argumentative practices. Taking place among peers, these practices are usually more conscious of their foundation upon

---

3 This classification proposal can be seen as a refinement of Toulmin's distinction between *warrant-using* and *warrant-establishing* arguments (2003 [1958]: 111) and Toulmin et at.'s distinction between *regular* and *critical* arguments in each argumentative field or practice (Toulmin et al. 1984 [1978]: 67).

(hinging upon) explicit warrants and of the norms and limits of their applicability.
3. Finally, argumentative normativity may (also argumentatively) be *assessed and discussed*. That is what (over)-reflective cultural communities (or particular practices within them) do. They problematize their own standards of reasonability, try to discuss and asses them. Sometimes they even change them, appealing to epistemic, ethical or empirical grounds. This is done through also argumentative (even) higher-level practices. Either within specialized or professional fields as in Toulmin et al.'s (1984 [1978]: 67) field-related *critical arguments*. Or in philosophical discussions about the justification of usually *deep warrants*.

An argumentative approach to these levels results in their characterization through different types of response in argumentative practice. Thus, if particularist analogy and counter-analogy (Alhambra, 2022) is the typical argumentative sign of the first level, regular warrant-using arguments, objections and refutations are the sign of the second level and the exploration of backings, determination of warrant-limits or exceptions, and presentation of rebuttals, an argumentative sign of the third one.

7. Conclusions

Wittgenstein's "hinge propositions" understood as "rules of testing", "rules of evidential significance" allegedly both make arguing and justifying (within a certain argumentative practice) possible and cannot be justified or argued for *within* than practice.

Toulmin's notion of "warrant" (rules designating on what kind of content *one may take it* that another content is justified) inherits many of the characteristics of Wittgenstein's hinges (through his own reflections on both "legal laws" and "laws of nature").

The most conspicuous feature of Wittgenstein's hinges, namely *unchallengeability*, applies to Toulmin's warrants only "locally", when talking about "regular warrants", but may apply to *very deep and very basic warrants to a much greater extent*, without being an absolute or metaphysically necessary feature.

These are the ones that we must (and usually do) follow in order take part in actual human justifying practices, but could have been different and could conceivably change. Wittgenstein's river-bed metaphor (*OC* § 97) is as important as the spade metaphor (*PI* §217).

These are the ones we follow in our common behavior and the ones in following which we raise our descendants. What I have called deep warrants in this paper might be a good example of them.

References

Alhambra, J. (2022). Argumentation by Analogy and Weighing of Reasons, *Informal Logic,* 42(4), 749-785.
Coliva, A. (2010). *Moore and Wittgenstein. Scepticism, Certainty and Common Sense.* Basingstoke: Palgrave Macmillan.
Coliva, A. (2016). Which Hinge Epistemology? *International Journal for the Study of Skepticism,* 6, 79-96.
Conant, J. (1998). Wittgenstein on Meaning and Use. *Philosophical Investigations,* 21, 222–250.
Fisher, R. W. (2014). Why it doesn't matter whether the virtues are truth-conducive, *Synthese,* 191, 1059-1073.
Fogelin, R. (2005 [1985]). The Logic of Deep Disagreements. *Informal Logic,* 25 (1), 3-11.
Josephson, J. R. (2001). On the proof dynamics of inference to the best explanation, *Cardozo Law Review,* 22, 1621-1643.
Marraud, H. (2020). *En buena lógica.* Guadalajara: Universidad de Guadalajara.
Moore, G.E. (1993 [1939]). Proof of an External World. In *G. E. Moore: Selected Writings* (pp. 147–170). New York: Routledge.
Moyal-Sharrock, D. (2004). *Understanding Wittgenstein's On Certainty.* Basingstoke: Palgrave Macmillan.
Olmos, P. (2021). Metaphilosophy and Argument: The case of the justification of Abduction, *Informal Logic,* 41(2), 131-164.
Siegel, H. (2013). Argumentation and the Epistemology of Disagreement. *Cogency* 5 (1), 135-170.
Siegel, H. (2021) Hinges, Disagreements, and Arguments: (Rationally) Believing Hinge Propositions and Arguing across Deep Disagreements. *Topoi* 40, 1107–1116.
Strawson, P. (1985). *Skepticism and Naturalism. Some Varieties.* London: Methuen.
Tindale, C. (2015). *The Philosophy of Argument and Audience Reception.* Cambridge: CUP.
Toulmin, S.E. (1953). *The Philosophy of Science.* London: Hutchinson University Library
Toulmin, S.E. (1958). *The Uses of Argument.* Cambridge: C.U.P.
Toulmin, S., R. Rieke y A. Janik (1984 [1978]). *An Introduction to Reasoning.* New York / London: Macmillan Publishing.
Williams, M. (2004). Wittgenstein, Truth and Certainty. In M. Koelbel and B. Weiss (eds.), *Wittgenstein's Lasting Significance* (pp. 249–284). London-New York: Routledge.
Wittgenstein, L. *Philosophical Investigations*
Wittgenstein, L. *On Certainty*
Woods, J. & B. Hudak (1989). By Parity of Reasoning, *Informal Logic* 9, 125-139.
Wright, C. (2004). Warrant for Nothing (and Foundations for Free)? *Aristotelian Society Supplementary Volume* 78: 167–212.

The Third Party in Shared Decision-Making: The Role of Patient Companions in Discussions Between Patients and Healthcare Professionals

Roosmaryn Pilgram
Leiden University Centre for Linguistics
r.pilgram@hum.leidenuniv.nl

Lotte van Poppel
Center for Language and Cognition Groningen
l.van.poppel@rug.nl

Abstract

Patients often bring along a companion to medical consultations, which may result in three-party argumentative interactions. While consultations ideally involve shared decision-making (SDM), the way in which SDM proceeds in consultations with three parties has, nonetheless, so far received little attention. In this study, we analyse how the presence of a third party can affect the decision-making process. To do so, we specify the roles that this party can fulfil, and discuss, using the pragma-dialectical framework, how these roles relate to discussion roles. Lastly, based on a qualitative analysis of examples, we illustrate how the roles that a third party could fulfil can be expressed in actual medical decision-making.

1. Introduction

In medical consultation, it is important that the patient and healthcare professional reach agreement on which care is necessary and desirable. Ideally, patients and professionals together take part in the decision-making process, resulting in so-called shared decision-making (SDM). In SDM, both the patient and the healthcare professional provide information and make choices together, such as the choice for a certain treatment (Stiggelbout et al., 2015). Each participant has relevant expertise: the healthcare professional brings in medical knowledge about the course of the disease and possible treatment options, while the patient has knowledge about their preferences, the experienced symptoms and their impact.

Patients often take someone along to a consultation, such as a partner or family member (see Bracher et al., 2020). Patient federations indeed advise patients to bring a companion to a medical consultation for practical or moral support. Following up on this advice means that there are (at least) three parties in the consultation: the patient, the healthcare professional and the patient's companion. Such a third party can influence SDM (see Hodgson et al., 2016), as they can bring in specific expertise, for example, about the impact a treatment has on the patient, and they can initiate questions or answer questions on behalf of the patient (Ekberg et al., 2015; Huber et al., 2016). Yet, so far, protocols and the extant literature on SDM have focused very little on what this means for the decision-making process.

In fact, for SDM, the exact consequences of the presence of a third party are unknown. However, when regarding SDM from an argumentation theoretical perspective, there are already some indications as to how a third party can impact the decision-making process. For example, the third party can act as an antagonist in disagreements about treatment with the healthcare professional or even the patient (Coe & Prendergast, 1985; Huber et al., 2016). To gain more insight into SDM with three parties, this study will examine the way in which third parties can affect the decision-making process from an argumentation theoretical perspective. We will limit ourselves to in person face-to-face interactions and informal companions (i.e., partners, family members or friends), thereby excluding e-consultations and official companions (e.g., assistance, counsellors and interpreters).

More specifically, based on the extant literature, we will describe the roles that an informal patient companion can perform in a consultation. We will additionally discuss the argumentative dimensions of SDM using the pragma-dialectical argumentation theory (Van Eemeren & Grootendorst, 1992). Subsequently, the roles that third parties can take upon themselves in a consultation will be related to the pragma-dialectical discussion roles. Lastly, based on excerpts from actual medical consultations, we will illustrate how taking an argumentation theoretical perspective on SDM can further our understanding of this process involving three parties.

2. Patient companions as participants

2.1. Patient companions: participants in medical consultation

When a medical decision has to be made, the healthcare professional is obliged to obtain the patient's informed consent (see, e.g., Art.338 of the *Wet op de geneeskundige behandelingsovereenkomst* from the Dutch Civil Code): the healthcare professional may only proceed with the proposed medical procedure if the patient consents after understanding what the procedure entails, what its pros and cons are and which alternative treatments (including refraining from treatment) are available.

A patient companion can aid the decision-making process, for example, as moral support or memory aid. The percentage of patients who bring along someone to a medical consultation varies widely, and seems to depend on factors such as consultation type, age and physical condition of the patient (cf. Greene et al., 1994). For example, 16-25% of patients in primary care and outpatient clinics are accompanied by an informal companion. Older patients and oncology patients take someone along in 63-86% of their consultations (Laidsaar-Powell et al., 2013; Pel-Littel, 2019).

Conversations with three (or more) participants, also known as 'polylogues', proceed differently than conversations with only two participants (Bruxelles & Kerbrat-Orecchioni, 2004). In polylogues, turn-taking is, for instance, more complex, since there are two interlocutors who could potentially respond after every turn. Furthermore, a polylogue also allows for one of the parties bringing in information that would otherwise be left unaddressed, and for a party to answer on behalf of another party. Parties can even build coalitions, in which they act together as one party (Coe & Prendergast, 1985). In consultations with patient companions as third parties, this can result in three distinct coalitions: a coalition between the patient and the companion, the patient and the professional, and the companion and the professional (Rosow, 1981).

It should be noted that a patient companion, who might be a partner or caregiver, may have an interest in the consultation outcome. This would, for example, be the case if treatment options have consequences for the relationship between the patient and the third party, or for actions by the third party, such as administering medication (Bracher et al., 2020; Huber et al., 2016). In the literature, different roles are described that a patient companion can perform in a consultation, ranging from more to less active (Street & Gordon, 2008; Adelman, Greene, and Charon, 1987). Adelman, Greene, and Charon (1987), for instance, introduce three main categories: the third party as a passive participant, an advocate, or an opponent.

However, these categories are not directly linked to particular roles in SDM.

2.2. Patient companions: participants in shared decision-making

The exact contribution of patient companions to SDM is not well known. In Shields et al. (2005) and Pel-Littel et al. (2019), no difference was found between two-way conversations and three-way conversations in terms of patient-centred communication and patient participation in the decision-making process. However, exploratory observations in real-life consultations show that a third party can in fact be influential.

Excerpt 1A, for example, illustrates a patient companion significantly contributing to the decision-making process. This example was taken from a corpus of over 700 recorded consultations in Isala Hospital in Zwolle, the Netherlands (Driever, 2022). The excerpt is just a fragment from a longer consultation in which a 73-year-old female patient (P) and her husband (companion C) talk with the patient's doctor (D). The patient suffers from lung problems and is often out of breath, so the doctor investigates possible causes and solutions.

Excerpt 1A Medical consultation between a doctor (D), patient (P) with lung disease and companion (C), the patient's husband (Isala hospital corpus, our translation from Dutch)

252	D	no, uhm I'm just thinking
253		Do you snore?
254	P	no.
255	C	Well
256	P	Well no [really]
257	C	[well]

In the excerpt, the physician assesses the patient's sleep and any potential sleeping disorders, including apnoea. As a result, he inquires if the patient has a habit of snoring (in line 253). The patient firmly replies with a "no" in response to this. However, the patient's partner expresses disagreement with this answer by uttering the interjection "well" ("nou" in Dutch). This prompts the patient to reinforce her initial reply (in 256).

What is interesting about *Excerpt 1A* is that the companion can provide the information needed for making decisions about the patient's health, as only he knows whether the patient snores. In this consultation, the information from the husband, in fact, supports the doctor's proposal to carry out a sleep study and test for apnoea. Therefore, the companion, with only a minimal contribution, seems to play a crucial role in the deliberation process.

3. Argumentation in consultations with three parties

3.1. Argumentation in medical consultations

Although medical consultations are largely informative in nature, argumentation can play an important role in them. One reason for this is that the ideal of SDM dictates healthcare professionals and patients to jointly go through the stages of the decision-making process, discuss different options and come to a shared decision, and it is precisely this discussion and weighing of options that is inherently argumentative. According to Stiggelbout et al. (2015, p. 2), the healthcare professional and patient need to go through the following four steps in SDM:

1. The professional informs the patient that a decision is to be made and that the patient's opinion is important;
2. The professional explains the options and the pros and cons of each relevant option;
3. The professional and patient discuss the patient's preferences; the professional supports the patient in deliberation;
4. The professional and patient discuss patient's decisional role preference, make or defer the decision, and discuss possible follow-up.

Step three is particularly interesting from an argumentative point of view: in this step, the patient and the healthcare professional do not simply make an inventory of the patient's preferences, but have to *deliberate* about the consequences of these preferences for the decision-making process. Already in classical antiquity, deliberation is linked to practical reasoning, where arguments in favour or against a particular proposal must be presented (see Aristotle's *Nicomachean Ethics*).

Various argumentation theorists see SDM as a form of argumentative discussion (see Van Eemeren, Garssen & Labrie, 2021; Goodnight, 2006; Labrie, 2012; Labrie & Schulz, 2014, 2015a, 2015b; Pilgram, 2015; Pilgram & Snoeck Henkemans, 2018; Snoeck Henkemans & Mohammed, 2012). Snoeck Henkemans and Mohammed (2012) emphasise that the goal of SDM is in line with that of the pragma-dialectical critical discussion: both aim to resolve a difference of opinion by critically testing the acceptability of a position. More specifically, during SDM, an argumentative discussion can arise about (parts of) the diagnosis, the treatment plan, the prognosis, and even procedural matters such as making a follow-up appointment. As such, medical consultation is seen as an argumentative activity type (Bigi, 2018; Pilgram, 2009; 2015).

In argumentative discussions in medical consultation, different types of differences of opinion can be distinguished. If a party only expresses doubts about a particular course of action ("I'm not sure whether I want to take medication"), the situation can be seen as a 'non-mixed' dispute. In a non-mixed dispute, one party has the role of protagonist, who has the burden of proof for a particular standpoint, and the other party has the role of a doubtful antagonist. A party can also advance an opposing standpoint in addition to expressing doubting. For example, a healthcare professional could recommend a certain treatment ("You should use an inhaler") and the patient could object to that with an alternative proposal ("I thought I could maybe get some antibiotics"). Such a situation qualifies as a 'mixed' dispute between the healthcare professional and patient, as both act as protagonist of their own standpoint and antagonist of the other party's standpoint (Van Eemeren & Grootendorst, 1992).

3.2. The third party's argumentative roles

As suggested in section 2.2, a third party can play an important role in argumentative discussions in a medical consultation. A third party could support the patient's or healthcare professional's view, but could also give rise to additional differences of opinion with each of the other parties, resulting in argumentative 'polylogues' (Lewinski & Aakhus, 2014). Disagreements in which the third party takes on an argumentative role are intriguing from the perspective that a medical consultation can be seen as a 'meeting between experts' – i.e., that the healthcare professional is the medical expert and the patient is the expert from experience (Tuckett et al., 1985). The third party does not necessarily have a certain expertise, institutionally speaking. How can this party then position itself in the discussion?

To answer this question, it is first of all important to determine whether the interaction can be understood as an argumentative discussion between the patient and healthcare professional, with the companion as additional intervening party in this discussion, or whether the discussion is between the other parties separately (healthcare professional vs. third party, or third party vs. patient). In the first case, where the third party joins the discussion between healthcare professional and patient, the involvement of the third party could result in coalition forming, or a combination of both mixed and non-mixed disagreements. Nonetheless, the same rules and conventions apply that govern such argumentative discussions without any participation of the third party. After all, the healthcare professional must still comply with the legal requirement of informed consent and will probably also attempt to reach a shared decision with the patient.

In the second case, in which a discussion arises between the third party and either the health professional or the patient, the regular rules and conventions for such discussions no longer immediately play a role. For example, a healthcare professional does not need to engage in SDM with the third party or get their explicit agreement to particular treatment. Nevertheless, these kinds of discussions might still serve the decision-making process between the patient and healthcare professional, for example, by clarifying the consequences of a diagnosis or the preferences of the patient. These discussions might therefore serve as sub-discussions in the main dispute between the patient and healthcare professional.

In discussions between the patient, healthcare professional *and* the third party, there is an additional point to take into account: the discussion situation in question. With 'discussion situation', we mean the combination of the type of disagreement between the parties (non-mixed, mixed or a combination of both), the discussion role that the parties take upon themselves (protagonist or antagonist), and any coalition formation (no coalition or possible coalition). From a theoretical point of view, twelve different discussion situations can be distinguished for disputes between patients, healthcare professionals and third parties (see *Table 1*).

The least complex discussion situations in *Table 1* are 1, 2 and 3. In these situations, there is only a non-mixed dispute, in which no coalition can be formed. This would, for example, be the case when the health professional proposes a treatment and the patient and third party simply ask why the professional proposes it (discussion situation 1). Discussion situations 1 and 2 differ relatively little from 'traditional' medical discussions between the patient and the healthcare professional, since their starting points and argumentative means are similar.

On the other hand, discussion situation 3 deviates significantly from typical two-party consultations: here, the third party acts as the protagonist of a standpoint, while the patient and healthcare professional question this standpoint. This situation does not fully correspond with any of the roles assigned to the third party in the existing literature (see section 2.2). It remains nonetheless a viable option, for example, when a third party suggests a treatment and both the patient and healthcare professional are uncertain about it.

Table I. Possible discussion situations for three-party disputes in medical consultation

Type of disagreement	Coalition	Nr.	Protagonist	Antagonist: doubt	Antagonist: opposition
Non-mixed	No	1	Health professional	Patient and 3rd party	-
		2	Patient	Health professional and 3rd party	-
		3	3rd party	Health professional and patient	-
	Possible	4	Health professional and patient	3rd party	-
		5	Health professional and 3rd party	Patient	-
		6	Patient and 3rd party	Health professional	-
Mixed	Possible	7a	Health professional	-	Patient and 3rd party
		7b	Patient and 3rd party	-	Health professional
		8a	3rd party	-	Health professional and patient
		8b	Health professional and patient	-	3rd party
		9a	Patient	-	Health professional and 3rd party
		9b	Health professional and 3rd party	-	Patient
Partly mixed, partly non-mixed	No	10a	Health professional	3rd party	Patient
		10b	Patient	3rd party	Health professional
		11a	Health professional	Patient	3rd party
		11b	3rd party	Patient	Health professional
		12a	3rd party	Health professional	Patient
		12b	Patient	Health professional	3rd party

The fact that the presence of a third party gives rise to a polylogical discussion becomes more clearly visible in discussion situations 4 to 12. These situations are more complex than the first three, because the parties involved form a coalition or because different types of disputes arise between these parties. Discussion situations 4 to 9 illustrate the theoretically possible coalitions. For example, a third party could act as the patient's advocate and together with the patient oppose a treatment plan of the doctor (discussion situation 7a), or a third party and patient can jointly propose an alternative plan (7b).

Discussion situations 10 to 12 show the possible differences of opinion between the three parties. These situations indicate the possibility that two parties have a mixed dispute with each other, while the other party doubts the positions of those parties and therefore has a non-mixed dispute with both of them.

4. The argumentative role of the third party in shared decision-making

Based on *Table 1*, we can identify different discussion situations and indicate what argumentative role patient companions can play in SDM. For example, recall that in *Excerpt 1A* (presented in section 2.2), the patient and third party have a mixed difference of opinion, because they adopt opposing standpoints about the patient's snoring, while the doctor takes the position of the doubting antagonist. In other words, the dispute in this excerpt is partly mixed and partly non-mixed, without any coalitions, resulting in situation 12b in *Table 1*.

Earlier on in the very same consultation, we also encountered a different discussion situation (see *Excerpt 1B*). In this preceding part, the doctor proposes to the patient to attend a COPD rehabilitation centre, so that she can improve her physical fitness. In lines 133 and 135, we can see that the patient does not accept the proposal ("not all that great") and gives an argument for her non-acceptance: "I'm not ready for that yet". Subsequently, it becomes evident that the third party and the doctor form a coalition, trying to convince the patient that the rehabilitation centre represents a favourable option for her due to the potential health benefits it offers her in the long run.

In contrast to the patient, the third party explicitly endorses the physician's proposal by acknowledging the problematic nature of the current situation (line 136: "No but it doesn't get any better this way of course") and by affirming the benefits of seeking treatment at a rehabilitation centre (line 142: "Yes"). Despite this, the patient remains unconvinced and declines the proposal. The consultation ends with the doctor advising the patient to consider this option. This particular segment thus displays the formation of a coalition between healthcare professional and third party, as they both advocate for the patient to go to a rehabilitation centre, while the patient opposes it. The discussion situation in this part of the consultation constitutes a mixed difference of opinion of type 9b.

Excerpt 1B Medical consultation between a doctor (D), patient (P) with lung disease and companion (C), the patient's husband (Isala hospital corpus, our translation from Dutch)

130	D	I don't know if that would be something
131		That you would go to some sort of rehabilitation centre?
132		Where you would only have to focus on the body?
133	P	Yeah yeah not all that great.
134	D	Hm?
135	P	Actually, I'm not ready for that yet.
136	C	No but it doesn't get any better this way of course.
137	D	Well yeah look that's it you know,
138		You know you are now 73
139		And if you would and if you go out for a few weeks,
140		huh you can see that as an investment in the rest of eh of your
141		[life.]
142	C	[life] Yes.
143	D	Yes
144		no it is of course eh yes quite a task.
145	P	Yes it sure is.

As opposed to a situation in which a coalition is clearly formed, such a coalition can also be merely suggested for strategic purposes. This appears to occur in *Excerpt 2*, in which the patient's companion (his wife) engages in an argumentative exchange with the doctor.

Excerpt 2 starts immediately after the patient has reported that his little finger and ring finger are causing discomfort, characterised by a tingling sensation that prevents him from sleeping at night. The doctor therefore suggests hand surgery (lines 1-12). In response to this proposal, the patient's wife expresses her concerns by saying: "Yes, well, that actually bothers us" (line 13). She continues by referring to colleagues with the same complaints as her husband who have had similar operations multiple times (lines 15-16). Her statement that those colleagues have had surgery "two three times" implies that such surgery is not often successful. In addition, she ends her turn with "but" (line 16), which also suggests a negative continuation. Indeed, the doctor seems to interpret the companion's contribution in this way and tries to disprove her assumption in lines 17 and 19 by indicating that multiple operations are an exception and by pointing to the success of such operations ("that operation is successful in 85 out of 100 cases").

The argumentative situation in the beginning of the fragment can be understood as a mixed dispute between the doctor serving as protagonist and the third party acting as antagonist: the doctor proposes considering surgery and the companion does not explicitly reject this proposal, but she does express a negative stance towards surgery. While the companion's statement could be interpreted as merely an expression of doubt, her support for her aversion to surgery suggests resistance to the doctor's proposal.

Excerpt 2: Medical consultation between a doctor (D), patient (P) with a pinched nerve and companion (C), the patient's wife (Den Hartog, 2019, pp. 41-42, our translation from Dutch)

```
1   D   Well, we- what can be done about it is surgery by the plastic surgeon
2       [...]. So the question is sort of how much you are bothered by it and
3       whether that's enough to possibly undergo the surgery for [...]
4   C   Yes well that actually does bother us like
5   D   Hmhm
6   C   Yeah I have many colleagues who've had the same syndrome well
7       they went under the knife two three times but
8   D   Hmm that does happen but is an exception
9   C   Yes
10  D   Yes, that surgery is in 85 out of 100 cases successful
11  D   But I can of course not predict whether you belong to that 85 or that
12      you
13  P   No    [no no
14  D         [or that you belong to the 15
15  C         [yeah that's true
```

It is noteworthy that the third party expresses her resistance through the first-person plural 'we', presenting her worries as a shared concern between herself and the patient. However, nowhere in the consultation does the patient seem to concur with that worry. In fact, the patient agrees with the doctor's judgment (in line 23) and asserts that it is crucial to address the issue with his fingers (in lines 27 and 34). Therefore, the fact that the third party presents her standpoint as a shared one does not seem entirely justified. The strategic use of a suggested shared perspective can be advantageous to the companion. According to the principles of SDM, the physician is obliged to consider the patient's preferences. By presenting herself as part of a coalition with the patient, the third party ensures that the doctor also takes her concerns seriously. Additionally, the use the pronoun 'we' suggests a majority, and under the guise of "safety in numbers", this could improve the third party's position.

5. Conclusion and discussion

In this contribution, we have demonstrated that a third party accompanying a patient to a medical consultation, such as a partner or adult child, can impact shared decision-making (SDM) with a healthcare professional. Since the ideal of SDM requires weighing the pros and cons of different options, argumentation can play an important role in a consultation. However, until now it is unknown how the presence of a third party exactly influences this argumentative part of a consultation.

Based on the different positions that the patient, healthcare professional and third party can take in a consultation, we have determined which different discussion situations are theoretically possible if the third party participates in the discussion. We have shown that more, and more complex, discussion situations can arise in a three-party consultation than in a dialogical consultation. The complexity of these situations depends on whether the type of disagreement is mixed, non-mixed or a combination thereof, whether the parties position themselves as protagonists or antagonists, and whether or not a coalition is formed.

In a number of the possible discussion situations, the third party can speak on behalf of the patient as their representative, advocating for their interests in disputes with the healthcare professional. Given the asymmetrical relationship between the professional and the patient, it can be challenging for a patient to explicitly express their doubts or objections to a doctor's treatment proposal. In such cases, a third party can act as an antagonist and voice their concerns on behalf of the patient. The third party could also form a coalition with the patient in order to strengthen their position. The parties can then supplement each other's arguments and express their position in the 'we'-form. In this way, the third party can further SDM.

However, we also showed that the third party could play a role in the decision-making process without necessarily benefitting the patient: the third party could (intentionally or not) wrongly suggest a coalition with the patient by putting forward the position using the first-person plural. Because of this possibility, the healthcare professional should carefully assess the patient's actual stance in order to arrive at a decision on which the patient and the healthcare professional can both agree.

Our analysis also revealed that a third party accompanying the patient to a medical consultation could bring in specific expertise relevant to SDM.

This expertise could include insight into the patient's daily life, the severity and frequency of symptoms, difficulties with medication compliance or the potential consequences of undergoing surgery. In fact, this expertise provides a third perspective on treatment options that can be valuable in the decision-making process.

In sum, in this contribution, we have argued that the presence of a third party can lead to all kinds of complex discussion situations and thus influence decision-making in different ways. This study therefore offers new insights that can be used to improve communication and decision-making between the patient, healthcare professional and third party. Although the inclusion of a third party is encouraged by patient organisations, and in many cases is unavoidable anyway due to the patient's condition, the guidelines for SDM do not explicitly consider such parties. Given the importance of SDM for patient satisfaction and better health outcomes, it is advisable to pay more attention to the influence of the third party on this process.

References

Adelman, R.D., Greene, M.G., & Charon, R. (1987). The physician-elderly patient-companion triad in the medical encounter: the development of a conceptual framework and research agenda. *The Gerontologist* 27(6), 729-34.

Aristoteles (1999). *The Nicomachean ethics*. Kitchener: Batoche Books.

Bigi, S. (2018). The role of argumentative practices within advice-seeking activity types. The case of the medical consultation. *Rivista Italiana di Filosofia del Linguaggio* 12(1), 42-52. https://doi.org/10.4396/20180602.

Bracher, M., Stewart, S., Reidy, C., Allen, C, Townsend, K., & Brindle, L. (2020). Partner involvement in treatment-related decision making in triadic clinical consultations – A systematic review of qualitative and quantitative studies. *Patient Education and Counseling* 103(2), 245-253. https://doi.org/10.1016/j.pec.2019.08.031.

Bruxelles, S., & Kerbrat-Orecchioni, C. (2004). Coalitions in polylogues. *Journal of Pragmatics* 36(1): 75–113.

Coe, R.M., & Prendergast, C.G. (1985). The formation of coalitions: interaction strategies in triads. *Sociology of Health & Illness* 7(2), 236-247. https://doi.org/10.1111/1467-9566.ep10949087.

Driever, E. (2022). Shared decision making in hospital care: what happens in practice. *Dissertation*. University of Groningen. https://doi.org/10.33612/diss.232457161

Eemeren, F.H., Garssen, B.J., & Labrie, N.H.M. (2021). *Argumentation between doctors and patients: Understanding clinical argumentative discourse*. Amsterdam / Philadelphia: John Benjamins.

Eemeren, F.H. van, & Grootendorst, R. (1992). *Argumentation, communication, and fallacies: A pragma-dialectical perspective*. Hillsdale: Lawrence Erlbaum Associates.

Ekberg, K., Meyer, C., Scarinci, N., Grenness, C., & Hickson, L. (2015). Family member involvement in audiology appointments with older people with hearing impairment. International Journal of Audiology 54(2), 70-76. https://doi.org/10.3109/14992027.2014.948218

Goodnight, G.T. (2006). When reasons matter most: Pragma-dialectics and the problem of informed consent. In: Houtlosser, P. & Rees, A. van (Eds.), Considering pragma-dialectics, Mahwah / London: Lawrence Erlbaum Associates, pp. 75-85.

Greene, M.G., Majerovitz, S.D., Adelman, R.D., & Rizzo, C. (1994). The effects of the presence of a third person on the physician-older patient medical interview. Journal of the American Geriatrics Society 42(4), 413-419.

Hartog, R. den (2019). De invloed van lage deontische stance op shared decision making. Een conversatie-analytisch onderzoek naar lage deontische stance in behandelvoorstellen tijdens arts-patiënt gesprekken. Masterscriptie. Rijksuniversiteit Groningen.

Hodgson, J., Pitt, P., Metcalfe, S., Halliday, J., Menezes, M., Fisher, J., Hickerton, C., Petersen, K., & McClaren, B. (2016). Experiences of prenatal diagnosis and decision-making about termination of pregnancy: a qualitative study. Australian and New Zealand Journal of Obstetrics and Gynaecology 56 (6), 605-613. https://doi.org/10.1111/ajo.12501.

Huber, J., Streuli, J.C., Lozankovski, N., Stredele, R.J.F., Moll, P., Hohenfellner, M., Huber, C.G., Ihrig, A., & Peters, T. (2016). The complex interplay of physician, patient, and spouse in preoperative counseling for radical prostatectomy: a comparative mixed-method analysis of 30 videotaped consultations. Psycho-Oncology 25(8), 949-956. https://doi.org/10.1002/pon.4041.

Labrie, N.H.M. (2012). Strategic maneuvering in treatment decision-making discussions: Two cases in point. Argumentation 26, 171-199. https://doi.org/10.1007/s10503-011-9228-5

Labrie, N.H.M. & Schulz, P.J. (2014). Does argumentation matter? A systematic literature review on the role of argumentation in doctor-patient communication. Health Communication 29(10), 996-1008. https://doi.org/10.1080/10410236.2013.829018.

Labrie, N.H.M. & Schulz, P.J. (2015a). Exploring the relationships between participatory decision-making, visit duration, and general practitioners' provision of argumentation to support their medical advice: Results from a content analysis. Patient Education and Counseling 98(5), 572-577. https://doi.org/10.1016/j.pec.2015.01.017.

Labrie, N.H.M. & Schulz, P.J. (2015b). Quantifying doctors' argumentation in general practice consultation through content analysis: Measurement development and preliminary results. Argumentation 29, 33-55. https://doi.org/10.1007/s10503-014-9331-5.

Laidsaar-Powell, R.C., Butow, P.N., Charles, C., Lam, W.W.T., Jansen, J., McCaffery, K.J., Shepherd, H.L., Tattersall, M.H., & Juraskova, I. (2013). Physician-patient-companion communication and decision-making: A systematic review of triadic medical consultations. Patient Education and Counseling 91(1), 3-13. https://doi.org/10.1016/j.pec.2012.11.007.

Lewinski, M., & Aakhus, M. (2014). Argumentative polylogues in a dialectical framework: A methodological inquiry. Argumentation 28(2), 161-185.

Pel-Littel, R.E., Buurman, B.M., Pol, M.H. van de, Yilmaz, N.G., Tulner, L.R., Minkman, M.M., Scholte Op Reimer, W.J.M., Elwyn, G., & Weert, J.C.M. van (2019). Measuring triadic decision making in older patients with multiple chronic conditions: Observer OPTIONMCC. *Patient Education and Counseling* 102(11), 1969-1976. https://doi.org/10.1016/j.pec.2019.06.020.

Pilgram, R. (2009). Argumentation in doctor-patient interaction: Medical consultation as a pragma-dialectical communicative activity type. *Studies in Communication Sciences* 9(2), 153-169.

Pilgram, R. (2015). *A doctor's argument by authority: An analytical and empirical study of strategic manoeuvring in medical consultation.* Dissertation. University of Amsterdam.

Pilgram, R., & Snoeck Henkemans, A.F. (2018). A pragma-dialectical perspective on obstacles to shared decision-making. *Journal of Argumentation in Context* 7(2), 161-176. https://doi.org/10.1075/jaic.18027.pil

Rosow, I. (1981). Coalitions in geriatric medicine. In M. Haug (Ed.), *Elderly patients and their doctors.* New York: Springer.

Shields, C.G., Epstein, R.M., Fiscella, K., Franks, P., McCann, R., McCormick, K., & Mallinger, J.B. (2005). Influence of accompanied encounters on patient-centeredness with older patients. *Journal of the American Board of Family Practice* 18(5), 344-354. https://doi.org/10.3122/jabfm.18.5.344

Snoeck Henkemans, A.F., & Mohammed, D. (2012). Institutional constraints on strategic maneuvering in shared medical decision-making. *Journal of Argumentation in Context* 1 (1), 19-32. https://doi.org/10.1075/jaic.1.1.03moh.

Stiggelbout, A.M., Pieterse, A.H., & Haes, J.C.J.M de (2015) Shared decision making: Concepts, evidence, and practice. *Patient Education and Counselling* 98(10), 1172-1179. http://dx.doi.org/10.1016/j.pec.2015.06.022.

Street, R.L., & Gordon, H.S. (2008). Companion participation in cancer consultations. *Psycho-oncology* 17, 244-251. https://doi.org/10.1002/pon.1225.

Tuckett, D., Boulton, M., Olson, C. & Williams, A. (1985). *Meetings between experts: An approach to sharing ideas in medical consultations.* Routledge: Abingdon.

Rhetoric, Anthropological Models and the Law

Federico Puppo
University of Trento
federico.puppo@unitn.it

Abstract

Nowadays rhetoric is mostly conceived in a negative way, as a technique used to gain persuasion, in addition to argumentation and/or against reason. The aim of my paper is to offer a completely different view on rhetoric, in the wake of the Aristotelian conception, with the basic idea is that rhetoric could have anything other than a pejorative sense and that rhetoric, despite his coming from the past, could still play a decisive role for the foundation of a highly valid account of law, coherently with the basic human right to a fair trial. An account based on the role played by the third, since rhetoric, as law and justice, is always ad alterum. From this point of view, rhetoric, as law and justice, is given within a social practice and it is more than a repertoire of means of persuasion or a theory of speaking. In the Aristotelian account, rhetoric is a way of thinking and a way of being. To be more precise, it is the zoon politikon's way of being, i.e. the way of being of man as living in the co-original linguistic and political dimension (Heidegger 2009). This is the same frame in which legal and political deliberations take place, since «we deliberate about things that are in our power and can be done» (Arist. Nic. Eth., 3, 1112), such as what is good or evil, or just or unjust.

1. Introduction

Nowadays rhetoric is mostly conceived in a negative way, as a technique used to gain persuasion, in addition to argumentation and/or against reason. The aim of my paper is to offer a completely different view on rhetoric, in the wake of the Aristotelian conception discussed, among others, by Martin Heidegger (2009): the basic idea is that rhetoric could have anything other than a pejorative sense (Tindale 2004, 2019; Danblon 2013; Rocci 2016; Piazza 2015) and that rhetoric, despite his coming from the past, could still play a decisive role for the foundation of a highly valid account of law.

In my view, rhetoric, as law and justice, is given within a social practice and it is more than a repertoire of means of persuasion or a theory of speaking well. In the Aristotelian account, which I share, rhetoric is a way of thinking and a way of being. To be more precise, it is the *zoon politikon*'s way of being, i.e. the way of being of man as living in the co-original linguistic and political dimensions (Heidegger 2009). This is the same frame in which legal and political deliberations take place. This account on rhetoric deals with a precise conception of the nature of law and I will try to explain why it suits with some basic legal principles, such as the due process of law.

My starting point is to outline some main characteristic of the legal context from an argumentative point of view, which reveals its rhetorical nature. Then, I will see how these features characterize the nature of judges in relation with some basic legal principles.

2. The nature of the legal context

The legal context, mainly exemplified by the trial, is at least characterized by the following features:

1) From an argumentative point of view, the trial is more than dialogical. It is clearly a trilogical (Plantin 1996) or polylogical situation (Lewiński, Aakhus 2014; Manzin 2014): we have, at least, two parties disputing in front of a judge and/or a jury (I will tell more on that in the last section). From a legal point of view, there is no trial without three parties: *Processus est actus trium personarum, actoris, rei, iudicis, in iudicio contendentium*. But, this is not the ultimate dimension of the trial: there are, from a certain point of view, more parties involved in the trial. It is true that, from a strict legal point of view, the decision binds only the parties to the case: but from another point of view, the decision can be matter of discussion for other judges, for legal scholars, even for society. It is not by chance that (at least in the Italian system, but it is a common rule) "justice shall be administered in the name of the people" (art. 101 Italian Constitution): it is not a private affair of the three parties involved in the trial, it has a wider range of interest.

2) From a phenomenological point of view, the dispute to be decided by the judge is caused by a practical matter. The judge is required to find out a practicable solution on the matter proposed by the parties. And s/he must do that on the basis of the law, which is also the institutional frame in which the parties speech: they are not free to say what they want when and how they want, but they must respect procedural legal rules. The communicative frame is institutionalized in nature.

3) Finally, the judge and/or the jury is the one who must be persuaded by the parties: immediately, in the trial, lawyers have no interest in persuading the other party. S/he has an interest in persuading the judge: the lawyer's speech is directed to the judge.

So: we are in a context in which the decision depends not on the parties who argue, but on a third party, the judge, who decides the case. The judge deliberates on a practical matter, which must be solved. The judge is the one we look at when we argue in a courtroom. The judge must be persuaded by the parties. To act as a lawyer means to take care of it.

All these features tell us about the nature of the context we are looking at: since the beginning of our legal-philosophical history, this has been described as a rhetorical situation. It is not argumentation, it is rhetoric. But the fact is that in our legal-philosophical history, as it is well known, rhetoric gained no more than a pejorative sense: is has been a gradual movement, which met his final passage with the modern model of law and knowledge, with an unavoidable loss of value and consideration for rhetoric.

3. Coming back to rhetoric

Already with the Roman Age and Middle Ages rhetoric became more and more relative to the art of beautiful writing (Barthes 1979), something pertaining stylistic aspects of our discourses, while, in the secular journey of knowledge, rhetoric earned the deep prejudice we perfectly know. Rhetoric has gradually lost his value, while scientific knowledge, with its typical anti-metaphysical and anti-Aristotelian characterization (Agazzi 2014), has earned it.

However, the weakening of the methodological monism of scientific and positivist knowledge of the last century did not lead to a full rediscovery of rhetoric, but rather to the development of studies on argumentation, in which the persuasiveness of a discourse deals with some kind of components to be addicted to its reasonableness. In the light of argumentative approach, the rhetorical components are not part of discourse's rationality: they came 'after' it, not 'with' it.

Rocci (2016), following Jacobs (2006) (2000), explains that this way of conceiving rhetoric is typical of the weak-compatibilist view, according to which the pursuit of rhetorical effectiveness, which includes persuasive effectiveness, can be compatible with the dialectical soundness of arguments but does not add to it. In the light of these approaches, in other words, rhetorical persuasion, or effectiveness, is not a necessary condition for the reasonableness of argumentation and/or for the inferential nature of the argument in itself. If I may, I would say that it is something that we cannot avoid, but maybe it would be better to not have it. But, there is another possibility: I mean, to adopt, as Rocci (2016) does, a strong-compatibilist view, whereby a certain kind of persuasive effectiveness is necessary to the pursuit of reasonable argument.

In other words, we must avoid the anti-rhetorical prejudice, typical of certain philosophical thought, typical of modernity but still present in Plato – for his distrust of passions (Tindale 2021, 177), that tries to 'tame' the phenomenon of persuasion, by considering it dangerous and against rationality or demonstration (Piazza, 2015).

What we have here is, again, the distinction between two different ways of conceiving rhetoric, due to the philosophical and humanistic models of argumentation (Danblon 2013). I believe that Danblon, but also Tindale or Rocci or Piazza, are right in defending the humanistic or strong-compatibilist views on rhetoric, which is more Aristotelian than Platonic. This requires to recognize to Aristotle's Rhetoric its strong philosophical value, by regaining, at the same time, the entailed anthropological model of *zoon politikon* (Heidegger 2009; Piazza 2015, 2019). In this conception, rhetoric is always an activity of logos, based on a reasoning together which has in the audience the most important party: in this classical and Aristotelian conception, to speak is always to speak ad alterum. We are linguistic and political animals not only in a biological or phenomenological point of view, but from an essential point of view, which means (as Aristotle pointed out in its Nicomachean Ethics) that, by talking, we do not give information about the world, but we evaluate, discuss and dispute the values of our actions and belief. Everyone do that inside the polis: rhetoric, from this point of view, it is not only a technique, but it becomes a way of being. The way of being of human beings in society (*zoon politikon*). The main concern of this kind of human being is taking care of the relationship between him and the others, mainly the one to be persuaded.

In this light, law can assume a very different value compared to the modern one, which identifies law with a set of rules to settling disputes or with a technique of social regulation: these can be some functions of law, but they cannot explain its essence. Because, whatever the function of law, it could not disregard the presence of the relationship: basically, a legal rule makes no sense without thinking to society and relations to be ruled. And these relations are the philosophical fundamental justification for law: law, as human beings, do not create the relation, which exists in itself before any kind of regulation. This is what Aristotle means by speaking, in its Politics, of the being by nature of the state. So, in this rhetorical view, law finds its value in maintain relations alive, by making them possible. And jurists must take care of it.

So, rhetoric mainly deals with a certain way of being of human beings and, at the same time, with a certain way of being of law, jurists, judges and lawyers. To argue in a rhetorical context, such as the judicial one, means to mainly look at the judge: it is not only because s/he is the one to be persuaded, but because s/he is the "alterum" essential for the relation. At the same time, the judge her/himself can assume a rhetorical role, and this is what I am going to see in the next section.

4. The rhetorical judge

If it could be quite easy to understand or accept the idea that lawyers are rhetoricians because they need to take care of the judge, we need to understand what it means to imply a rhetorical conception of law for the judiciary. Piazza (2015) helps me in this task: she explains that the rhetorical man, as the *zoon politikon*, is the one who has logos, because, at the same time, he persuades and lets himself be persuaded.

To persuade and to be persuaded are the two faces of the same rational coin: reasoning it is not only persuading, but it also to be persuaded (by using ethos, pathos and logos). We usually believe, by thinking to rhetoric as a technique, that judges are part of the rhetorical process because they must be persuaded by lawyers: they are conceived like the passive 'target' of lawyers' discourses. To conceive rhetoric as a way of being, without any pejorative view, allows us to add the component of logos in relation to persuasion: judges take part in the rhetorical relation and so, in their turn, they reason. Being persuaded, s/he must persuade. And this is what s/he does in giving reasons for her/his decision, talking to the parties to the case and to all the other potential parties: other judges, legal scholars, society.

This conception of judging gives me the possibility to highlight another reason for which rhetoric better fits with legal experience then argumentation. In a nutshell, I believe that the basic argumentative scheme is: two parties (proponent and opponent) having dialogue, or arguing, one to each other, for different aims. Two parties, without a third one. I know that it can be an oversimplification, but this is what can explain the weak compatibilist view: the third party is added later, to expand the model to polylogical situations, but it is not conceived since the beginning as a constitutive part of it. In a strong compatibilist or humanistic view, the situation is different, because the third is immediately and by essence part of the context. But, and this is the point, as it is very well known, what Aristotle claims in his Rhetoric is that rhetorician does not argue as a dialectical does.

In the rhetorical domains, the orator does not have a dialogue made of questioning and answering with the other party: what we have is more similar to a monologue, pronounced to get a third party to make a decision or to change her/his mind. In the typical rhetorical situation, the hearer cannot really interact during the speech, even if s/he has the great power to judge on it and to take a decision (Piazza 2015, 112). This 'monological' point can strike the attention of legal scholars, because we usually describe the trial as a typical dialogical situation, arguing one to each other in front of a judge. But it is a very typical way of arguing: it is not like in common dialogues, because the parties (even the judge) must respect procedural rules, which provide when and how talking or questioning. More than that, it is pretty clear that the party who makes questions or gives answers is making a speech directed to the judge: questions and answers are made by looking at her/him. As I said, the third is really part of the communicative process made during the trial.

But s/he cannot be conceived as a 'dialectical' judge, very active, the one who searches for truth making questions and asking for answers. This type of judge is the one the inquisitorial system, typical of the modern conception of law, not by chance a conception which was born in context really against rhetoric. The 'rhetorical' judge is completely different: s/he does not speak, s/he listens. *Audiatur et altera pars* is the basic principle of each trial, since the classical model embodied by Athena in Aeschylus' Eumenides, nowadays institutionalized by the human right to a fair trial. This principle has been defined by the Council of Europe as a fair and public hearing, within a reasonable time, by an independent and impartial court. The principle of independence and impartiality of the judiciary has been considered part of the "general principles of law recognised by civilized nations" (Article 38 of the International Court of Justice Statute). As to impartiality, the UN Human Rights Committee stated that it "implies that judges must not harbour any preconceptions about the matter put before them, and that they must not act in ways that promote the interests of one of the parties" (Human Rights Committee, Communication No. 387/1989 (Karttunen v. Finland), UN Doc. CCPR/C/46/D/387/1989, para. 7.2.).

The argumentative model seems to me not really capable of grasping the essence of a fair trial. Not only because the judge can be conceived as the 'target' of the persuasion, in a passive way; but because to describe the activity of the judge as 'argumentation' could bring her/him towards a dialectical model, in contrast with the right to a fair trial. From this point of view, it seems to me the rhetorical conception of law based on the third could also strength the value of the right to a fair trial, which perfectly fits with the rhetorical model.

5. Conclusion

In this paper, I tried to explain why we need, from a certain point of view, to rediscover rhetoric: it is time to overcome the modern prejudice. To look at law from a rhetorical perspective gives us the possibility to shape the unavoidable condition of relation which must be supposed by any kind of legislation. The being 'ad alterum' of rhetorical persuasion runs parallel to the model of human being as *zoon politikon*, together with a precise understanding of the mutual rational relation between persuading and be persuaded. In this way, it has been possible find a legal-philosophical ground or justification for the right to a fair trial principle on the rhetorical model of the judge.

In conclusion, the rhetorical understanding of law, inspired by the strong compatibilist views and by the humanist models, by searching for a deep philosophical understanding of Aristotle's Rhetoric, leads us at conceiving law not as a set of norms produced to solve disputes, but as the living conditions for making a dispute possible. Law, as rhetoric, is more than a technique: it is the unavoidable condition of living together, by protecting the relation which is already among us. This relation depends on us not because we create it, but because we must safeguard it.

References

Agazzi, E. (2014). Scientific Objectivity and Its Contexts. Berlin: Springer Cham.
Aristotle (2018). *The Art of Rhetoric*, Trans. by R. Waterfield with an Introduction and Notes by H. Yunis. Oxford: Oxford University Press.
Barthes, R. (1979). *La retorica antica. Alle origini del linguaggio letterario e delle tecniche di comunicazione*. It. trans., Milano, Bompiani.
Danblon, E. (2013). The Reason of Rhetoric. *Philosophy & Rhetoric, 46.4*, 493-507.
Heidegger, M. (2009). *Basic Concepts of Aristotelian Philosophy*. Bloomington: Indiana University Press.
Jacobs, S. (2006). Nonfallacious Rhetorical Strategies: Lyndon Johnson's Daisy Ad. *Argumentation, 20(4)*, 421–442.
Jacobs, S. (2000). Rhetoric and Dialectic from the Standpoint of Normative Pragmatics. *Argumentation, 14(3)*, 261–286.
Lewiński, M., Aakhus, M. (2014). Argumentative polylogues in a dialectical framework: a methodological inquiry. *Argumentation, 28 (2)*, 161-185.
Manzin, M. (2014). *Argomentazione giuridica e retorica forense: dieci riletture sul ragionamento processuale*. Torino: Giappichelli.
Mazzarese, T. (2008). Towards a Positivist Reading of Neo-constitutionalism. *Jura Gentium. Rivista di filosofia del diritto internazionale e della politica globale, 1/2008* (https://www.juragentium.org/topics/rights/en/mazzares.htm).
Piazza, F. (2019). Retorica e vita quotidiana. Che cosa ha ancora da dirci Heidegger sulla Retorica di Aristotele. *Giornale di Metafisica, 1/2109*, 261-277
Piazza, F. (2015). *La Retorica di Aristotele. Introduzione alla lettura*. Roma: Carocci.
Plantin, C. (1996). Le trilogue argumentatif. Présentation de modèle, analyse de cas. *Langue française, 112*, 9-30.
Rocci, A. (2016). Ragionevolezza dell'impegno persuasivo. In: P. Nanni, E. Rigotti & C. Wolfsgruber (Eds.), *Argomentare: per un rapporto ragionevole con la realtà* (pp. , 88-120), Milano: Fondazione per la Sussidiarietà.
Tindale, C. (2021), *The Anthropology of Argument. Cultural Foundation of Rhetoric and Reason*. New York: Routledge.
Tindale, C.W. (2019). Informal Logic and the Nature of Argument. In F. Puppo (Ed.), *Informal Logic: A 'Canadian' Approach to Argument* (pp. 375-401), Windsor: Windsor Studies in Argumentation.

Tindale, Christopher W. (2004). *Rhetorical Argumentation: Principles of Theory and Practice*. Thousand Oaks (CA): Sage.

LISTING AS AN ARGUMENTATIVE RESOURCE IN POLITICAL DISCOURSE

MENNO H. REIJVEN
University of Amsterdam
m.h.reijven@uva.nl

ALINA DURRANI
University of Massachusetts Amherst

GONEN DORI-HACOHEN
University of Massachusetts Amherst

Abstract

Lists are discursive resources with a wide variety of purposes. They can be used to manage the interaction and to advance various speech acts. There is also a line of research looking at lists as a rhetorical resource, having identified it as a claptrap. In this paper, we look at ways in which lists can contribute to argumentative discourse. When looking at lists which present one with a number of different reasons for a standpoint, some lists can be characterized as itemizing lists, as they simply enumerate a couple of arguments. Yet, some lists can be seen as directional, as there is a deliberate ordering of the items in order to communicate something more about the principle which brings the list items together. Then, third, there are also lists which are used to establish a generalized position. As it is not about the separate elements on the list, but the general principle, these lists can be called generalizing lists. These are three ways in which lists can contribute to argumentation.

1. Introduction

Lists can be an important resource in political discourse. Atkinson (1984) has identified the three-part list as an important rhetorical tool to generate applause. Namely, the three-part list signals a completion point towards which the audience can orient during a speech. This is a research finding which has been confirmed in recent years (Bull & Waddle, 2021; Bull & Miskinis, 2015). Moreover, the three-part list seems to be fundamental throughout language use, being identified as a basic structure in conversation as well (e.g., Jefferson, 1990; Lerner, 1994; Selting, 2007).

Yet, lists are more than tools to communicate a completion point. Lists, as a discursive structure, have been shown to generate more meaning than could have been communicated through other linguistic means. Through listing, speakers can refer to an organizational principle connecting elements beyond those overtly mentioned in the list (Barotto & Mauri, 2018) or connect different items together as belonging to the same category (Karlsson, 2020). Important to note is that these organizational principles are subjective selections by a speaker (Dori-Hacohen, 2020). This subjectivity of lists becomes also clear by noticing that lists which are longer than three parts are a recurring phenomenon (Dori-Hacohen, 2020). Hence, while lists carry an aura of objectivity as a discursive tool, in contrast to narratives (see Schiffrin, 1994), they are useful tools for a speaker to establish meaning and are therefore useful to argumentative discourse. In this paper, we investigate how lists can contribute to argumentative discourse.

2. Literature review

Lists are a discursive structure, and bring together pieces of information as belonging to a same set. While some scholars refer to 'listing' of the same information as reduplication (e.g., Karlsson, 2020), we follow Jefferson's (1990) perspective that lists can consist of the repetition of the same item. Lists as a structure bringing together (potentially different) items can therefore besides connecting and reformulating information, also resolve disfluencies in discourse (Kahane, Pietandrea & Gerdes, 2019). The discursive structure of a list enables each of these processes.

Interactionally, this means that lists can be used purposefully by a speaker. Lerner (1994), for example, has noticed that by establishing a category beyond a concrete referent can be used to manage interactional demands. When a speaker may anticipate a rejection, they can employ a list to open up the potential elements which would constitute agreement, resulting in a softening of the rejection (Lerner, 1994). According to Selting (2007), lists are often part of a larger three-part structure in which the lists proper do, for instance, "elaborating, explicating, exemplifying [and] illustrating" (p. 491). The opening up of the category of referred to elements can clarify the issue at hand. The category invoked through the list can thus fulfill a variety of interactional functions.

Two main types of lists are typically distinguished. Sánchez-Ayala (2003) distinguishes between lists which frame and lists which demonstrate. With regards to the former, common ground is incrementally built to coordinate and monitor the orientation towards the topic at hand.

As to the latter, a list can offer evidence or support for a position. Similarly, Schiffrin (1995) claims that there are descriptive and argumentative lists. While argumentative lists contain examples to support a position, descriptive lists sketch a situation and often appear within a narrative structure.

With respect to argumentative lists, rhetorical strength can be accomplished in different ways. One way in which lists can contribute to persuasion is through the sheer number of elements. Larson (2019) argued that #metoo was effective as an enormous list was created which overwhelmed people and thereby it could accomplish political persuasion. Rains (1992) stresses that it does not matter that later items to the list become strictly spoken meaningless. Each item still contributes to additional persuasiveness as there is a suggestion that more arguments could be given (Rains, 1992).

Also important to note is that the order of items in lists in discourse is also meaningful. Kahane, Pietandrea and Gerdes (2019) have claimed that the meaning of a list is defined through the first item, to which all other items have a special (i.e, paradigmatic) relationship. This item is namely the basis for identification of the underlying categorization principle which is communicated through the list. Similarly, Jefferson (1990) has claimed that the last item may sometimes appear as an afterthought rather than something important (p. 81). While sometimes this can be exploited strategically, filling up the list can also just seem like a discursive chore, as only with three items the list would be complete (Jefferson, 1990). In sum, the order of the items has been shown to be important, with special attention having been devoted to the first and last items on a list. Rhetorically, the order of items on a list can also matter. Lists can develop with increasing intensity (e.g., Atkinson, 1984), meaning that the order of the elements is not interchangeable (Inbar, 2020).

Both the function of and ordering within the list are considerations in the analysis below.

3. Data

We used two corpora to create a collection of lists for analysis. First, we used the first 2016 U.S.A. presidential debate. Second, we used interactions by presidential candidates on U.S.A. broadcast late-night talk shows during the 2016 elections and the 2020 Democratic primaries. We use these two types of data because they complement each other, while in both contexts candidates have to argue for their position. On presidential debates, candidates receive longer turns and have to directly respond to each other. During interviews on late-night talk shows, candidates talk conversationally with the host about their ideas without this possible challenge.

4. Listing reasons

One way in which lists are used in argumentative discourse is to share multiple reasons for a position. Consider the following list by Hillary Clinton during the 2016 U.S. presidential debate. Just before this argument by Clinton, Trump has accused her of not being trustworthy or capable to negotiate good trade deals for the U.S. because her husband, former President Bill Clinton, approved NAFTA during his term in office. According to Trump, NAFTA has been the worst trade deal ever negotiated, and he believes that it is problematic that Hillary Clinton favors this deal. In response, Clinton reflects on her time as a senator, and raises, in the form of a list, the criteria she used to evaluate trade deals she had to vote on.

> (1) Hillary Clinton: When I was in the Senate, I had a number of trade deals that came before me. And I held them all to the same test. <u>Will they create jobs in America? Will they raise incomes in America? And are they good for our national security?</u> Some of them I voted for, the biggest one, a multinational one known as CAFTA, I voted against, and because I hold the same standards as I look at all of these trade deals.

In this quote, to rebut Trump, Clinton shares that in her time as a Senator, she had to vote for "a number of trade deals." First, Clinton claims that she "held them to the same test." At last, she claims that the items on the list are "the same standards" she has used over and over again to "look at all of these trade deals." The reference to "test" and "standards" suggest that the three list elements are an exhaustive representation of her criteria for good trade deals. Specifically, any deal should "create jobs," "raise incomes," and enhance "our national security." Argumentatively, each item on the list is equally important as only taken together they constitute the complete test for a good trade deal. She always considered each specific element.

The ordering of the list is relevant to consider here. The items seem to lack direction and appear to be interchangeable in terms of order. The order is just tied to the topic of the talk. At the beginning of this Q&A-section, moderator of the debate Lester Holt asked the candidates how they plan to "create the kinds of jobs that will put more money into the pockets of American workers" (not shown). The first two items of the list fit in well with the topic as introduced by the moderator and are also salient topics within this segment of the debate. The structure of the list is used to add another consideration as argument into the discussion, one which Clinton considers to be potentially relevant too. Argumentatively, the list establishes a categorization principle for a good test, which goes beyond *simply* considering economic benefits, further defending her ability and character to be president. While the list presents three co-equal arguments, the order allows Clinton to broaden the topic beyond the question of the moderator through the last item to complete the list.

Earlier in the debate, Clinton also offered a couple of reasons within a list structure to defend a position. There is a danger that a last item appears as an afterthought (Jefferson, 1990), particularly when the list becomes longer than the typical three parts. Yet, this can be counteracted within such lists by stressing the importance of the last item. That is, the last item may receive further subordinative argumentation, as can be seen in the next example from Clinton from the 2016 U.S. presidential debate.

(2) Hillary Clinton: I want us to invest in you. I want us to invest in your future. That means jobs <u>in infrastructure, in advanced manufacturing, innovation and technology, clean renewable energy, and small business</u>, because most of the new jobs will come from small business.

In this case, Clinton explicates through a list what she thinks "investing in you" means. She defends her position that she will invest in the American people by listing where she wants to create jobs: in "infrastructure, advanced manufacturing, innovation and technology, clean renewable energy, and small business." This last item is then raised in importance through an extension. Clinton continues by claiming that "most of the new jobs will come from small business," reducing the importance of the other items on her list. In this case, the last-but-not-least strategy results in stressing the last item at the expense of the other items. She marks this as her main argument, while the other items can be seen as alternative arguments for her future success. In this example and the previous one, each item is in principle a similar premise defending the conclusion – each being a separate reason – without there being much of a structure besides broadening the categorization principle beyond the topic under discussion or marking one reason as the main argument.

Compare this to the following list advanced by Elizabeth Warren at *The Late Show with Stephen Colbert* in 2019. Just before the next excerpt, Warren was asked by Colbert what she thinks the most egregious thing is which was done by President Trump and the Republicans in Congress during Trump's term as President. Specifically, the question concerns asking about what they did to help the "one tenth of the one percent" to get richer. In contrast to Clinton's lists above, this list has a fixed internal organization: the items are not interchangeable. Instead, there is a progression from the first to the last.

(3) Elizabeth Warren: Well, so, look at this fundamental question about how America works. When Donald Trump and the Republican Congress wanted to give away a trillion plus dollars to <u>millionaires,</u>

billionaires and giant corporations, they got their act together in just a few weeks and managed to pass a bunch of tax breaks for them.

Warren starts her response to this question by noticing that helping the rich is "how America works." She explains that "Donald Trump and the Republican Congress" can, when that is the case, "get their act together" really fast, in contrast to when they need to help ordinary citizens. When explaining the Republican actions, Warren uses a list to explain who were the recipients of "a bunch of tax breaks:" "millionaires, billionaires and giant corporations." There is an increase in intensity from "just" millionaires up to the "giant corporations." She is establishing the lower and upper bounds of entities Republicans are tempted to help. This list establishes the underlying principle of categorization precisely through the ordering of the items. Still, each element on the list could act as a separate reason, as each element is supposed to prompt Republicans to act: any wealthy group of people is presumed to be supported by them.

These lists which outline a number of premises in a fixed order can also be used to manage the discussion topic. In the next excerpt, Clinton was asked the question by moderator Lester Holt during the 2016 presidential debate how she is planning to "heal the divide" caused by bad "race relations" in the country. In his question, Holt referred to recent events like the police shootings on African-Americans in Tulsa and Charlotte. Clinton starts her answer as follows.

(4) Hillary Clinton: Well, you're right. Race remains a significant challenge in our country. Unfortunately, race still determines too much, often determines where people live, determines what kind of education in their public schools they can get, and yes, it determines how they're treated in the criminal justice system. We've just seen those two tragic examples in both Tulsa and Charlotte. And we've got to do several things at the same time.

When receiving the floor, Clinton affirms that she agrees with Holt that there is a "bitter gap" to be bridged by the next president. Clinton continues this thought explicitly by stating that "race remains a significant challenge in our country." Then, she provides three reasons showing that race is still a challenge in the U.S., which she does in the form of a list. Race is a significant challenge, as it "still determines too much," meaning that it determines "where people live," their "education" and "how they're treated in the criminal justice system." The listing is done here by repeating the word "determines" and a wh-question word after it. The listing device connects the list to the main claim by replacing its central element ("too much").

While the list offers three reasons for why race is still a challenge in U.S. society, there is an order to the items listed. People grow up in a neighborhood, and enter the school system there, and may face the criminal justice system later on – a sequence representing interconnected problems with structural racism. In addition, the third item is also the only item linked to the question originally asked by the moderator, as this pertained to police violence towards people of color. Clinton marks this by adding "yes" to her utterance, affirming the words by moderator Lester Holt again. Thus, what she is doing through this list is embedding the issue within the larger problematic state of affairs, acknowledging that the topic of the discussion is only a small part of the overall problem with many interconnected components.

This is then further acknowledged in the argumentation. In the subsequent utterances in her turn, Clinton just continues to discuss the last item of the list, as requested by the question. The list has contextualized the topic of discussion, but by working towards the last element, it becomes relevant to further discuss and defend the last element, similarly as was done in example 2, which is done by reiterating the two examples introduced by Lester Holt in the question: "two tragic examples in both Tulsa and Charlotte."

Next, Clinton outlines her plans to improve "race relations," which she announces by saying "we've got to do several things at the same time." The list showed her awareness of the breadth of the topic, but the direction of the list enabled Clinton to still respond to the question and share her ideas on how to improve the criminal justice system. The list enabled her to manage the topic: as a discursive structure it allowed her to both deviate and return to the topic at hand, and maintain coherence in her talk. Still, she gave three separate arguments for why race is still a challenge in the U.S.

In these three cases, we saw lists being used to list *reasons*. All elements act as a reason for a position. In excerpt (1), the test/standard used had to be represented and this judgment had to be comprehensively communicated. In excerpt (2), the argument is about the breath of investments to create jobs. In excerpt (3), it is about the range of people Republicans are willing to support with their policies. In excerpt (4), it is about the full context of the problem under discussion. Overall, we observed two types of lists. On the one hand, there are itemizing lists where different items are brought together. Then, the list is used to advance a number of premises for a position. On the other hand, as in excerpt (3) and (4), there are directional lists where there is a clear progression from the first through the last item. Besides communicating a number of premises, the underlying coherence of the list is communicated as well, enhancing the argumentative strength of the talk.

5. Establishing premises

Lists can also be used to deflect a position advanced by the opponent. Namely, a list can enlarge the number of referents, providing alternatives to the claim made by the other party. Trump, during the first U.S. presidential debate of 2016, used a list to undermine the claim that Russia hacked the servers of the Democratic party. There is a clear progression in this list to open up the possible culprits to ultimately make the claim that "we don't know" who is responsible for this. This was particularly relevant for Trump to challenge as Clinton used this as a reason to call out Trump for taking a friendly stance towards president Putin from Russia.

> (5) Donald Trump: Now, <u>whether that was Russia, whether that was China, whether that was another country</u>, we don't know, because the truth is, under President Obama we've lost control of things that we used to have control over.

With this list, Trump rebuts Clinton's assertion that Russia was behind the cyberattacks on the servers of the Democratic National Party. First, Trump repeats Clinton's suggestion (that it "was Russia"), but then opens it up to "China" as well, and at last even that it could be "another country." The items on the list are not equal and their order matters. The first item is a repetition of Clinton's point and is the claim to be refuted. It could be seen as a concession, admitting that it *could* be Russia who did it. However, then, Trump continues that there is an alternative culprit, China, starting to enlarge the set of possibilities. Last, Trump opens up this set beyond these two countries to any country to make the point that "we don't know." Hence, the direction of the list is designed to undermine Clinton's position and support a strong counterclaim. In this way, Trump is able to deflect the attack on his stance on Russia.

Interestingly, in this case, Trump defends his list. While the list defends the claim that we do not know who launched the cyberattacks by establishing that any country could have done this, Trump then continues by claiming that this is the case "because" "under Obama we've lost control of things that we used to have control over." It could be any country because the U.S. has no way of knowing the answer. Hence, rather than being three separate arguments, the list is used to defuse Clinton's position (the first item on the list) and move towards the claim that any country could have done this, which is then a claim being defended by subordinative argumentation.

A similar process can be observed in the next excerpt, although Clinton in the first U.S. presidential debate is not criticizing a position but simply establishing one. In the excerpt below, Clinton responds to Trump's claim that ISIS is "beating us at our own game," the internet, suggesting it is embarrassing that the U.S. cannot stop it. She advances a list which is letting her move beyond the threat of ISIS on U.S. soil alone.

(6) Hillary Clinton: I think we need to do much more with our tech companies to prevent ISIS and their operatives from being able to use the internet to radicalize, even direct people <u>in our country, and Europe, and elsewhere</u>. But we also have to intensify our air strikes against ISIS ...

Clinton starts with claiming that she has crafted plans to fight ISIS, which involve dealing with them online. She believes that the U.S. government can utilize American "tech companies" to get a hold of ISIS on the internet, which Trump suggested the U.S. is supposed to be able to control. Thus, then, according to Clinton, it is possible to prevent ISIS to "use the internet to radicalize [and] direct people." Where ISIS is able to instruct people, is established through a list: "in our country, and Europe and elsewhere." Clinton establishes that the danger of ISIS is everywhere and that their activity on the internet should also be tackled everywhere in order to curb ISIS.

There is some ambiguity regarding the referent of "elsewhere." It completes the three-part list, but does not clearly designate the exact set of entities being referenced. Clinton suggests that our "tech companies" have to become more involved because ISIS is directing people everywhere in *the West*, as this is the generalization of the two meaningful items mentioned in this list. The list is used to establish a category without explicitly naming it, simply creating a suggestion of inclusion. In this case, as in the previous case with Trump, thanks to the empty and open list completer (elsewhere or anywhere), a broad and open premise is established through the list.

6. Conclusion

Lists can have various argumentative functions, and in this paper, we only surveyed a few of them. Important to note is that lists are not just advanced to provide multiple reasons for a standpoint or multiple examples carrying support for a position. Of course, there can be lists which simply list a number of reasons in defense for a standpoint. Sometimes, the reasons given are interchangeable – the order is not set in stone. In example 1, each criterion carries the same argumentative weight and is simply a co-equal item on the list. Similarly in example 2, each item on the list is a source of jobs and the order of the items is not necessarily like this. In principle, they are co-equal items, each being a premise in defense of the standpoint. Still, given the discourse at hand, the organization can be explained. Example 1 is structured by the topic of the talk, with the first two items tied to the topic at hand while the third deviates from it. In example 2, Clinton marked the last item as most important, making the ordering of the items relevant herself. These lists, in which in principle each of the items are co-equal, can be referred to as itemizing lists. They are a resource through which speakers can communicate a number of reasons within a discursive structure.

Lists providing multiple reasons can also have a more fixed structure with more dependency of the separate items on each other. In example 3, we saw that Warren listed three categories of "rich entities," ordered progressively. Thereby, the first item established the lower boundary, while the last item marked the upper one. While each of those separately was shown to be a reason for Republicans to get to work, together they mark the spectrum of entities they are willing to work for. Similarly, the list in example 4 was structured in a comparable way. While each item on the list constituted an argument for the fact that "race relations" are a problem in the U.S., they were ordered in a progressive way, reflecting the complex racist state of affairs. We refer to these lists as directional lists, as there is a progression form the first to the last item which communicates meaning in addition to each of the items as a premise in the argument.

Third, we also saw that lists can be used to establish a premise. Rather than communicating a number of reasons, the list is advanced to communicate a categorization principle. In example 5, Trump used a list to undermine a point made by Clinton. The first item of the list was the point Clinton made in her turn. The second item opened up the class of referents, and the last general list completer accomplished establishing as categorization principle for the list "any country", which was diametrically opposed to Clinton's position that it certainly was Russia which was behind the cyberattacks on the Democratic National Party. In excerpt 6, Clinton used a list to establish the premise that ISIS uses the internet to radicalize and direct people in Western countries, without having to state it in that way. She did not advance three separate arguments, but invoked a premise through this list regarding being inclusive and comprehensive. These are generalizing lists, as they aim at establishing a general category to be used in the reasoning.

In this paper, we just discussed three ways in which lists can be used in argumentation. For sure, this topic needs future attention to understand these dynamics in more detail and investigate further the various ways in which lists can help convince an addressee.

References

Atkinson, Max (1984). *Our Master's Voice. The language and body language of politics.* London: Methuen.

Barotto, A., & Mauri, C. (2018). Constructing lists to construct categories. *Italian Journal of Linguistics, 30*(1), 95–134. https://doi.org/10.26346/1120-2726-117

Bull, P. & Miskinis K. (2015). Whipping it up! An analysis of audience responses to political rhetoric in speeches from the 2012 American presidential elections. *Journal of Language and Social Psychology, 34*(5), 521–538. doi:10.1177/0261927X14564466.

Bull, P. & Waddle M. (2021). Speaker-audience intercommunication in political speeches: A contrast of cultures. *Journal of Pragmatics, 186,* 167–178. https://doi.org/10.1016/j.pragma.2021.10.001.

Dori-Hacohen, G. (2020). The "Long List" in oral interactions: Definition, examples, context, and some of its achievements. *Pragmatics. Quarterly Publication of the International Pragmatics Association (IPrA), 30*(3), 303–325. https://doi.org/10.1075/prag.19007.dor

Inbar, A. (2020). List constructions. In R. A. Berman & E. Dattner (Eds.), *Studies in Language Companion Series* (Vol. 210, pp. 623–658). John Benjamins Publishing Company. https://doi.org/10.1075/slcs.210.18inb

Jefferson, G. (1990). List-construction as a task and as a resource. In G. Psathas (Ed.), *Interaction Competence* (pp. 63–92). University Press of America.

Kahane, S., Pietrandrea, P., & Gerdes, K. (2019). The annotation of list structures. In A. Lacheret-Dujour, S. Kahane, & P. Pietrandrea (Eds.), *Rhapsodie: A prosodic and syntactic treebank for spoken French* (pp. 69–95). John Benjamins Publishing Company. https://doi.org/10.1075/scl.89.06kah

Karlsson, S. (2020). The Meanings of List Constructions: Explicating Interactional Polysemy. In: K. Mullan, B. Peeters, & L. Sadow (eds.), *Studies in Ethnopragmatics, Cultural Semantics, and Intercultural Communication* (pp. 223-238). Singapore: Springer Singapore. https://doi.org/10.1007/978-981-32-9983-2_12.

Larson, S. R. (2019). "Just let this sink in": Feminist Megethos and the Role of Lists in #MeToo. *Rhetoric Review, 38*(4), 432–444.

Lerner, G. H. (1994). Responsive list construction: A conversational resource for accomplishing multifaceted social action. *Journal of Language and Social Psychology, 13*(1), 20–33.

Rains, C. (1992). "You die for life": On the use of poetic devices in argumentation. *Language in Society, 21*(2), 253–276.

Sánchez-Ayala, I. (2003). Constructions as Resources for Interaction: Lists in English and Spanish Conversation. *Discourse Studies, 5*(3), 323–349. https://doi.org/10.1177/14614456030053003

Schiffrin, D. (1994). Making a list. *Discourse Processes, 17*(3), 377–406. https://doi.org/10.1080/01638539409544875

Schiffrin, D. (1995). *Approaches to discourse.* Cambridge: Blackwell.

Selting, M. (2007). Lists as embedded structures and the prosody of list construction as an interactional resource. *Journal of Pragmatics, 39*(3), 483–526. https://doi.org/10.1016/j.pragma.2006.07.008

CONSPIRACY AND NON-CONSPIRACY ARGUMENTS IN ANTI-VACCINE DISCOURSE

THÉOPHILE ROBINEAU
Université Paris Cité
theophile.robineau@gmail.com

Abstract

In recent years, we have witnessed an increase in the visibility of certain discourses described by the media and a number of academic observers as 'conspiracy theories'. Taguieff (2021) even goes so far as to claim that we have entered the 'era of conspiracy theories'. The Internet has played a decisive role in the multiplication, diversification and massive dissemination of these discourses, and their labelling as 'conspiracy theories' has undoubtedly contributed to the phenomenon, creating a single category out of a heterogeneous nebula of discursive productions, which gives it greater visibility and makes it an object of debate.

There is now a significant amount of multidisciplinary work (Danblon, 2010) aimed at characterizing so-called 'conspiratorial' discourses, a characterization often conceived as a prerequisite for combating their dissemination. However, a reading of this work reveals a double vision of the object.

On the one hand, conspiratorial speeches are seen as marked by a pathetic rhetorical dimension, with emotions (fear, anger, resentment) being considered obstacles to the deployment of critical reason. They are manifestations of the 'paranoid style' (Hofstader 2012). This pathologizing view of conspiratorial thinking sees those who engage in it as psychologically fragile individuals. Although discussed since the 1990s, this approach has left its mark on contemporary work on conspiratorial discourse.

Since then, some have sought to recognize a common logic in the discourses they take as their object, which would allow them to be characterized as conspiratorial. The idea is not to situate them outside the realm of reason (Taguieff (2021) even asserts that conspiratorial discourse obeys the principle according to which 'Everything that is officially held to be true [must] be subjected to ruthless critical scrutiny'), but to see a rogue logic at work, marked by epistemological assumptions that nothing happens by chance, that the world can be made readable by reconstructing hidden agendas that link events together, based on misinformation and bad epistemological arguments (Oswald, Lewinski, Greco & Villata, 2022).

The fact is that these characteristics do not always seem to be sufficient to define a 'conspiracy discourse' and can be applied to a large number of discourses circulating in the public sphere, which presents the risk of disqualifying them by lumping them together under this label.

By comparing three interviews conducted with people opposed to the French government's vaccine policy, this paper proposes ways of distinguishing between discourses that one might be tempted to call conspiracy theories and others that are not. In particular, by showing that some of the features considered characteristic of conspiracy speeches are not always sufficient to distinguish these two types of speeches.

1. Introduction

Even if history is not lacking in theses proposing a reading of the world's development based on a global causal system, and referring to the will of a hidden and malevolent entity (we can think, for example, of the 'conspiracies of lepers' in the Middle Ages (Chapoutot, 2021), or the attribution of the causes of the French Revolution to the action of the Jews and Freemasonry (Nicolas, 2015)), these theories seem to have multiplied, to the point that some authors consider that we have entered the 'age of conspiracy theories' since the 2010s (Taguieff, 2021) or describe our era as the 'era of conspiracy' (Peltier, 2021). For its part, the French government has multiplied initiatives to fight against what it now considers a 'major issue' and a danger to democracy (Bronner, 2022).

The term 'conspiracy theorist' has therefore been circulating more and more in the French public arena for several years. Three aspects stand out in relation to the use of this term: on the one hand, it is systematically disqualifying; on the other hand, beyond its negative axiology, it is difficult to find a common semantic core to its different uses. Finally, this variation in usage goes hand in hand with a low level of definitional activity for the term, which is mostly used as if it were self-evident.

In view of this vagueness, I wondered whether it was possible, within the framework of an analysis of argumentative discourse, to bring to light discursive characteristics common to these discourses that some people describe as 'conspiracy'.

2. Approach

In order to do this, I chose to focus on a specific position expressed in the context of the debate on vaccination: that of speakers expressing an anti-vaccine position. This debate has the advantage of having given rise to numerous antagonistic positions based on diverse positioning strategies, some of which could be qualified as conspiracy theories, which makes it possible, in a thematically homogeneous corpus, to compare arguments that could be described as conspiratorial with arguments that do not invite such a characterization.

As a methodology, I chose to conduct interviews with people opposed to the covid-19 vaccine.

One of the advantages of conducting interviews is that they offer the respondents a space in which they can develop their thoughts, without any immediate horizon of confrontation or judgement. This interactional dimension seemed to me to be essential in order to be able to get the interviewee to make his or her position explicit and to get him or her to react on points that he or she would not have addressed spontaneously, which would not necessarily have been possible without the physical presence of the researcher (Beaud, S., & Weber, F., 2010).

Adopting a comprehensive interview approach seemed to me to be a way of gathering their words in a context of trust, so that the anti-vaccine respondents could develop a discourse that was not marked by a defensive position against an antagonist.

I will rely on a definition of argumentation as "a mode of constructing discourse aimed at making it more resistant to contestation" (Doury, 2003). Also, during the interviews collected, I invited the anti-vaccine respondents to react to the reported pro-vaccine speeches, in order to lead them to consolidate the support of their position. I then conducted a comparison of the way in which the various respondents proceeded in the face of this confrontation with the counter-discourse.

Beyond these considerations linked to the methodology of the investigation, one of the first difficulties I faced in my work was the risk of having a 'circular approach', which would amount to collecting conspiracy speeches in order to search for and find what had made me intuitively choose them in the first place. To avoid this bias, I tried never to presuppose that a discourse would be conspiracist or not. I conducted several interviews with anti-vaccine people who were anti-vaccine for multiple reasons.

Once this data had been collected, I sought to identify, based on objective criteria, the main anti-vaccine positioning strategies of my interviewees. Firstly, I will see whether the discourses whose speakers claim to be conspiracy theorists themselves, or who say they have already been labelled as conspiracy theorists by others, present specific discursive characteristics. In parallel (and this phase of the work is still in the programmatic stage), I would like to set up a reception study to identify the speeches that are actually perceived as conspiracy theorists; I will then see whether or not these speeches have specific discursive-argumentative characteristics.

3. Presentation of the three interviews

The work presented here is a first application of my approach on a corpus of three interviews collected during the year 2022. All three are interviews with anti-vaccine people, each for different reasons.

The first interview was conducted with a doctor who claims not to be a conspiracy theorist - indeed, he mentions certain "conspiracy theories" from which he distances himself. Nevertheless, when asked what he thinks of the vaccine policy and why he opposes it, he explains that for him the RNA vaccine is not intended to protect against covid-19 but simply to allow "big pharma" - the world's big pharmaceutical lobby - to make money from people's health. Big pharma being controlled by "a few families" which he would soon identify as Jewish families.

Respondent 1 *"to fill their pockets only [...] big pharma's money and there are some families who hold this."*

The belief in a worldwide Jewish conspiracy, which is behind most of the events in history, is one of the fundamental historical elements of the conspiratorial imagination (Poliakov, 2006). Later in the interview, the respondent will explain that behind each of the camps of the Second World War, or behind the assassination of Kennedy, was the Rotschild family to profit from it.

The second interview shows almost none of the characteristics that one would associate with conspiracy. The respondent simply expressed doubts about the validity of the government's health policy and her wish not to be vaccinated. Like the first respondent, she does not define herself as a conspiracy theorist or as anti-vaccine. Her reticence towards vaccines stems first of all from a more general feeling of distrust towards medicine, based on her values and personal experience.

Respondent 2 *"I don't feel in danger from this disease and in general in my life, so I'm not anti-vaccine or anything, but in general in my life I avoid taking as many things as possible that don't belong to my body. In everything that is medicine and all that, if I have to take it, I take it, but I avoid it as much as possible."*

Finally, the third interview sometimes contains statements that one would spontaneously want to associate with conspiracy, but in a less marked way than in the first interview. In particular, she justifies her rejection of the vaccine by associating it with the government, which she does not trust to conduct a good vaccination policy. Although at times she uses phraseology that one might be tempted to call conspiracy, in this case by explaining that a mask factory was deliberately closed a year before the start of the pandemic and that the RG - the French secret service - knew about it, which would make some of the government's decisions part of a preplanned plan hidden from the public.

Respondent 3 *"Normally the first thing to do is to put the factory back in place. Now they're closing the factory the year before, the last factory that makes masks, it's not normal."*

Interviewer *"For you, it was voluntary?"*

Respondent 3 *"How do you want me to explain it? The guys who are capable of knowing exactly everything, they know everything! The RG, you can't be at that level of responsibility and not know things. The guys know. I mean, so you don't choose to [close]!"*

However, most of the time, the respondent focuses her criticism on the simple "way of managing" the health crisis without presupposing any malicious or hidden intentions. It is more a case of the government being accused of amateurism than a denunciation of a real desire to do harm.

Respondent 3 *"I understood that there was an epidemic, for me it was not at all propaganda. I think that, really, it could have been dangerous. However, as the speech went on, I said to myself No! what? I don't believe in this way of managing things! I think there were deaths because they didn't manage the health crisis, quite the contrary."*

These three interviews were therefore selected for an initial exploratory study in order to obtain anti-vaccine speeches illustrating various anti-vaccine positions with regard to their potential conspiracy nature.

4. Different types of modalisations

The interviews were analyzed using the NVivo annotation software, and the same analysis grid was applied to each of the three interviews. This analysis grid is built around the main enunciative, and discursive categories usually considered central to the construction of an argumentative discourse.

We first tried to identify all the discursive mechanisms generally involved in the construction of an argumentative discourse. Positive or negative axiology, the expression of degrees of certainty with regard to the statements, enunciative heterogeneity, concessive forms, types of arguments such as testimony, arguments by precedent, comparisons, as well as the presence or absence of connectors traditionally regarded as argumentative.

Table I. Argumentative discursive mechanisms

Discursive mechanism	Example
Positive axiology	There are some MPs who for me have been great at speaking out. (#2)
Negative axiology	I think it's despicable how they handled it! (#3)
Expression of degrees of certainty	I don't know where it is at the last time I checked. It hasn't been renewed if I'm not mistaken. I'll have to check. (#1)
Concessive forms	They should have tried, they should have tried, they're right but I think they didn't go for the right population. (#2)
Enunciative heterogeneity	Even Raoult said that it is a statistical error to think that multiple sclerosis is linked to the hepatitis vaccine. (#1)

To the language phenomena traditionally considered relevant for the description of any argumentative sequence, we have added themes described in the scientific literature as characteristic of conspiracy discourse.

The elements identified are: belief in the existence of a global Jewish conspiracy, existence of malicious groups with hidden interests (Romero Reche, A., 2020)., high intelligence of the malicious actors (Oswald, Lewinski, Greco & Villata, 2022), conception of history as being influenced by these groups, questioning the causes of negative events, stupidity or ignorance of the masses about the existence of malicious interests against them (Hofstadter, R., 1964)

Table II. Characteristic themes of conspiracy theories

Themes	Example
Existence of a Jewish lobby	When there is a war there are war debts. So, on each side you have the generals, the officers, and some politicians from Germany and the Axis forces, and on the other side the Allies, and at the end of the table, because we're discussing money, on one side there's Mr Rothschild! And on the other side there's Mr Rothschild's cousin, so he's the one who's collect the jackpot. (#1)
Existence of malicious and hidden groups	The same people who are descended from the communists of the time of the Russian revolution, these people hold the power. (#1)
Intelligence or intentionality of malicious actors	The Rothschilds and their cousins are excessively intelligent people, excessively businesslike. (#1)
Conspiracy phraseology	It's true that there are some dubious things about graphemes and all that... There are questions to be asked about marking people. (#1)
Unifying world view	[Macron] who comes to give him the badge of the super businessman [to the boss of Pfizer], but it's crazy, it's crazy this level. (#3)
Transmission of insider knowledge	What I propose is to send you some links to documentation or interviews with certain people. (#1)

Finally, we have created a category that we had not planned to activate at the beginning, but which seemed necessary to us after listening to the interviews, which struck us as having a strong emotional dimension. The emotions in question are mainly negative: indignation, fear, anger, but also sometimes positive, especially when expressing the pleasure of resisting the government and a form of heuristic joy when the interviewees explain that they have discovered the 'hidden' truth or pierced the lies of official discourse.

Table III. Emotional dimension

Emotions	Example
Negative emotions	When you have your xx year old daughter crying on the phone one morning and she says: I'm going to lose my job! Do you realize, Dad? That what they are doing is serious! It's serious! (#2)
Positive emotions	For the time being, it was nice, it's that there are people I think who opened their eyes to a certain manipulation of the media. (#2)

5. Comparisons

5.1. Critique of the public debate

Each of the three interviews was conducted using the same interview grid, and therefore follows a similar procedure, which allows us to compare the way in which each respondent answers certain specific questions.

After collecting their general feelings about the covid-19 crisis and French vaccination policy, they were asked more specifically why they had chosen not to be vaccinated. The three respondents then all began by recounting the circumstances in which their initial doubts arose. This stage is characterized by a strong negative axiology attached to their background and the way in which the respondents feel their point of view has been treated, which makes it possible to identify the angles of attack that make sense to them.

Even before entering into the discussion of the benefits or harms of the vaccine, the respondents expressed their views on the media treatment of the health crisis and stated that their critical thinking on the vaccine had difficulty finding an echo in the media, as the latter hindered the anti-vaccine positions of which they refused to be respectful relays. This difficulty in being heard and recognized leads the respondents to formulate criticisms of the way the public debate works and, thanks to the use of statements with a deontic value or criticizing the modalities of the public discussion, which gives a glimpse of what it should be.

Respondent 1 *"All these renowned professors, Fourtillan, well we don't put medical professors in psychiatry because they have a different opinion anyway!"*

Respondent 2 *"I'm not a doctor, so I don't have an opinion on the matter. But my opinion as a citizen is that everyone has the right to express themselves."*

Respondent 3 *"Raoult said it: it had become almost a religion of the thing [of the vaccine], that is to say there is a dogma and it is that one and no other. You could no longer question anything about it."*

In these three examples, the respondents do not explicitly seek to defend an opinion but criticize the way in which opinions that differ from the pro-vaccine line are disqualified in the media.

The first example uses the status and recognition of the individuals concerned ("renowned professors"; "medical professors") to support the conclusion that they should not be put into psychiatry - disqualified as mad - simply because they have a different opinion. The "anyway" can be read here as a marker of indignation.

In the second example, the respondent uses a concessive form "I'm not a doctor [...] But" to establish the assertiveness of her statement by continuing "my opinion as a citizen is that everyone has the right to express themselves". Here too, she takes a firm stance on the way in which public debate should be conducted by reminding us that, as a citizen, she has the legitimacy to give her opinion.

The third example uses an argumentative structure somewhat similar to that of the first: she first quotes Professor Raoult, a controversial French doctor but considered by her as an expert, who would denounce religion and medical "dogma" in relation to the vaccine. The use of polyphony "it is that one and no other" suggests the voice of a censor who would reject any opinion different from his own without bothering to justify himself. The next sentence, which serves as a conclusion, uses a generalizing "you" to seek the approval of the interviewer in denouncing the obstacles to questioning, which are necessary for the debate of ideas.

5.2. Denunciation of an official discourse

The importance given by the respondents to respecting the right to express themselves and the plurality of opinions, while at the same time demonstrating a form of hypercriticism towards the government's word, is a perfect reflection of the phenomenon of 'cognitive asymmetry' (Wagner-Egger, 2020, cited in Taguieff, 2021) identified as characteristic of the conspiracy fighter. This can be observed through the use of the word 'official', which is a recurrent qualifier for thinking that is considered dogmatic. This criticism of 'official' discourse is also accompanied by a negative axiology or qualifiers considered disqualifying in the respondents' discourse:

Respondent 1 *"I think it's several [Jewish] families because when you look at the official media in France, they are all run by the same cousins."*

Respondent 3 *"anyway if you listen to all the official propaganda, it's basically that: everything is fine!"*

The first interview will talk about the media and official figures, which it contrasts with alternative media and a reading of the statistics that give different conclusions about the danger of the vaccine. The fact that the owners of the official media are supposedly members of a few Jewish families justifies the fact that their speeches are all in the same direction.

The third interview also uses the term official media, going so far as to speak of official propaganda, which could be linked to the malice of the government, as the propaganda was the work of the state.

Respondent 2 *"there are people I think that they have opened their eyes to some manipulation of the media."*

Only the second interview does not use the term 'official', the respondent makes the negative axiology of the word "manipulation" attributed to the media. However, she does not elaborate on the causes of this manipulation, which is mainly related to the bad way of managing the public debate.

5.3. Ethos of resistance

In this French media context where pro-vaccine speech is hegemonic, defending an anti-vaccine position goes hand in hand, among the respondents, with the construction of an ethos of resistance in the face of the government's vaccine policy, at once victimized and revolted. This ethos has been identified in the literature as characteristic of conspiracy discourses (Uscinski & Parent, 2014).

Respondent 1 *" I compare that to my parents who were in the resistance during the war. [...] you have the resistance fighters, there's the Gestapo, you obviously have the Vichy regime, but you also have, and we mustn't forget this, 30% of the population who are collaborators [with the German occupier] and who would denounce you for this or that [...] opportunists and cowards."*

Respondent 3 *"I feel like I'm fighting a little war, a little resistance struggle: I won't go to the restaurant anymore."*

In the first and third respondents, the reference to the imaginary of resistance - which is very important in France - is assumed. The use of the word "war" allows the respondents to oppose very easily identifiable camps: in example number 1, the use of the deictic "war" explicitly refers to the Second World War, which allows the respondent to identify with his parents, resistance fighters against the oppressor. Comparing the situation experienced during the health crisis with that experienced under occupied France makes it possible to associate the current government negatively with that of Vichy, which takes on an additional dimension in the first interview where the pro-vaccine people are described as collaborators, suggesting the existence of external influences.

This idea was developed later in the interview with the help of a strongly negative axiology "opportunists and cowards", which underlined the radical nature of the respondent's comments.

The absence of the deictic in the third example leaves the reference to the Second World War vaguer, and the introductory use of 'at the limit' shows that the respondent is aware of the radical nature of her feelings about the situation and seeks to qualify them with the help of a modalization. The use of the adjective "little" for struggle and war is in line with this and should be linked to the fact that her act of resistance was ultimately only to stop going to the restaurant.

Respondent 2 *"I felt very unfairly attacked: I was going to lose my job! So I spent nights watching the National Assembly."*

In comparizon, the rhetoric associated with war is not found in the second respondent, or at least not as explicitly. Spending nights listening to the National Assembly may evoke the clandestine listening of Radio London (the name given to the BBC's French-language broadcasts for the resistance and its supporters).

It is more the cost of her resistance that is thematized here: the respondent explains that she considers herself to be a victim of the injustice she suffered, while seeking to fight against it.

5.4. Cohabitation of ethos

The rejection of an official discourse in the name of respect for the plurality of opinions, combined with the construction of an ethos of resistance, allows us to observe the cohabitation of two seemingly conflicting ethos: on the one hand, an 'open' ethos and, on the other, a 'dogmatic' ethos. This takes the form of questioning and a humble attitude towards knowledge, coupled with the assertion of strong positions that leave no room for doubt or negotiation.

Respondent 1 *"I don't hold the truth, but I do hold the doubt."*

Respondent 1 *"There is no reason to vaccinate children. No reason at all!"*

Respondent 2 *"I find it difficult to have a thought when you are not from this [medical] background, to have a serene and calm thought."*

Respondent 2 *"I was convinced that I had to stick to my position, to what I thought from the start in fact. And that I wouldn't give in under pressure."*

Respondent 3 *"On the WHO, on the criteria of the pandemic, I think... I don't know. I don't know. I... no, I don't know, I'm not familiar enough with the subject to have an opinion on the matter."*

Respondent 3 *"I can't! It's not possible! It's too big if you want, it's obvious and I know it hurts, I swear I'd love to think something else! I'd love to think that they've found a great vaccine..."*

In each of these three examples, the surveys explicitly express either a position of ignorance "I hold the doubt" "I don't know", or their difficulties in forming a satisfactory opinion "a serene and calm thought". This attitude towards the truth echoes the denunciations of the "official" word and calls for respect for the plurality of everyone's opinions.

Nevertheless, it is also accompanied by a very firm rejection of the government's policies, which resonates with the ethos of resistance. This rejection is justified by the expression of forms of categorical indignation in respondents 1 and 3: "no reason to vaccinate children", "it's not possible!". Finally, in respondent 2, the use of the term "convinced" testifies to the strength of her position.

This cohabitation of two contradictory ethos has also been identified as a rhetorical device characteristic of conspiracy theories (Dominicy, 2010).

6. Discussion

These comparisons allow us to make an initial analysis: on the one hand, the characteristics formally identified as being able to be used to define a conspiracy discourse are found in each of the three discourses: criticism of the media, denunciation of a false official word, ethos of resistance, cohabitation of conflicting ethos.

However, the presence of these characteristics does not allow us to conclude without discussion that the speeches analyzed are conspiracies. Thus, the respondents are objectively part of a minority, whose point of view is little represented - or in any case, largely disqualified when it is - in most of the major media. Thus, the ethos of victimhood and resistance that they construct is consistent with their objective position in the public debate. In the same way, calling for the practice of systematic doubt can be as much a form of critical thinking (Danblon, 2020) as an attitude of generalized suspicion characteristic of conspiracy.

However, a more detailed analysis of each of these characteristics allows us to identify a form of gradation between the interviews.

For example, we find different modalizations in the assertiveness of the respondents' comments. When they expressed their indignation about the media's treatment of the anti-vaccine discourse, the first respondent was more assertive "I compare" than respondents 2 and 3 who put forward their personal and sensitive experience "I felt" "I have the impression".

This gradation is also found in their criticism of the government. While respondent 1 did not hesitate to explicitly compare the government and the pro-vaccine people to the Vichy government, the Gestapo and the collaborators during the occupation, respondent 3 was more measured and made more use of modalizations to express her nuance.

Finally, respondent 2 is the one who is the most charitable towards the government throughout the interview:

Respondent 2 *"On the first vaccine they went in thinking... well it was a bit of a panic, we had to find a solution. So they said, well, let's go for it! Let's try it! [...] We had to try! They had to try, they were right, but I think they went to the wrong people."*

In this example, the use of reported speech shows a form of empathy, or at least the possibility of putting oneself in the shoes of one's political opponents. The respondent goes so far as to take on their own arguments "They had to..." with a very marked assertive dimension used to find excuses for them. The final concessive structure ultimately makes it possible to qualify his degree of agreement with the vaccine policy that has been put in place. It is therefore not a matter of categorical opposition as in interviews 1 and 3.

This gradation is also found in the tolerance shown by each of the respondents towards pro-vaccine or pro-government discourse. When asked about pro-vaccine authority figures, the then Minister of Health, Olivier Veran, and the French medical regulatory body, the Ordre des Médecins, respondent 1 immediately disqualified their words.

Respondent 1 *"in France, thanks to crooked people like Veran, well all those who were in power at the time, had the opportunity to say there is no other treatment. [...] The Ordre des Medecins was created under the Vichy regime with all the good intentions that they had in Vichy of course! And so it's a wart that has remained [...] all at the mercy of money."*

Respondent 1 frequently uses the insults "crooked", "wart" and associates his opponents with collaborators. He also questions the sincerity of the latter, whose word would be subject to financial interests (the insult 'crooked' refers to being corrupt) and not to those of the French. The use of 'all' here indicates that the respondent is making a generalization and does not bother to distinguish between the words of the doctors.

Respondent 2, on the other hand, is again more charitable. She maintains her opposition to the vaccination policy but attributes what she considers to be failures to simple errors, to a lack of intelligence.

Respondent 2 *"I'll tell Olivier Veran that he was wrong. That it's not complicated in fact, you don't need to be a doctor to say that there are facts that go against his analysis. The first thing I would say to him, you know, if I had him in front of me? I would say frankly I thought he was smarter than that."*

The axiology here is much less negative, the respondent assumes that being stupid is less serious for a politician than being dishonest, but above all she is engaged in a dialogue with her opponents. The use of speech verbs and reported speech allows respondent 2 to project herself into a situation of oral exchange.

Moreover, the criticism she addresses to the minister here is based less on a disqualification of his person than on external facts, which would contradict her speech. Respondent 2 therefore attacks the speech rather than the person.

Finally, respondent 3 takes care to distinguish between the government's words on the one hand and those of other doctors (including the medical association) on the other. Here, she mentions three successive cases of doctors, two of whom (Olivier Veran and Agnès Buzin) were successively Health Ministers during the epidemic. She compares them to Professor Didier Raoult, who is more critical of the government's health policy.

Respondent 3 *" Veran, doctor, podiatrist or whatever. Is he a neurologist? Or I don't know. I mean, what does the guy know about virology? I mean, he is not a doctor!"*

Respondent 3 *"I listened to Raoult [...] the day I was convinced of what he was saying was when he appeared before the Senate committee. He really stayed there for three hours! He deconstructed everything apart. Yeah, he did. From A to Z. For three hours I listened and I listened. Just after Buzin who is half crying, she doesn't know what she's saying, she's babbling, she's having papers passed to her because she doesn't know anything. I mean, that's not work!"*

Here, the disqualification of the government's word is based on the questioning of the ministers' competence, compared to that of other organizations or doctors that the respondent considers more legitimate. Olivier Veran is disqualified because he is not a virologist, while Agnès Buzin's comments are deemed inadmissible because they do not give any guarantees of seriousness. He was too emotional: "half crying"; "she's babbling".

The respondent's conclusion "that's not work" clearly indicates that the criterion that she considers crucial to distinguish between doctors' words is the quality of their appearance before the Senate. The use of the verb "to deconstruct", as if one were deconstructing an argument, shows the importance attached to the quality of the demonstration by each of the protagonists in the debate.

In other words, the respondent is willing to listen to the government's word. Nevertheless, she demands that this speech be legitimate (that the speaker has competence in the field he is talking about) but also that the speech be well constructed and rational. There is no a priori refusal as with respondent 1.

The three respondents thus share the fact of disqualifying the words of their opponents, although their attacks do not carry the same radical disqualification.

Respondents 2 and 3 each accept to give contradictory discourses a chance. Respondent 2 initially thought the Minister was 'smarter than that', which indicates that she gradually became convinced of the opposite. Respondent 3 also admits to a process that led to her being convinced 'the day'.

Respondent 1, on the other hand, disqualified her opponents a priori, on the basis of their moral qualities or historical affiliation, the medical order being the heir to Vichy. Moreover, the reading grid he applies to the current situation seems to pre-exist him and to be part of a broader conception of history in which the government's current vaccination policy is only one event among others.

7. Conclusion

The three respondents whose speeches I analysed all testified that they had been described at one time or another as conspiracy theorists. All of them, however, reject this description as disqualifying or even insulting.

A detailed analysis of their speeches nonetheless brings out features that have been identified in the literature as characteristic of conspiracy theories. The denunciation of an 'official' word or government 'propaganda', a criticism of the media judged to be biased in their treatment of information, or even liars. We also find calls for caution, doubt and a critical spirit with regard to the government's vaccine policy, which coexist with a radical denunciation of pro-vaccine discourse and the affirmation of certain certainties by the respondents.

Nevertheless, despite the presence of these characteristics usually associated with conspiracy speeches, the interviews differ from one another in the nature and intensity of the criticisms addressed to their opponent. Some base their attacks more on the evocation of historical references, associating pro-vaccine discourses with collaboration during the Second World War. Others, on the other hand, focus on defending the rules of the debate and consider that it is not well conducted in France.

It is these distinctions, more than certain other characteristics associated with conspiracy theories, that seem to allow us to distinguish the discourses that would be more readily qualified as conspiracy theories from the others.

However, these initial observations should be confirmed in a larger corpus and compared with a survey of the reception of the material, in order to verify whether these characteristics do indeed allow us to associate a type of discourse with a category that could be qualified as "conspiracy theories".

References

Beaud, S., & Weber, F. (2010). Guide de l'enquête de terrain: Produire et analyser des données ethnographiques (4e éd. augmentée). la Découverte.

Chapoutot, J. (2021). Le grand récit: Introduction à l'histoire de notre temps. PUF.

Danblon, E. (2020). Régimes de rationalité, post-vérité et conspirationnisme: A-t-on perdu le goût du vrai ? Argumentation et Analyse Du Discours, 25. https://doi.org/10.4000/aad.4528

Dominicy, M. 2010. Les sources cognitives de la théorie du complot : La causalité et les faits. In Danblon, E., & Nicolas, L. (Eds.), Les rhétoriques de la conspiration. CNRS Éditions. doi :10.4000/books.editionscnrs.16262

Doury, M. (2003). L'évaluation des arguments dans les discours ordinaires: Le cas de l'accusation d'amalgame. Langage et société, 105, 9-37. https://doi.org/10.3917/ls.105.0009

Hofstadter, R., Raynaud, P., & Charnay, J. (2012). Le style paranoïaque: Théories du complot et droite radicale en Amérique. F. Bourin.

Nicolas, L. (2015). Jesuits, Jews, and Freemasons: Rhetorics of conspiracy. Diogenes, 249-250, 75-87. https://www.cairn-int.info/journal--2015-1-page-75.htm.

Oswald, S., Lewinski, M., Greco, S., & Villata, S. (2022). The Pandemic of Argumentation.

Peltier, M. (2021). L'ère du complotisme : La maladie d'une société fracturée.

Poliakov, L., & Taguieff, P.-A. (2006). La causalité diabolique (Reproduction en fac-similé). Calmann-Lévy.

Romero Reche, A. (2020). Conspiracy Theory: Secrecy and Transparency. Journal of the CIPH, 98, 81-102. https://doi.org/10.3917/rdes.098.0081

Stephan, G., & Vauchez, Y. (2021). Dévoiler les «bobards» des médias dominants: Les stratégies de (dé)légitimation de la réinformation. RESET, 10. https://doi.org/10.4000/reset.3180

Taguieff, P. 2021. Les théories du complot, Que sais-je ? Paris

Uscinski, J. E., & Parent, J. M. (2014). American conspiracy theories. Oxford Univ. Press.

EXPLORING THE ARGUMENTATIVE POTENTIAL OF DOUBT IN MEDICAL CONSULTATIONS

MARIA GRAZIA ROSSI
Institute of Philosophy (IFILNOVA), Faculty of Social and Human Sciences, NOVA University of Lisbon, Lisbon, Portugal
mgrazia.rossi@fcsh.unl.pt

DIMA MOHAMMED
Institute of Philosophy (IFILNOVA), Faculty of Social and Human Sciences, NOVA University of Lisbon, Lisbon, Portugal

SARAH BIGI
Department of Linguistic Sciences and Foreign Literatures, Catholic University of the Sacred Heart, Milan, Italy

Abstract

In this paper, we explore a particular normative dimension of the strategies used by healthcare professionals to manage uncertainty in medical consultations. We analyze medical consultations focusing on the argumentative potential of doubts expressed by patients and their families by examining the inferences activated beyond what is explicitly stated in the sequences exhibiting doubt. Using the distinction between three argumentative potentials of doubt (ambivalent doubt, skeptic doubt, and denialism), we run an exploratory analysis on an Italian corpus of 52 medical consultations between health professionals and patients with Type 2 diabetes. The study sheds light on the nature of the doubts expressed in medical consultations, how healthcare professionals discursively manage uncertainty and using which strategies. Based on the analysis, we argue that when expressed in medical consultations, skeptical doubts should be explicitly acknowledged and adequately addressed to prevent the doubt from acquiring a denialist potential, which can undermine patient adherence and the possibility of favorable health outcomes.

1. Introduction

Uncertainty in healthcare is a multifaceted phenomenon that appears in diverse forms and requires different management strategies (Dahm & Crock, 2022; Eachempati et al., 2022; P. K. J. Han, Klein, & Arora, 2011; Paul K.J. Han et al., 2021; Kalke, Studd, & Scherr, 2021). Considering the context of chronic care and diabetes, we focus on a specific dimension of uncertainty, mainly related to how patients' doubts appear in the dialogical interactions. That is, we look at how doubts about the management and treatment of diabetes are expressed (by patients and their families) and addressed (by healthcare professionals).

From a discoursive perspective, the expression of doubt represents what Candlin (2000) calls a *crucial site*, a potential indicator of a critical moment in medical interactions that can lead to significant direct and indirect effects if not appropriately handled. On a relational and epistemic level, doubts can lead to misunderstandings and mistrust of the medical system and authorities. Moreover, doubts not adequately recognized and managed can also provoke indirect clinical effects such as, for example, suboptimal outcomes (including non-adherence of patients, treatment dropout, and requests for a second opinion), contributing to increasing healthcare costs and affecting the sustainability of the healthcare system.

Previous studies in clinical communication pointed out that minimizing and discursively downplaying expressed concerns can reinforce patients' doubts and potentially impact their treatment adherence (Stevens, 2018). However, discursive analyses of doubts are still missing in clinical communication and diabetes care. In this respect, Han and colleagues (2021) clarified why existing taxonomies classifying different doubts and strategies to manage uncertainty are insufficient to avoid suboptimal outcomes. As Han and colleagues (2021, p. 288) put it: "Yet the taxonomy is purely descriptive; it offers no definitive answers to the normative question of how physicians ought to manage different uncertainties." While this issue is recognized as central in healthcare, an argumentative framework has not been used so far to address this challenge. Indeed, an argumentative framework has the explanatory potential to provide normative criteria to distinguish between different uncertainties and to recommend discursive practices to manage them. The main aim of this study is to contribute to facing this challenge in healthcare communication through the adoption of argumentative lenses and tools (cf. also Bigi, 2018; Eemeren, Garssen, & Labrie, 2021; Jackson, 2020; Rossi, Macagno, & Bigi, 2022; Snoeck Henkemans & Mohammed, 2012). More specifically, we offer an analysis of the argumentative potentials of doubt expressed by patients and their families (Mohammed, 2019a, 2019b; Mohammed & Rossi, 2022), looking at how healthcare professionals interpret and manage doubts and through which discursive practices.

2. Argumentative potentials of doubt in interpersonal healthcare communication

The concept of argumentative potential (Mohammed, 2019b; see also Kjeldsen, 2007; Serafis, 2022) refers to the implicit argumentative dimension that may be attributed to a certain discourse or to parts of it in function of some argumentativity that may be inherent in language (Anscombre & Ducrot, 1983) or in the context of its use (see for example Amossy, 2009). This can be analyzed, for example, by identifying possible argumentative inferences that can be activated beyond what is explicitly said by the speakers. This concept has been used in previous research to analyze public controversies, such as the controversy between the #MeToo movement and Anti-#MeToo Manifesto (Mohammed, 2019a), political controversies (Mohammed, 2019b), and health controversies about the COVID-19 vaccine (Mohammed & Rossi, 2022). In this latter case, Mohammed & Rossi (2022) looked at common argumentative patterns of anti-vaccination proponents about COVID-19 vaccine safety. We distinguished three categories of argumentative potentials that a doubt can have: ambivalence, skepticism, and denialism.

For example, the doubt about the safety of the Oxford-AstraZeneca vaccine motivated by the widely circulated information that "Several people have died from unusual blood clots after getting the Oxford-AstraZeneca Vaccine" (EMA, 2021, April 7) can activate different argumentative potentials and support three inferences. An ambivalent doubt inference would have the minimum argumentative potential, or the minimum degree of doubt about the safety of vaccines (e.g., "I am not sure if the vaccine is safe or not"). A skeptic doubt would have a stronger potential, i.e. activate an inference that can support a skeptical position against vaccine safety (e.g., "I do not think that the vaccine is safe"). Finally, denialism would be the result of an inference that puts doubt about vaccine safety in the reasoning typical of conspiracy theories: doubt about safety is evidence against the official stories about vaccines and their benefit altogether, discrediting the whole system and healthcare authorities (e.g., "The official story about the vaccine is not credible"). In this contribution, we show the usefulness of these three categories of argumentative potentials to analyze how doubts are handled in a corpus of patient-professional interactions.

Doubts in interpersonal communication might be expressed and managed differently than public health controversies. These two forms of communication are indeed diverse in many ways. Public controversies are more and more characterized by polarized debates (Flores et al., 2022), while the preference for agreement in ordinary interpersonal exchanges is well-documented (e.g., Pomerantz, 1985; Schegloff, Jefferson, & Sacks, 1977). As van Eemeren and Grootendorst (2004, p. 98) noted, raising doubts is "contrary to the preference for agreement that predominates in ordinary exchanges."

Preference for agreement is even more common in expert/non-expert communication, with all its asymmetries of experience, knowledge, power, and role. Scholars have thoroughly analyzed asymmetry in medical interactions (Ainsworth-Vaughn, 1998; Beisecker, 1990; Bigi & Rossi, 2020, 2023; Maynard, 1991; Todd, 1989), pointing out how patients generally agree and rarely say when they have not understood (Graham & Brookey, 2008; Rossi & Macagno, 2020). Moreover, Andrade (2020, p. 6) suggested that "little conspiracy mongering surrounds" the disease when there are good prospects for treatment, as in the case of diabetes.[1] Therefore, some defining characteristics of chronic care make this context much less likely to contain the attribution of denialist argumentative potentials to doubts expressed by patients and their families.[2]

3. Methodological note

We conducted an exploratory study on a corpus of medical interviews between Type 2 diabetes patients and healthcare professionals recorded from March 2012 through March 2014 in the diabetes outpatient clinic of the Azienda Ospedaliera Istituti Clinici di Perfezionamento di Milano (A.O.I.C.P.), Italy (Bigi, 2014). The Ethical Committee of the I.C.P approved the protocol in January 2012.

We analyzed 52 verbatim transcripts with 16 Italian patients and six healthcare professionals by identifying different types of doubt patients and their families expressed. In this first exploratory study, we focused only on doubts expressed in the context of proposals made by healthcare professionals. We codified doubts about the proposal itself or about a premise supporting the proposal, that is, doubts about one or more preconditions for the acceptability of the proposal. Doubts about the efficiency of the proposal or about the interpretation of clinical data used to support the proposal fall in this category. Our analysis does not include other doubts, such as doubts followed by purely explanatory sequences (e.g., request of information); doubts followed by argumentative sequences but not about a proposal; explicit refusals.

We used an incremental locality principle to account for how patients and their families doubt a proposal and determine which of the three categories of argumentative potential cases belonged to. We used the incremental locality principle inspired by previous philosophical analyses

---

[1] Raising doubts practice and conflicts might be more frequent in other medical contexts, such as mental health (McCabe, 2021) or vaccination consultations (Rentmeester, 2013).
[2] In our analysis, we detected only two cases of denialist doubt.

of the practice of doubting, such as – among others – Ludwig Wittgenstein (1969; see § 24, 27, 115, 120, 450, 625). In a Wittgensteinian framework, "Meaningful doubt is always local and presupposes prior certainty" (Rummens & De Mesel, 2022, p. 136). For our analysis, we translated this principle by characterizing each doubt as ambivalent, skeptical, or denialist based on the degree of locality presupposed by its interpretation. We looked at how doubts are interpreted in the data, considering the degree of locality expressed through linguistic evidence and discursive markers (e.g., repetition as an indicator of skepticism; cf. § 4).

Ambivalent doubts are those open to the highest degree of local interpretation and incorporate the minimum degree of argumentative potential. They amount to just not being sure whether the proposal made by the healthcare provider is the best one for the patient or not. Skeptic doubts are those available to a medium degree of local interpretation; they are doubts with a stronger argumentative potential that seem to support a skeptical position about a proposal, something like "I do not think that the proposed treatment is the best for my case." Finally, doubts with a denialist argumentative potential are open to the minimum degree of local interpretation and incorporate the maximum degree of argumentative potential. Denialist doubts can incorporate the refutational narratives typical of conspirational thinking and attitudes, having the potential to discredit not only the proposal itself but also the institutions behind it – mainly in the specific context under consideration, healthcare systems, healthcare authorities, and pharmaceutical companies.

4. Handling the skeptic argumentative potential: a double-edged strategy

In our corpus, the most representative cases of argumentative doubts about a proposal or its preconditions are doubts that can be interpreted according to the skeptical argumentative potential. In what follows, we analyze two prototypical examples to illustrate the most frequent discursive strategy healthcare professionals use to handle this kind of doubt.

The first example concerns the efficacy of generic drugs as clinically equivalent to brand-name drugs. Upon passing specific bioequivalence studies, generic medicines are approved as pharmaceutical equivalents or alternatives (WHO, 2016). Generic drugs have been shown to represent a cost-effective solution for removing financial barriers, thus promoting equity of access to care and sustainability of health systems (Godman et al., 2021). That is also why they are offered to patients by healthcare professionals. However, patients often have negative perceptions of generic medicines. The literature shows that patients resist using them due to various concerns and misconceptions (Alrasheedy et al., 2014;

Mostafa, Mohammad, & Ebrahim, 2021), which sometimes they bring as a topic of discussion in medical consultations. The example below can be regarded as an exemplification of this more general trend.

In lines 1 and 3, the patient (P) introduces the topic by asking the doctor (D) if the generic metformin, an oral hypoglycemic drug used to treat non-insulin-dependent diabetes, is the same as the brand-name one.

Example 1 | P19_2

We used the following conventions for all the examples reproduced in this article. The original transcript is in Italian; a translation in italics follows each speech turn in English. We marked the patients' turns expressing doubts in bold; we underlined in bold the healthcare professionals' speech turns where doubts are acknowledged, dismissed, or discussed. Finally, we inserted comments between brackets to make it visible when a turn was particularly relevant for our analysis.

1. P: **a volte mi danno questa metformina**
   Sometimes they give me this metformin

2. D: sì
   Yes

3. P: **eh ma è uguale?**
   Is it the same?

4. D: **è una bella domanda**
   Good question

   (doubt acknowledged)

5. P: **io avevo visto che dentro- anche la forma delle pastiglia (.) son diverse (.) lo lo chiedo al farmacista e mi dice è la stessa cosa**
   I saw that inside- also the shape of the pills are different. I asked the pharmacist and he tells me it's the same thing

6. D: allora sono due generici. Quindi sono uguali eh::: come tipologia di farmaco, come sostanza che c'è dentro e dosaggio
   So, they are two generics. Therefore they are the same eh the type of medicine, the substance inside and dosage

   (D confirms the information P got from the pharmacist)

7.	P:	**ma però-**
		But

8.	D:	ci siamo accorti
		We realized

9.	P:	**io ho letto un bigliettino che ci son su delle cose diverse nell'altro**
		I read a note that there are different things in the other

10.	D:	**[vediamo]**
		Let's see

(doubt acknowledged)

11.	P:	[ed è per questo] che l'ho portato oggi
		which is why I brought it today

12.	D:	**vediamo, mi faccia vedere che così poi glielo spiego**
		Let's see, let me see so I can explain it to you

(doubt acknowledged and partnership building)

Until line 4, when the doctor acknowledges the doubt as a legitimate concern, the patient's request might be interpreted as an ambivalent doubt about the efficacy of the generic drug. The doctor's move is particularly interesting since it opens the dialogical space needed to explore better what is behind the patient's concern. The patient is not missing any information: he already asked the pharmacist, who confirmed that the brand-name metformin and the generic one are the same (line 5). However, the patient is somewhat skeptical about it, as is made clear if we look at the specific formulation he uses in line 1 (i.e., "they give me this metformin"). Such formulation expresses a sense of not having control over the decision, a sense of external and not autonomous agency, instead attributed to the pharmacist (line 5) and (perhaps) to the dietician. Indeed, in the previous visit, the dietician also explained that the generic drugs are equivalent, thus reinforcing our attribution of a skeptical interpretation to the patient's doubt.

Moreover, in lines 7 and 9, the patient reacts skeptically after the doctor confirms the same information again. The skeptic potential of the patient's doubt is therefore evident in consideration of the repetitive questions made by the patient, who, in spite of the information already shared by various healthcare professionals, insists on the same topic of concern.

The doctor's move strategically curbs the skeptic argumentative potential of the patient's requests in two interrelated ways: strengthening the partnership with the patient and acknowledging the doubt as an ambivalent one, that is, downgrading it (lines 10 and 12). Similar to what we argued in relation to doubts emerging in the context of public health controversies (Mohammed & Rossi, 2022), downgrading skeptic potential to ambivalence might be an effective strategy to deal with doubts. By taking the time to look for the differences between generic and brand-name drugs, the doctor can explain why the differences detected by the patient do not impact the efficacy of the generic drug and are, therefore, not relevant.

However, our analysis also shows that downgrading doubt with a skeptical potential and treating it as an ambivalent doubt can sometimes be problematic. Our second example provides an interesting case in this regard.

The second excerpt comes from a visit between a nurse and a patient under insulin treatment with poor glycaemic control. During the visit, the nurse discovers that the patient is changing the therapy without consulting the doctor, inserting an oral hypoglycemic drug typically used to treat non-insulin-dependent diabetes, and lowering the prescribed insulin units. Thus, understanding the danger for the patient's health, the nurse calls the doctor, who yells at the patient insisting on the importance of adherence to treatments and providing information on how to administer insulin. The excerpt starts after all this has happened, with the patient asking again for further details on what to do if the blood sugar levels drop and go too low (line 1). Once again, repeating the exact request looks like a discursive practice used by patients to express doubt and show skeptical attitudes toward a proposal that has already been justified (see also Regina Wu, 2009).

Example 2 | P6_4

1. P: se io:::- perché poi le misuro no, **se io vedo che scendo troppo,**
    cosa faccio?
    If I... because then I measure them, right? If I see that I'm going down too low, what do I do?

2. D: **innanzitutto ci chiama. numero uno. prima di mettere mano alla terapia voi ci dovete sempre chiamare**
    Number one you call us. Before you modify the therapy you should always call us

3. P: **e chi chiamo?**
    So who do I call?

| 4. | D: | **l'infermeria qua, noi. che sappiamo poi cosa fare con cosa dire. NUMERO o NUMERO. dalle dodici alle tredici è l'orario migliore** |

The nurse here, us. We know what to do, what to say. [READS OUT TEL. NUMBER] or [READS OUT TEL. NUMBER]. Between noon and one pm it's the best time to call

| 5. | P: | sì |
| | | Yes |

(sounds like an agreement)

| 6. | D: | **per contattarci.** nel momento in cui vede che sono mo:::lto basse, allora lì mangi subito qualcosa, si rasserena un attimo sulla terapia. Ma prima di andarla a ritoccare, ci dia un colpo di telefono. così la sistemiamo [via telefonica] |

This is to contact us. When you see they are really low, then in that moment eat something right away, you relax a bit about the therapy. But before you really change it, call us. So we can change it over the phone

| 7. | P: | [va bene] | ok |

(sounds like an agreement)

| 8. | D: | mh? |
| | | mmh? |

| 9. | P: | **eh no perché può capitare** |
| | | *Right, because, I mean, it can happen* |

| 10. | D: | non prenda altre pastiglie però. non associ altre pastiglie all'insulina perché se no ovviamente l'effetto è maggiorato e poi [succedono danni] |

But don't take other pills. Don't add other pills to your insulin becauseotherwise obviously, the effect is increased and then there's trouble

We can describe what is happening by taking the context into account, both in terms of the patient's history in previous consultations and verbal exchanges with the doctor in this same consultation. We can interpret the doctor's strategy (lines 2, 4, and 6) as an attempt to attribute an ambivalent argumentative potential that ignores the skeptical argumentative potential behind the patient's requests. Indeed, the doctor's answers (lines 2, 4, and 6) seem to generate an information-giving exchange or an explanation dialogue: that is, the doctor provides practical instructions on what the patient should do if the blood sugar levels drop. Yet, simultaneously, the doctor reproaches the patient restating the authority of healthcare professionals as the only ones who can modify and authorize the change of pharmacological treatments (e.g., "Before you modify the therapy you should always call us," line 2). Or in other words, emphasizing the authority of healthcare professionals in choosing the more appropriate therapy, the doctor is dismissive and only apparently legitimizes the patient's demands. In this respect, it is interesting that the dialogue finishes with an apparent agreement (lines 5 and 7), and nonetheless, the patient dropped out and stopped going to the healthcare facility.

In addition to the aggressive tone, we may speculate that the doctor's treatment proposal did not convince the patient, as the doctor increased the insulin units again. The patient is not adhering to insulin, probably due to the fear of hypoglycemia that might be associated with it.

Insulin refusal, poor adherence to insulin therapy, or incorrect insulin administration are well-known problems in diabetes care (Negash & Mekonen, 2023). Such problems are associated with different psychological, social, and economic factors (e.g., anxiety and fear of injection-related discomfort or hypoglycemia, limitations in daily activities, costs, etc.) and some misconceptions and myths about insulin injections and side effects (Singh & Jain, 2020). Therefore, when patients have doubts about insulin treatment, it is crucial to explore their perspectives to address their concerns and correct wrong presuppositions where appropriate. Unfortunately, the patient's perspective is not explored in this second example and, more generally, in the visit from which it is drawn. This is why treating the patient's requests as ambivalent by repeating the same information to address his doubts is not enough to curb their skeptic potential. It is what is behind that doubts (i.e., the fear of hypoglycaemia) that should be explored, legitimized, and possibly engaged with.

5. Concluding remarks

Analyzing how patients and their families express doubt represents the first step to facing the normative challenge of providing recommendations to healthcare professionals on managing doubt and uncertainty in healthcare communication. In this contribution, we took this first step by examining the dialogical paths generated by sequences expressing doubts and the discursive practices used by healthcare professionals to manage them. More specifically, we looked at how healthcare professionals curbed the argumentative potential of skeptic doubts, which are the most representative in our corpus but also the most dangerous. Indeed, a skeptic doubt always has the potential to become less and less local, losing its epistemic meaningfulness, acquiring a denialist argumentative potential, questioning the healthcare system and authorities, and putting patients' health in danger. Similar to what we discussed in the context of public health controversies, also in interpersonal healthcare communication, it is crucial to acknowledge doubt as legitimate. This is especially important in chronic care, where patients have a long history of illness and often strong feelings that they know what to do and how to deal with it correctly. In our corpus of diabetes consultations, we found patients that challenged the doctor's proposals based on the argument that they have had diabetes for ages and, therefore, they know how to modify the treatment and reduce insulin.

Our exploratory analysis showed that downgrading the argumentative potential of doubts is a common discursive practice used by healthcare professionals to cope with skepticism. Our discussion also revealed that sometimes this strategy could be problematic (as discussed in example 2). However, what is problematic is not the downgrading strategy *per se* but how it is discoursively realized. Our analysis suggests that the main task in implementing the process of *doubt legitimization* is to explore what is behind the doubt, have the chance to understand patients' concerns and correct potential misunderstandings or misconceptions that often hide behind the expression of doubt. Indeed, simply offering the same information does not allow healthcare professionals to determine the causes behind the doubt and erase it.

More systematic analyses of consultations, better if conducted in different medical contexts and languages, are needed to recommend strategies to manage the argumentative potential of doubt and modify problematic discoursive practices.

Acknowledgments This work was supported by the Fundação para a Ciência e a Tecnologia (research grant no. SFRH/BPD/115073/2016).

References

Ainsworth-Vaughn, N. (1998). *Claiming power in doctor-patient talk*. Oxford: Oxford University Press.

Alrasheedy, A. A., Hassali, M. A., Stewart, K., Kong, D. C., Aljadhey, H., Mohamed Ibrahim, M., & Al-Tamimi, S. K. (2014). Patient knowledge, perceptions, and acceptance of generic medicines: a comprehensive review of the current literature. *Patient Intell, 6*, 1–29.

Amossy, R. (2009). Argumentation in Discourse: A Socio-discursive Approach to Arguments. *Informal Logic, 29*(3), 252. https://doi.org/10.22329/il.v29i3.2843

Andrade, G. (2020). Medical conspiracy theories: cognitive science and implications for ethics. *Medicine, Health Care and Philosophy, 23*(3), 505–518. https://doi.org/10.1007/s11019-020-09951-6

Anscombre, J.-C., & Ducrot, O. (1983). *L'argumentation dans la langue*. Bruxelles: Pierre Mardaga.

Beisecker, A. E. (1990). Patient Power in Doctor-Patient Communication: What Do We Know? *Health Communication*, *2*(2), 105–122. https://doi.org/10.1207/s15327027hc0202_4

Bigi, S. (2014). Healthy Reasoning: The Role of Effective Argumentation for Enhancing Elderly Patients' Self-management Abilities in Chronic Care. *Studies in Health Technology and Informatics*, *203*, 193–203.

Bigi, S. (2018). The role of argumentative practices within advice-seeking activity types. The case of the medical consultation. *Rivista Italiana Di Filosofia Del Linguaggio*, *12*(1), 42–52.

Bigi, S., & Rossi, M. G. (2020). Considering Mono- and Multilingual Interactions on a Continuum: An Analysis of Interactions in Medical Settings. In C. Hohenstein & M. Lévy-Tödter (Eds.), *Multilingual Healthcare: A Global View on Communicative Challenges* (pp. 11–37). https://doi.org/10.1007/978-3-658-27120-6_2

Bigi, S., & Rossi, M. G. (2023). Fostering interdisciplinary knowledge translation at the interface between healthcare communication and pragmatics. In *A pragmatic agenda for healthcare: fostering inclusion and active participation through shared understanding*.

Candlin, C. N. (2000). *The Cardiff lecture 2000: Reinventing the patient/client: New challenges to healthcare communication (Cardiff Papers on Healthcare Discourse 2)*. (Cardiff Un). Cardiff.

Dahm, M. R., & Crock, C. (2022). Understanding and Communicating Uncertainty in Achieving Diagnostic Excellence. *JAMA*, *327*(12), 1127–1128. https://doi.org/10.1001/jama.2022.2141

Eachempati, P., Büchter, R. B., Ks, K. K., Hanks, S., Martin, J., & Nasser, M. (2022). Developing an integrated multilevel model of uncertainty in health care: A qualitative systematic review and thematic synthesis. *BMJ Global Health*, *7*(5). https://doi.org/10.1136/bmjgh-2021-008113

Eemeren, F. H., Garssen, B., & Labrie, N. (2021). *Argumentation between Doctors and Patients: Understanding clinical argumentative discourse*. Retrieved from https://www.jbe-platform.com/content/books/9789027260109

EMA. (2021). COVID-19 Vaccine Janssen: assessment of very rare cases of unusual blood clots with low platelets continues. In *European Medicines Agency*. Retrieved from https://www.ema.europa.eu/en/news/covid-19-vaccine-janssen-assessment-very-rare-cases-unusual-blood-clots-low-platelets-continues

Flores, A., Cole, J. C., Dickert, S., Eom, K., Jiga-Boy, G. M., Kogut, T., ... Van Boven, L. (2022). Politicians polarize and experts depolarize public support for COVID-19 management policies across countries. *Proceedings of the National Academy of Sciences*, *119*(3), e2117543119. https://doi.org/10.1073/pnas.2117543119

Godman, B., Massele, A., Fadare, J., Kwon, H.-Y., Kurdi, A., Kalemeera, F., ... Meyer, J. C. (2021). Generic drugs—essential for the sustainability of healthcare systems with numerous strategies to enhance their use. *Pharmaceutical Sciences and Biomedical Analysis Journal*, *4*(1).

Graham, S., & Brookey, J. (2008). Do patients understand? *The Permanente Journal*, *12*(3), 67–69.

Han, P. K. J., Klein, W. M. P., & Arora, N. K. (2011). Varieties of Uncertainty in Health Care: A Conceptual Taxonomy. *Medical Decision Making, 31*(6), 828–838. https://doi.org/10.1177/0272989x11393976

Han, Paul K.J., Strout, T. D., Gutheil, C., Germann, C., King, B., Ofstad, E., ... Trowbridge, R. (2021). How Physicians Manage Medical Uncertainty: A Qualitative Study and Conceptual Taxonomy. *Medical Decision Making, 41*(3), 275–291. https://doi.org/10.1177/0272989X21992340

Jackson, S. (2020). Evidence in Health Controversies. *OSSA Conference Archive, 15*. Retrieved from https://scholar.uwindsor.ca/ossaarchive/OSSA12/Friday/15

Kalke, K., Studd, H., & Scherr, C. L. (2021). The communication of uncertainty in health: A scoping review. *Patient Education and Counseling, 104*(8), 1945–1961. https://doi.org/10.1016/j.pec.2021.01.034

Kjeldsen, J. E. (2007). Visual Argumentation in Scandinavian Political Advertising: A Cognitive, Contextual, and Reception Oriented Approach. *Argumentation and Advocacy, 43*(3–4), 124–132. https://doi.org/10.1080/00028533.2007.11821668

Maynard, D. W. (1991). Interaction and Asymmetry in Clinical Discourse. *American Journal of Sociology, 97*(2), 448–495. Retrieved from http://www.jstor.org/stable/2781383

McCabe, R. (2021). When patients and clinician (dis)agree about the nature of the problem: The role of displays of shared understanding in acceptance of treatment. *Social Science & Medicine (1982), 290*, 114208. https://doi.org/10.1016/j.socscimed.2021.114208

Mohammed, D. (2019a). Managing Argumentative Potential in the Networked Public Sphere : The Anti- # MeToo Manifesto as a Case in Point. In B. Garssen, D. Godden, G. R. Mitchell, & Wagemans J.H.M. (Eds.), *Proceedings of the 9th conference of the International Society for the Study of Argumentation* (pp. 813–822). Amsterdam: Sic Sat.

Mohammed, D. (2019b). Standing Standpoints and Argumentative Associates: What is at Stake in a Public Political Argument? *Argumentation, 33*(3), 307–322. https://doi.org/10.1007/s10503-018-9473-y

Mohammed, D., & Rossi, M. G. (2022). *The Argumentative Potential of Doubt: From Legitimate Concerns to Conspiracy Theories About COVID-19 Vaccines - The Pandemic of Argumentation* (S. Oswald, M. Lewiński, S. Greco, & S. Villata, eds.). https://doi.org/10.1007/978-3-030-91017-4_7

Mostafa, S., Mohammad, M. A., & Ebrahim, J. (2021). Policies and practices catalyzing the use of generic medicines: a systematic search and review. *Ethiopian Journal of Health Sciences, 31*(1).

Negash, Z., & Mekonen, T. (2023). Patient perception towards shifting oral antihyperglycemic agents to injectable insulin and associated factors in the diabetes clinic of Tikur Anbessa specialized hospital: Cross-sectional study. *Metabolism Open, 17*, 100228. https://doi.org/https://doi.org/10.1016/j.metop.2022.100228

Pomerantz, A. (1985). Agreeing and disagreeing with assessments: some features of preferred/dispreferred turn shapes. In J. M. Atkinson (Ed.), *Structures of Social Action* (pp. 57–101). https://doi.org/10.1017/CBO9780511665868.008

Regina Wu, R.-J. (2009). Repetition in the initiation of repair. In J. Sidnell (Ed.), *Conversation Analysis: Comparative Perspectives* (pp. 31–59). https://doi.org/DOI: 10.1017/CBO9780511635670.003

Rentmeester, C. A. (2013). Professionalism, fidelity and relationship-preservation. *Human Vaccines & Immunotherapeutics*, *9*(8), 1812–1814. https://doi.org/10.4161/hv.24432

Rossi, M. G., & Macagno, F. (2020). Coding Problematic Understanding in Patient–provider Interactions. *Health Communication*, *35*(12), 1487–1496. https://doi.org/10.1080/10410236.2019.1652384

Rossi, M. G., Macagno, F., & Bigi, S. (2022). Dialogical functions of metaphors in medical interactions. *Text and Talk*, *42*(1), 77–103. https://doi.org/10.1515/text-2019-0166

Rummens, S., & De Mesel, B. (2022). A Wittgensteinian Account of Free Will and Moral Responsibility. In *Philosophical Perspectives on Moral Certainty* (pp. 132–155). Routledge.

Schegloff, E., Jefferson, G., & Sacks, H. (1977). The preference for self-correction in the organization of repair in conversation. *Language*, 361–382.

Serafis, D. (2022). Unveiling the rationale of soft hate speech in multimodal artefacts. *Journal of Language and Discrimination*. https://doi.org/10.1558/jld.22363

Singh, S. K., & Jain, R. (2020). Myths About Insulin Therapy. *RSSDI's Insulin Monograph: A Complete Guide to Insulin Therapy*, 245.

Snoeck Henkemans, A. F., & Mohammed, D. (2012). Institutional constraints on strategic maneuvering in shared medical decision-making. *Journal of Argumentation in Context*, *1*(1), 19–32. https://doi.org/https://doi.org/10.1075/jaic.1.1.03moh

Stevens, L. M. (2018). "We have to be mythbusters": Clinician attitudes about the legitimacy of patient concerns and dissatisfaction with contraception. *Social Science & Medicine (1982)*, *212*, 145–152. https://doi.org/10.1016/j.socscimed.2018.07.020

Todd, A. D. (1989). *Intimate Adversaries: Cultural Conflict Between Doctors and Women Patients*. Philadelphia: University of Pennsylvania Press.

WHO. (2016). Generic medicines: interchangeability of WHO-prequalified generics. *WHO Drug Information*, *30*(3), 370–375. Retrieved from https://apps.who.int/iris/handle/10665/331014

Wittgenstein, L., Anscombe, G. E. M., von Wright, G. H., Paul, D., & Anscombe, G. E. M. (1969). *On certainty* (Vol. 174). Blackwell Oxford.

The Types of Appeal to Religious Authorities. Kairos and Interreligious Communication

Lucia Salvato
Università Cattolica del Sacro Cuore
lucia.salvato@unicatt.it

Abstract

The aim of this study is to analyze the different uses of appeal to religious authorities in argumentative monologues on theological matters. The focus is on the main speeches – homilies, greetings, and discourses during lay or religious meetings – given by Joseph Ratzinger during his fourth and final apostolic voyage to Germany (2011) as Pope Benedict XVI. Depending on the nature of the interlocutors – lay or religious – Ratzinger appeals to different types of religious authorities. The paper will analyze the different argumentative roles that each authority plays in Ratzinger's speeches, considering Toulmin's model. Additionally using argumentation schemes by Walton, it aims to reconstruct the argumentative conclusions that Ratzinger draws from the words of each religious authority and their function within his arguments.

1. Introduction

This study is part of a recent and still under-researched tradition of studies on a particular kind of critical discussion, the so-called "monolectical reasoning" (Walton 1990: 405), or "argumentative monologue" (Dascal 2005; Rigotti 2005; Rocci 2005; Rigotti and Rocci 2005; van Eemeren and Peng 2017). The study examines a series of speeches – homilies, greetings, and discourses during religious or lay meetings – given by Joseph Ratzinger during his fourth and final apostolic voyage to Germany (2011) as Pope Benedict XVI. In the analysis, I will focus on the different types of religious authorities Ratzinger appeals to in his speeches, depending on the lay or religious nature (Catholic, Evangelical, Orthodox, Jewish, and Muslim) of his interlocutors. In fact, Ratzinger's application to different authorities depending on his interlocutors corresponds to a rhetorical strategy grounded in a sort of "epistemological relativism" (Macagno/Damele 2013: 379) that is involved in the concept of kairos (Untersteiner 1954).

The term *kairos* denotes one of four meanings with which the ancient Greeks indicated time: 'the opportune moment', 'the right time' to do or say something to someone. Kairos has thus a qualitative nature and represents an essential rhetorical dimension which is strictly bound to the speaker's ability (Macagno/Damele 2013: 378-380). Ratzinger can certainly be considered among those 'educated people' with the requisite ability to adapt to variable, contingent circumstances. His way of appealing to different types of religious authorities – using authoritative texts and persons – depends on the interlocutors he is addressing. For instance, with seminarians he mainly invokes the saints and the Gospels; with a Muslim community, he openly aspires to a "fruitful collaboration" with Christians, quoting the German Basic Law to underline the need for due recognition of the public dimension of religious adherence based on agreement on certain inalienable values.

2. Types of authority: Walton's and Toulmin's models

The pragmatic aim of my analysis is to assess the strength of Ratzinger's arguments by focusing on those which contain an appeal to a religious authority. Ratzinger usually states his claims to different audiences using different kinds of argument: those based *on example* (Walton 2006:40), *on position to know* (Walton 2006:85, 86, 132), *from popular opinion* (Walton 2006:91-95, 132), *from popular practice* (Walton 2006:93-94), *from values* (Macagno, and Walton 2008), and by *appeal to expert opinion* (Walton 2006:72, 84-91, 133, 144, 188).

The aim of the following paragraphs is therefore to highlight the argumentative role that the main authorities cited by Ratzinger play in his speeches and the conclusions he draws from their words. In this evaluation, I will use Toulmin's argumentative model (1958). Together with Walton's argumentation scheme (Walton 1989, 1997, 2006), Toulmin's model (1958) can help focus Ratzinger's claims – which are linked to appeals to authorities or authoritative texts – by identifying the three main elements of an argument: the utterance/*claim, grounds,* and *warrant* used to state the claim. The *claim* is the main argument, the assertion that an author intends to prove. The *grounds* are the facts supporting the claim as explicit, incontrovertible data. The *warrant*, which can be implied or explicitly stated, expresses the guarantee that yields general support by linking the grounds (the evidence factors) to the claim (the communicated meaning). In his critical argumentation, Walton makes a fundamental distinction between "asserting a proposition and questioning a proposition" (Walton 2006:26). While questioning a proposition usually implies a neutral stance, to assert something means to claim that it is true. The assertion is therefore the expression of a pro point of view, meaning that the speaker must stand behind that claim and defend it if they want to maintain it. Walton's distinction is important because it concerns the way the different moves in a dialogue relate to previous ones.

According to Toulmin's framework, the analysis of arguments should include three other parts, the *backing, rebuttal,* and *qualifier*. The *backing* is an often-implied statement which a speaker relies on (Toulmin 1958: 103-104). It refers to additional support of the warrant, especially when the latter is implied. It is thus defined by Toulmin as "the *backing* of the warrants" (Toulmin 1958: 103; italics in the original). The *rebuttal* is a move that presents a "counter-argument" to the speaker's original argument (Walton 2006:27), thus acknowledging another perspective or interpretation of the situation or assumed statement. Including a rebuttal in an argument can help build credibility in an argumentation, allowing the speaker to present themselves as an unbiased thinker, which in turn can help the audience judge the situation objectively. The same can result from the introduction of a *qualifier* as this can show that the claim may not be true in all circumstances. The resulting scheme is as follows:

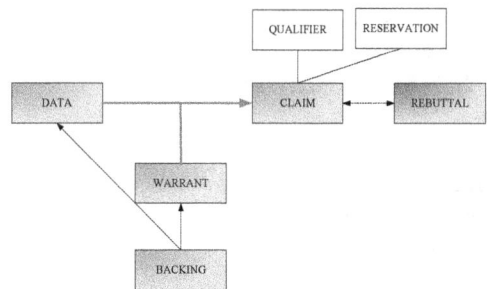

Figure 1. Toulmin's argumentation scheme

As is well-known, due to his long – mostly academic – teaching experience, Ratzinger's speeches are often a sort of didactic message, through which he seeks to reach his audience's mind and heart most clearly. Toulmin's model can thus help us follow his arguments and judge his way of presenting his claims.

3. The analysis. Argument from expert opinion

In this analysis I will take three questions as my starting point: *What is an appeal to an authority? What is a religious authority? Why do speakers stage other voices, especially in religious contexts?*

An appeal to an authority is part of a reasoning based on a source one has reason to think is reliable. Experts are presumably reliable and therefore authoritative to those who quote them because they are considered a "source of knowledge" or "a source of advice" (Walton 2006:85). By using appeals to authorities, speakers seek to make their discourse more impactful. Nowadays the term "authority" can have different connotations, as highlighted by *The New Oxford Dictionary of English* (1998:113; cf. Salvato 2023, in press).

In general, the argument from expert opinion is a type of reasoning in which it is assumed that the source is expert in a subject domain. Walton developed a scheme as a general structure to evaluate appeals to authority in argumentation, considering most of the previous interpretations (Walton 1997:199-210). Walton's "argumentation scheme" is a form of inference from premises to a conclusion and is summarized as follows:[1]

> E is an expert in domain D.
> E asserts that A is known to be true.
> A is within D.
> Therefore, A may (plausibly) be taken to be true

However, religious authorities' credibility and trustworthiness can be problematic, especially when their expertise in a specific area of religious faith is used to make statements closely linked to everyday life. Appeals to expert theologians, saints, or doctors and fathers of the church can introduce a 'critical' question on the relevance of their expertise to the different (social) issues introduced in an argumentation, with the audience considering the experts' field as only indirectly related to it.

Aside from an expert's general trustworthiness, citing religious experts introduces substantial problems, about the level of their authority, about why speakers stage other voices, especially in religious contexts, as well as about whether religious authorities' words stem from a *position to know* based on personal experience with appropriate but human knowledge, or whether they also convey divine illumination (cf. Salvato 2023, in press). Like any other authority but in the name of a supernatural order, a religious authority can decide (and should act, in word and deed, according to) what is permitted and what is forbidden. Staging other voices in

---

[1] D = Domain, E = Expert, A = Assertion.

religious contexts can therefore mean not only improving speakers' chances of making their discourse impactful. Particularly in the Christian Catholic church, when someone appeals to experts like doctors or fathers of the church, it is also intended to give religious believers the certainty that the utterance is grounded in rigorous Roman Catholic teaching. In his reflections, Ratzinger usually stages other experts' voices to gain official support for his argumentation and emphasize the intention of his words, allowing the audience to better comprehend the conclusion of his argumentation.

The following paragraphs will demonstrate how Ratzinger uses the support of religious experts in his argumentations.

4. Ratzinger's religious authorities

In Ratzinger's reflections, the religious authorities he quotes most frequently are authoritative persons or texts: saints (15 times) or theologians, among them doctors and fathers of the church, as well as official texts and documents of the Roman Catholic church like the New Testament, especially the Gospels (30 times), and encyclicals by his predecessors (seven times). With religious authorities, it is pertinent to consider whether the use of religious opinion is a matter of faith, that is, of theology, or whether it can also relate to everyday life. A starting point for argumentative analysis is hence the distinction between the types of authority that religious experts or official writings have and experts' different experiences and fields of knowledge.

An official text which provides important support for this evaluation is the *Compendium of the Social Doctrine of the Church*. According to this text, the church's heritage has its roots in the Sacred Scriptures, specifically the Gospels quoting Jesus' words and the apostolic writings. However, this heritage took shape and body through expert theologians, the so-called doctors and fathers of the church. Doctors of the church are defined as those who in any age have affirmed and defended Christian orthodoxy with their writings and whose role has been solemnly proclaimed by a pope or ecumenical council. The term fathers of the church designates the other group of authoritative ecclesiastical theologians and writers, who established the doctrinal foundations of Christianity.

While these church doctors and fathers are recognized theologians who defend Christian orthodoxy, saints' knowledge can be considered according to a *position to know*. Most of the 10,000 saints currently recognized by the Roman Catholic church were canonized because of their heroic lives rather than their teaching, having borne witness to Jesus Christ in everyday circumstances. The position of saints as 'authoritative sources' for the Christian church therefore stems mainly from having lived ordinary lives with a genuine religious consciousness. They live their lives in faith and reason, as *"the two cognitive paths of the Church's social doctrine"* are *"revelation and human nature"* (*Compendium*: Nr. 75). Therefore, their personal experience becomes "knowledge illuminated by faith" and in this sense they can be considered exemplary and authoritative people.

The saints Ratzinger quotes in his speeches are (in alphabetical order): Saint Augustine, Saint Bonaventure, Saint Boniface, Saint Elizabeth of Thuringia, Saint Kilian, Saint Paul, Saint Peter, and blessed[2] Mother Teresa of Calcutta. Among these, only Augustine and Bonaventure were proclaimed "doctors of the church", while Augustine – together with Saints Gregory, Jerome, and Ambrose – is considered a "father of the church".

As Ratzinger stated in his homily during the Holy Mass in Erfurt (24th Sept), "God's presence is clearly seen in the saints. Their witness to the faith can also give us the courage to begin afresh today." In this homily, Ratzinger takes the three patron saints of the Diocese of Erfurt – Saints Elizabeth of Thuringia, Boniface, and Kilian – as exemplary figures. However, he does not quote their words, instead telling of their Christian lives of genuine faith. Ratzinger recognizes them as figures bearing witness to Jesus Christ in many situations, thus his appeal to such exemplary saints can be considered as a general support to his claims. The three patron saints provided a living witness of the church "that ever endures", that is, "the witness of faith that makes all times fruitful and shows us the path of life".

Using three texts from the speeches, homilies, and discourses given by Ratzinger during his 2011 visit to Germany, I will now address the question *Who is quoted in religious texts?*

5. Uses of authorities: direct backing

The saint whom Ratzinger mainly quotes in his speeches and homilies on this German visit is Saint Paul. Paul is regularly cited in official texts as a saint and doctor of the Roman Catholic church, and his letters feature in its liturgy as the 'word of God'. I will now analyze Paul's cited words to see how Ratzinger appeals to him as an authority, that is, the role the saint plays in Ratzinger's texts and the conclusions he reaches.

The first quotation is taken from his *Letter to the Romans* (Rom 8:28) and forms part of Ratzinger's reasoning during the "Vespers of the Blessed Virgin Mary" in Etzelsbach (23rd Sept.):

"'We know that in everything God works for good with those who love him, who are called according to his purpose' (Rom 8:28), as we have just heard in the reading from the Letter to the Romans."

---

[2] Mother Teresa was later canonized by Pope Francis on 4 September 2016.

Toulmin's scheme allows us to better focus Ratzinger's argument and his way of appealing to Paul's authority.

5.1. Uses of quotations: theological matters

Paul's words are part of the reading from the Letter to the Romans that the congregation at Ratzinger's celebration have just heard. Consequently, they should be considered as the *grounds*, that is, the evidence validating Ratzinger's next words. However, as Toulmin underlines (1958: 98-99), the distinction between data and warrants is not always absolute and it is only possible "in *some* situations" (italics in the original) to distinguish clearly two different logical functions. Paul's words could then also be seen as a supporting guarantee, as the *warrant* authorizing Ratzinger's next step towards the claim of this argument.

However, before expressing this, Ratzinger's words are supported by the *grounds*, that is, the evidence for all Christians that the Virgin Mary is the interlocutor between God and humanity:

"With Mary, God has worked for good in everything, and he does not cease, through Mary, to cause good to spread further in the world."

Two further supporting statements are then introduced. Both can be considered *backing* parts establishing the reliability and relevance of the warrant. The first (backing1) refers to Christian knowledge taken for granted, based on John's Gospel (Joh 19:25-27). It regards the fact that on the cross, Jesus delivered Mary to humanity as their mother and humanity to Mary as her children:

"Looking down from the Cross, from the throne of grace and salvation, Jesus gave us his mother Mary to be our mother. At the moment of his self-offering for mankind, he makes Mary as it were the channel of the rivers of grace that flow from the Cross. At the foot of the Cross, Mary becomes our fellow traveler and protector on life's journey."

The second supporting statement (*backing2*) is an appeal to authority, a quotation from Lumen Gentium, the Dogmatic Constitution promulgated by Pope Paul VI in 1964 during the Second Vatican Council. It regards the certainty Christians must have of the Virgin Mary's presence in human life, interceding for "her son's sisters and brothers":

"By her motherly love she cares for her son's sisters and brothers who still journey on earth surrounded by dangers and difficulties, until they are led into their blessed home," as the Second Vatican Council expressed it (*Lumen Gentium*, 62)."

Having established these supports, Ratzinger can now make his own statement. He starts it with an affirmative *Ja/Yes*, accompanied in the English translation by the adverbial reinforcer *indeed,* emphasizing Ratzinger's personal view:

"Yes indeed, in life we pass through high-points and low-points, but Mary intercedes for us with her Son and helps us to discover the power of his divine love, and to open ourselves to that love."

With Toulmin's framework, the argument can now be restated as follows:

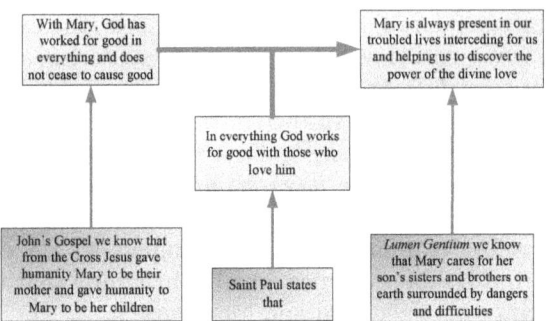

Figure 2. Ratzinger's first quotation according to Toulmin's scheme

As we have just seen, Ratzinger's argument contains two backing statements. One (backing1) is based on Ratzinger's confidence that his Christian audience (at the Vespers) knows the basic texts, specifically the Gospels. Without having to cite the source, he shares with them John's words. The other supporting statement (backing2) is based on another basic text, Lumen Gentium, whose knowledge Ratzinger cannot take for granted. Therefore, he makes the source of this quotation explicit and underlines that Mary cares for her son's sisters and brothers on earth.

Through these three supporting statements – Paul, John's Gospel, and *Lumen Gentium* – Ratzinger can conclude the Vespers celebrated in honor of the Blessed Virgin Mary with a sort of promise to the congregation: the mother of God never abandons humanity in its dangers and difficulties.

6. Authorities and religious life: faith

Another quotation by Saint Paul is used by Ratzinger during his meeting with seminarians in St Charles Borromeo Seminary Chapel in Freiburg (24th Sept). The appeal to Paul regards the second element that Ratzinger considers relevant in one's faith in God. The first element is the inner, personal life with Christ, that "constant inner journey with the word of God", expressed in a personal "being with Christ". The second element should be a personal union with the whole church and all fellow believers. As Saint Paul expressed this in a succinct expression, Ratzinger alludes to it as follows:

"[...] another is that we can only ever believe within the 'we'. I sometimes say that Saint Paul wrote: 'Faith comes from hearing' – not from reading."

Paul's words regard the proper origin of Christian faith, the "hearing" which constitutes listening to other Christian believers and thus in living within the "we" of the whole church. Hence the appeal to Paul should be considered as the grounds, that is, the evidence validating Ratzinger's next words "not from reading" and the paragraph that follows.

6.1. Authorities and religious life: Ratzinger's five statements

Ratzinger dedicates a long paragraph to this second part of the claim, which can be divided into five parts according to five statements:

"It needs reading as well, but it comes from hearing, that is to say from the living word, addressed to me by the other, whom I can hear, addressed to me by the Church throughout the ages, from her contemporary word, spoken to me by the priests, bishops and my fellow believers.

Faith must include a 'you' and [...] a 'we'. And it is very important to practice this mutual support, to learn how to accept the other as the other in his otherness, and to learn that he has to support me in my otherness, in order to become 'we', so that we can also build community in the parish, calling people into the community of the word, and journeying with one another towards the living God.

This requires the very particular 'we' that is the seminary, and also the parish, but it also requires us always to look beyond the particular, limited 'we' towards the great 'we' that is the Church of all times and places: [...]

When we say: 'We are Church' – well, it is true: that is what we are, we are not just anybody. [...] The 'we' is the whole community of believers, today and in all times and places. And so I always say: within the community of believers, yes, there is as it were the voice of the valid majority [...].

We are Church: let us be Church, let us be Church precisely by opening ourselves and stepping outside ourselves and being Church with others."

In the first part, Ratzinger's utterance also functions as the *warrant* because he underlines that a community involves both reading and hearing. However, the appeal to Paul's words is necessary to underline that "the living word" is addressed to Christians by others, that is, by the whole church, and must hence be heard. The other statements all focus on the importance of listening to "the voice of the valid majority". The second part emphasizes the importance of mutual support. The third underlines that Christians should look beyond the limited "we" towards the great "we" embodied by "the Church of all times and places". The fourth highlights that only the whole community of believers forms "the voice of the valid majority". In the fifth, he exhorts the community of seminarians to be open to others.

6.2. Authorities and religious life: rebuttal

Two further statements follow these five statements, functioning as a *rebuttal*:

"[...] it requires that we do not make ourselves the sole criterion.

[...] but there can never be a majority against the apostles or against the saints: that would be a false majority."

By including a *rebuttal* in his argument, Ratzinger acknowledges that his view can have a counterpart, thus helping the audience judge the situation objectively. The two statements sound like exceptions that might invalidate the claim if they occurred. Therefore, Ratzinger twice admonishes his interlocutors – young seminarians approaching both the secular and the Christian world – to avoid the dangerous attitude of considering oneself "the sole criterion", thus running the risk of going against the church's authorities. A fundamental aspect between both rebuttals starts with the conjunction *but*, which changes the standpoint by underlining that the "we" is more extensive than a small group asserting that "they are the church". St. Paul is a person in *position to know* for two reasons, his proximity to Christ, and his direct experience of the Christian faith.

With Toulmin's framework, the argument can now be summarized thus:

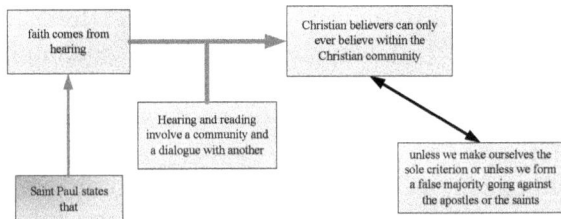

Figure 3. Ratzinger's second quotation using Toulmin's scheme

7. How to behave: the double warrant

In his homily during the Holy Mass at the airport in Freiburg (25th Sept), Ratzinger recalls Saint Paul's words six times, once from the Letter to the Hebrews, five times from the Letter to the Philippians. These quotations can be considered according to three different paragraphs, among which the third and last one can be divided into three argumentative steps. According to Toulmin's scheme, a warrant, claim, and backing can be recognized, but the figure of Saint Paul can function in this case as a double warrant. To his exhortation to unity, Paul adds a call to humility, saying:

"Do nothing from selfishness or conceit, but in humility count others better than yourselves. Let each of you look not only to his own interests, but also to the interests of others" (Phil 2:3-4).

Paul's words function as the warrant stated by a Christian authority according to two guarantees. First, Paul knows Jesus' words very well. Second, having devoted his life to others, he represents a model of Christian life. Ratzinger needs Paul's words to support his claim on the importance of both humility and service of others in Christian life, when he states:

"Christian life is a life for others: existing for others, humble service of neighbor and of the common good. Dear friends, humility is a virtue that does not enjoy great esteem in the world of today, or indeed of any time. But the Lord's disciples know that this virtue is, so to speak, the oil that makes the process of dialogue fruitful, cooperation possible and unity sincere. The Latin word for humility, *humilitas*, is derived from *humus* and indicates closeness to the earth. Those who are humble stand with their two feet on the ground, but above all they listen to Christ, the Word of God, who ceaselessly renews the Church and each of her members."

This part can be schematized as follows:

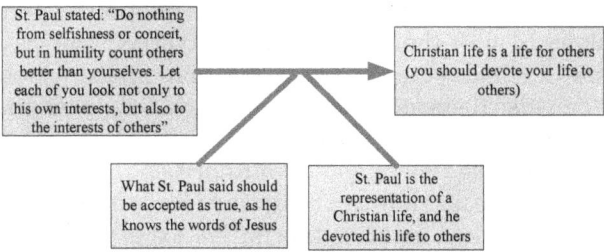

Figure 4. Ratzinger's second quotation using Toulmin's scheme

However, there is also a last statement that could function as a *backing*, in which Ratzinger can focus the main features of humility as a virtue, as they were experienced by further authoritative figures, the Lord's disciples. The argument can be outlined as follows:

> *warrant1*: (since) Saint Paul exhorts us to humility
> *warrant2*: from the Lord's disciples we know that this virtue makes the process of dialogue fruitful, cooperation possible and unity sincere
>   *claim*: Christian life is a life for others

The conclusion of the homily consists in an exhortation to ask God for the courage and humility "to walk the path of faith, to draw from the riches of his mercy, and to fix our gaze on Christ". Sustained by Saint Paul's authoritative words and the disciples' experience with Jesus, Ratzinger's conclusion receives official support, which can involve more deeply and persuade the audience.

8. Conclusions

In analyzing the proposed passages from Ratzinger's visit to Germany, I have used both Walton's method and Toulmin's analytical framework of argumentative structures. The latter is clear and flexible and can enable a comprehensive understanding of how information, statements, or experts' quotations are used in arguments. Used in conjunction with Walton's questions and analytical method, it becomes an even more complete and heuristic framework that can help students training in formal logic, that is, checking the logic in their drafts, especially by emphasizing the reader's or interlocutor's probable responses.

From the analysis the following significant elements emerged.

The appeal to others' words within arguments *from popular* or *from expert opinion* often entails a polyphonic perspective, in which monological speakers match different voices with their own, thus gaining a dialectical dimension. Through the polyphonic effect, quoting experts' words can allow speakers to shift their commitments to other sources and use them as premises for subsequent argumentation steps (cf. Ducrot 1984). Polyphony can also decrease the burden of proof on speakers, transferring it from themselves to their sources.

Moreover, the same religious authorities can have a multiple argumentative value, as they can be invoked for different reasons. St Paul, for example, is invoked as an authoritative expert from a perspective of position to know, as a witness of the word of God, and as a model of a life lived according to the Gospel and Jesus' words. This is relevant for the moral dimension of claims because it becomes a sort of multiple argumentation which enables distinction between the appeal to authorities as experts (Prof X says Y) and the citation of complex authorities (X, as witness/expert/model, says Y). At the same time, we can note that – depending on the conclusion – one or different warrants can be activated.

Further research could focus analysis on authoritative figures such as saints or blessed people using the following questions: Who is the example? Can the example of Christian life be considered a source of authority? In what sense? It is not uncommon for religious authorities to have this twofold argumentative function, where they are experts in the sense of position to know and as examples to be followed.

Lastly, because of their clarity, Toulmin's and Walton's frameworks can be used for didactic aims, for example by school and university teachers to aid the development of students' writing skills in composition classes. For example, having supervised hundreds of teaching assistants who had used Toulmin's model, Annette Rottenberg (1985) highlights in her book *Elements of Argument* that the model is useful for defending one's claims directly and efficiently; Locker and Keene (1983: 103-109) consider the model "easy-to-apply" whilst Karbach (Karbach 1987: 90, 81-91) found it valuable in business classes to develop business and technical writing programs.

References

Bakhtin, M. M. (1968). *Dostoevskij. Poetica e stilistica*. Torino: Einaudi.
Bakhtin, M. M. (1979). Das Wort im Roman. In R. Grübel (Ed.), *Die Ästhetik des Wortes* (pp. 154-300). Frankfurt a.M.: Suhrkamp Verlag.
Bakhtin, M. M. (1986). The problem of Speech Genres. In M. M. Bakhtin. C. Emerson and M. Holquist (Eds.), *Speech Genres and other late essays* (pp. 60-102). University of Texas Press: Austin.
Compendio della dottrina sociale della chiesa (2004). Città del Vaticano: LEV.
Dascal, M. (2005). Debating with Myself and Debating with Other". In P. Barrotta, and M. Dascal (Eds.), *Controversies and Subjectivity* (pp. 33-73). Amsterdam: John Benjamins.
Ducrot, O. (1984). *Le dire et le dit*. Paris: Les éditions de minuit.
Eemeren, F. H. van, and Wu Peng. (2017). Contextualizing Pragma-Dialectics. *Argumentation in Context, 12*. John Benjamins Publishing Company.
Karbach, J. (1987). Using Toulmin's Model of Argumentation. *The Journal of Teaching Writing 6(1) Spring 1987*, 81-91.
Locker, K. O. and M. L. Keene. (1983). Using Toulmin Logic in Business and Technical Writing Classes. In W. Keats Sparrow and N.A. Pickett (Eds.), *Technical and Business Communication in Two-Year Programs*. (pp. 103-109). Urbana: National Council of Teachers of English.
Macagno, F. & Damele, G. (2013). The Dialogical Force of Implicit Premises: Presumptions in Enthymemes. *Informal Logic*, Vol. 33, No. 3 (2013), 361-389.
Macagno, F. and Walton, D.N. (2008). Persuasive Definitions: Values, Meanings and Implicit Disagreements. *Informal Logic. 28(3)*, 203-228.
Rigotti, E. (2005). Congruity Theory and Argumentation. Argumentation in Dialogic Interaction. *Studies in Communication Sciences, Special Issue*, 75-96.
Rigotti, E. and Rocci A. (2005). From Argument Analysis to Cultural Keywords (and Back Again). In F.H. van Eemeren and P. Houtlosser (Eds.), *Argumentation in Practice*, (pp. 135-142). Amsterdam/Philadelphia: John Benjamins.
Rocci, A. (2005). Connective Predicates in Monologic and Dialogic Argumentation. Argumentation in Dialogic Interaction. *Studies in Communication Sciences, Special Issue*, 97-118.

Rottenberg, A. T. (1985). *Elements of Argument*. New York, St Martin's Press.
Salvato, L. (2005). *Polyphones Erzählen. Zum Phänomen der Erlebten Rede in deutschen Romanen der Jahrhundertwende*. Bern: Peter Lang.
Salvato, L. (2023, in press). The appeal to religious authority: a case-study. *Journal of Argumentation in Context*.
The New Oxford Dictionary of English (1998). Pearsall J. (Ed.). Oxford: Clarendon Press.
Toulmin, S. (1958). *The Uses of Argument*. Cambridge, UK: Cambridge University Press.
Untersteiner, M. (1954). *The sophists*. Oxford: Basil Blackwell.
Walton, D. (1989/2008). *Informal Logic. A pragmatic approach*. New York, NY: Cambridge University Press.
Walton, D. (1997). *Appeal to expert opinion. Arguments from Authority*. University Park/Pennsylvania: The Pennsylvania State University Press.
Walton, D. (2006). *Fundamentals of Critical Argumentation*. New York, NY: Cambridge University Press.
Walton, D., Reed C., and Macagno F. (2008). *Argumentation Schemes*. New York, NY: Cambridge University Press.

Websites

www.vatican.va
https://www.vatican.va/content/benedict-xvi/it.html
www.fondazioneratzinger.va

Compendium of the Social Doctrine of the Church
English version
https://www.vatican.va/roman_curia/pontifical_councils/justpeace/documents/rc_pc_justpeace_doc_20060526_compendio-dott-soc_en.html
Italian version
https://www.vatican.va/roman_curia/pontifical_councils/justpeace/documents/rc_pc_justpeace_doc_20060526_compendio-dott-soc_it.html

Ratzinger's speeches

Marian Vespers at the Wallfahrtskapelle (Etzelsbach, 23 September 2011).
Meeting with seminarians at St Charles Borromeo Seminary Chapel (Freiburg im Bresgau, 24 September 2011)
Homily during the Holy Mass at the Cathedral Square in Erfurt (24 September 2011).
Homily during the Holy Mass at the airport in Freiburg (25 September 2011).

ARGUMENTATION AND EPISTEMIC DISTRIBUTED VIGILANCE

CRISTIÁN SANTIBÁÑEZ[1]
Universidad Católica de la Santísima de Concepción, Chile
cristiansantibanezy@yahoo.com

Abstract

Epistemic vigilance (Sperber et al., 2010) refers to the natural individual competence to filter or scan information when the agent perceives that something in the message or in the source of the message may be misleading, or because the beliefs being communicated generate noise in her core beliefs. According to the authors, epistemic vigilance is part of our cognitive design to balance trust and vigilance. In this paper it is proposed that, from the point of view of argumentative activity, a better conceptual notion would be epistemic distributed vigilance, this is, an epistemic vigilance distributed in various sources that coherently emerge as a collective agency (groups with specific identities, value-based communities, corporations, institutions). In this proposal epistemic distributed vigilance refers to the natural tendency (or capacity) of group' members, or the group as a whole, to deposit both its beliefs and the reflective processes in other members of the group or in institutional semi-formal or formal mechanisms in order to decide what to believe and what courses of action to take. This paper argues that this epistemic collective agency is the result of several facts, both at the individual and the social level. For example, there is a social epistemic asymmetry, this is, some agents have faster and more reliable access to information, and there is a cognitive collective division of labor. Theoretical consequences of the proposed concept are that there is an endorsement of epistemic vigilance distributed among members of the group who do not necessarily speak for themselves but represent the group. An example of this would be the acquisition and formation of beliefs in children: they have an epistemic cognitive dependence (on the family or peers, Tomasello, 2014), they trust adults by default, they have access to beliefs through environmental physical beliefs distributed through repositories (i.e., objects, physical structures of schools).

---

[1] This paper is part of the Fondecyt 1200021 research project.

1. Introduction

The logician Johan van Bentham proposed some years ago that *logic is the immune system of the mind*. The sentence is part of the following paragraph: "What seems crucial about us is not the use of infallible methods, but reasoning with whatever means we have, plus an amazing facility for belief revision, i.e., coping with problems as they arise. It is this dynamic feature of human rationality which logicians and philosophers should try to understand better. Logic is not some vaccination campaign eradicating all diseases once and for all. It is rather the immune system of the mind! (van Bentham, 2007: 91). The idea would be that we have this natural equipment called logic to defend us against dangerous communication or beliefs. But do we really have this defense called logic or otherwise?

One of the most robust conceptual efforts, in my opinion, to address this question was captured in the concept of *epistemic vigilance* proposed by Sperber et al (2010). Years later, one of Sperber's students referred to the idea as *open vigilance* (Mercier, 2020). Both approaches assume that the mind is a well-designed cognitive mechanism that has reached its current efficiency after a long evolutionary path.

What the three ideas have in common (van Bentham's metaphor, Sperber et al.'s approach, and Mercier's turn) is that the efficiency of the mind to assess communication is placed in the individual's performance. Against this assumption, in this paper I will develop an approach that emphasizes that the assessment of communication is distributed in different minds that form groups, in the proximal and distant environment, and in the artefacts that we create to deposit some beliefs.

To this end, in this paper I use some ideas from the literature on distributed cognition, and social and communitarian epistemology. The paper is organized as follows: in section 2, I describe both the concept of epistemic vigilance, and the term open vigilance; in section 3, I briefly reflect on the notion of epistemic community using Kusch's (2002) approach, which considers knowledge as a social status; in section 4, I summarize some of the core ideas of distributed cognition that I will use in section 5; I develop, in section 5, my proposal termed as *epistemic distributed vigilance*. In the conclusions, I will only recall some of the main points of my proposal.

2. Epistemic Vigilance

Because some people would like to take advantage of us by communicating something that is not the case, we better watch their steps. This mental capacity of vigilance in the domain of communication (of information and beliefs) has evolved over time, in a back-and-forth game of strategies, both to deceive someone and to catch the deceiver. From the point of view of communication theory and the psychology of reasoning, the two main theories of this game are captured in the notion of *epistemic vigilance* proposed by Sperber et al. (2010), and in the concept of *open vigilance* recently developed by Mercier (2020). Both proposals rely heavily on evolutionary and cognitive theories. In the following, I will briefly describe both as the conceptual background from which my proposal emerges.

Sperber et al. (2010; Sperber, 2013) develop their concept of epistemic vigilance by assuming that since people have conflicting interests and they can use any means to achieve their interests, including lying, deceiving and manipulation (Mascaro & Sperber, 2009), it would be optimal for agents to be cognitively equipped with a mechanism to detect fraudulent communication.

The authors describe the problem very clearly: "How reliable are others as sources of information? In general, they are mistaken no more often than we are—after all, 'we' and 'they' refer to the same people—and they know things that we don't know. So, it should be advantageous to rely even blindly on the competence of others. Would it be more advantageous to modulate our trust by exercising some degree of vigilance towards the competence of others? That would depend on the cost and reliability of such vigilance. But in any case, the major problem posed by communicated information has to do not with the competence of others, but with their interests and their honesty. While the interests of others often overlap with our own, they rarely coincide with ours exactly. In a variety of situations, their interests are best served by misleading or deceiving us. It is because of the risk of deception that epistemic vigilance may be not merely advantageous but indispensable if communication itself is to remain advantageous. (Sperber et al., 2010: 359-360).

As can be deduced from this important quote, for Sperber et al. (2010) the mechanism of epistemic vigilance arises, in principle, from a calibration of trust between the interests of the communicator, the content communicated and the interests of the receiver(s). This calibration of trust has become crucial due to the need for valuable information that others may have. It seems that it does not matter if the communicator has a high competence, epistemic or otherwise, on the information communicated, but her honesty and interests are extremely relevant for the receiver. High competence in the information conveyed means that the communicator is an expert on the subject, has some experience of it, or has witnessed the events. But the problem, according to Sperber et al. (2010), is that given the fact that the communicator knows, consciously or unconsciously, that she can influence the receiver to achieve her goal, it is better to build some reputation of trustworthiness. But building a reputation of trustworthiness takes time and is not an easy task, so it is cheaper to lie. So the receiver, in turn, must be epistemically vigilant in order to avoid the damage of a potentially deceptive communication and to exercise constant control over the default trust. As we can immediately imagine from our daily experience, this cognitive task of being vigilant costs energy (and sometime a lot of it) and, more importantly, there is the problem of how reliable our own vigilance is.

In the Sperber et al.'s (2010) framework, vigilance in our communicative exchanges achieves a beneficial evolutionary stability of communication, a stability between potential harm and blind trust. However, the authors explicitly point out that vigilance does not have to be extremely efficient by trying to eliminate all the dishonest information sent by the communicator.

For some critical observers of the concept (Michaelian, 2013b; see also Michaelian, 2010, 2011, 2012a, 2012b, 2012c, 2013a; Michaelian & Sutton, 2013), this concession is not enough, because it seems that not even bare vigilance is on exercised among interlocutors, and this is why Sperber et al. point out that: "In order to gain a better grasp of the mechanisms for epistemic vigilance towards the source, what is most urgently needed is not more empirical work on lie detection or general judgments of trustworthiness, but research on how trust and mistrust are calibrated to the situation, the interlocutors and the topic of communication" (2010: 370–1). The author's emphasis on calibrating trust between participants assumes that communication is a cooperative endeavor from which a beneficial outcome is expected, otherwise no intentional communication between people would be possible. Some of these aspects are developed in the following important quote: "Most human communication is carried out intentionally and overtly: The communicator performs an action by which she not only conveys some information but also conveys that she is doing so intentionally (Grice, 1975; Sperber and Wilson, 1995). For communication of this type to succeed, both communicator and addressee must cooperate by investing some effort: in the communicator's case, the effort required to perform a communicative action, and in the addressee's case, the effort required to attend to it and interpret it. Neither is likely to invest this effort without expecting some benefit in return. For the addressee, the normally expected benefit is to acquire some true and relevant information. For the communicator, it is to produce some intended effect in the addressee. To fulfil the addressee's expectations, the communicator should do her best to communicate true information. To fulfil her own expectations, by contrast, she should choose to communicate the information most likely to produce the intended effect in the addressee, regardless of whether it is true or false." (360)

For Sperber et al. (2010), communication is essentially a cooperative coordination in which the parties have expectations that need to be met. In this endeavor, the most delicate part is the fact that the communicator, in principle, would not care about using fake information to achieve her goals. The concept of epistemic vigilance thus refers to the receiver's ability to assess the source of the content (is the speaker worthy of belief?), and the content itself (is what is being communicated useful to me?). Epistemic vigilance is not a cognitive mechanism against trust, but only against blind trust; it is a delicate balance between epistemic trust and epistemic vigilance.

There is a particular use of epistemic vigilance when we receive information that goes against our basic beliefs, or when we detect some kind of inconsistency (between the reputation of the communicator and the context of the communication, between what has been taken for granted by default and the unexpected new information); in these contexts, Sperber et al. (2010) seem to imply, our tendency is to assess the information more carefully and the sender gets its highest pick. Excessive trust is a danger in the face of free riders, those who parasitically take advantage of us without reciprocating. This would be one reason why we are wired with this cognitive tool (Kissine & Klein, 2014).

According to Sperber and co-authors, this cognitive tool can do several things at once. It can assess the reliability of the speaker (whether the speaker is competent, i.e. whether she is epistemically reliable, and whether she is benevolent, i.e. morally reliable); but the tool also assesses the degree of credibility of the incoming information, i.e. the tool analyzes the quality of the propositions independently of their source. Thus, epistemic vigilance provides general impressions of the sender's trustworthiness (considering all kinds of cues, such as facial expressions), and evaluates the sender's reliability on the basis of contextual factors, such as the convergence between the content of the proposition and the speaker's intentional effort (perhaps the sender wants us to believe something we know he does not believe).

The authors use an interesting metaphor to explain why we are collectively vigilant: "Our mutual trust in the street is largely based on our mutual vigilance. Similarly, in communication, it is not that we can generally be trustful and therefore need to be vigilant only in rare and special circumstances. We could not be mutually trustful unless we were mutually vigilant." (364). But is this a good analogy, namely, comparing walking and communicating?

3. Are We Vigilant? Questions from a communitarian perspective

The very fact that I am writing this complementary reflection on the concept under analyze is evidence that we are all exercising epistemic vigilance. But at the same time, I am doing so by using some vigilant remarks already made by other colleagues who are part of the community in which these ideas circulate. So I am vigilant because others have been before me. This is not mutualism, i.e. I am not doing this because others are doing the same to me, indeed I am not criticizing Sperber et al. (2010)

because I have been criticized by them (they do not even know that I exist!). So the argument here is that any epistemic vigilance is exercised within in a social design of distributed checks and balances in which agents acquire, hold and communicate beliefs. Furthermore, as socially constituted agents, we deposit beliefs in artefacts that are at our disposal for present and future vigilance, as we extend our cognitive skills and ideas into artefacts or other mechanisms located in the environment.

This argument will be explained in this section by considering a communitarian perspective on how we acquire and communicate beliefs, in particular using Kusch's (2002) approach. In the next section (4), I will address the problem of distributed epistemic construction.

Kusch supports his approach by pointing out some limitations of the individualistic approach to belief formation, namely, that there are natural and social asymmetries in the access to information, that individual cognition is limited (see also Mercier, 2020), that belief formation is a matter of training and use, and that there are very important epistemic social factors to consider (such as what are called *epistemic social injustices* (Fricker, 2007)). These facts add up to a communitarian view of the process of belief formation and, for our purposes, the communication of beliefs.

The communitarian view, according to Kusch, presupposes two important propositions: 1) that knowledge and its cognates, such as 'knowing' and 'knower', mark a social status, and 2) that the fundamental owners of knowledge are groups of people rather than the individuals. The agents within a community become epistemic in virtue of a systematic, direct, and indirect, dialogical training. To illustrate this, Kusch uses the way in which testimony works. In a nutshell, Kusch describes testimony as a performative act, i.e. it constitutes the facts it reports, only if there is a collective -agreement- that recognizes the transmission of information advanced by the testifier as such. Empirical belief formation via testimony is also a political move, in terms of public and collective authorizations.

4. Distributed Cognition: How Do We Surveille?

We have a kind of control over the beliefs communicated by checking the plausibility of these beliefs within the web of shared beliefs, that is, we control our peers by using preexisting beliefs (with confirmation bias being the most robust mechanism for filtering them (Mercier, 2020)). Another way to control communication in general, according to Mercier (2020: 53), is to use pre-existing inferential mechanisms. We set these mechanisms in motion when we argue. Argumentation is the most paradigmatic cognitive and cultural mechanism for exercising collective vigilance. In real argumentative settings, not everyone has the same ability to access and evaluate information, nor the same skills, and usually social asymmetries also play an important role in determining the outcome of disagreements. Social differences between arguers (such as reputation as communicators (Origgi, 2018), social status, epistemic asymmetries, power, among others) are facts that have organised cultures - and indeed animals - for hundreds of thousands of years (Diamond, 2012). We have evolved by incorporating these social experiences as cognitive patterns of adaptation to build niches (Sterelny, 2003).

My approach to dealing with such powerful evidence is to observe how a distributed cognitive perspective might shed light on the problem of epistemic vigilance from an argumentative point of view. In short, the distributed cognitive approach holds that all the resources necessary to solve a task are collectively located among the members of the group. Groups are made up of agents who are jointly and intentionally involved in solving a problem or task, and in which everyone contributes to the solution.

According to Steiner (1966), there are several ways in which members of a group can work together to solve a problem, such as:

1. Additive: Members can do something because the sum of their individual contributions generates the resolution.
2. Compensatory: Members can produce a result as a function of their individual solutions, for example by aggregating judgments under a majority rule.
3. Conjunctive: Members can do something that requires everyone to succeed at the same time in order to succeed as a group, e.g. mountaineers.
4. Disjunctive: Members can work independently and as a group to choose the best solution from individual efforts.
5. Complementary: Members can institute a division of labor in which a complex task is broken down into different parts assigned to different agents, thus distributing the problem to be solved among the agents and tools and the structural characteristics of their environment (teams, universities, corporations).

It seems that we use the 5 behaviors as arguers. Looking at these possibilities from the point of view of contextual patterns, they are also prototypical argumentative settings. Mediations could be an example of an additive argumentative setting. Parliaments could be an example of a compensatory setting. Team debates are examples of conjunctive contexts. Commercial companies tend to incorporate disjunctive mechanisms. As said, universities are prototypical examples of complementary settings. Certainly, these contexts are likely to combine the 5 behaviors during the solution of the task at hand and, merging with argumentation studies, these settings obviously make use of different types of dialogues during the exchange of opinions in order to solve the problem or achieve the goal.

5. Argumentation

Two questions need to be answered in order to see in what sense epistemic vigilance is a distributed practice and that argumentation is its natural mechanism. First, we need to determine, at least provisionally, which cognitive skills that are usually thought to operate at the individual level are manifested at the group level (Pettit, 2003). Memory and intentionality are two of them. Theiner (2013; Theiner, Allen & Goldstone, 2010) proposes the following parameters as group cognition (paradigmatic examples are transactive memory systems) that express memory and intentionality: (1) the system can adapt its behavior to a changing environment; (2) S can process information coming from its environment; (3) S can selectively and intentionally attend to its environment; (4) S can create internal representations of its environment; 5) S can modify its environment by creating artefacts; (6) S can realize that it is a cognitive agent; (7) S can have conscious experiences of itself and the world. (1) and (2) are behavioral, but (3)–(5) obviously involve intentionality; (6) and (7) are more problematic, because they involve the idea of consciousness.

The second question would be to determine in what sense a group is a cognizant in its own merit. It is one thing for cognition to be distributed, and another for a group in which cognition is distributed to be cognizant (the problem of consciousness being the biggest problem). The first step in answering this question is to consider that groups with members who know each other well, who have had a long time to adjust to each other, and trust each other, are candidates for transactive memory systems that go beyond what their individuals can do when considered independently (Theiner, 2013). In such groups, memories are stored in relation to experiences, there is access to this stored information, and it serves to solve problems more effectively. Memories need not be the exclusive property of individuals.

Although Kusch does not make use of the cognitive literature in his communitarian approach to belief formation, it is possible to complement it with the 5 behaviors identified, bearing in mind that within the groups there are likely to be the epistemic asymmetries that Kusch distinguishes.

It is worth to remembering that the first wave of cognitive studies claimed that certain cognitive or mental processes are extended beyond the limits of (biological) organisms. At that time, the functional principle of parity or similarity between extensions and knowing bodies was postulated. But with the second wave of cognitive studies, this principle (Clark, 2008; Clark & Chalmers, 1998) was replaced by the following idea: the transition from functional parity to complementarity between them as vehicles of cognition. It is about the integration between intra and extra mental entities. The third wave (Cash, 2013; Kirchhoff, 2012) of cognitive studies has added, based on the enactive perspective, that cognition often not only involves social interaction but that action depends on institutional forms of practices, norms and conventions that constitute individual cognition (e.g. Gallagher & Crisafi, 2009; De Jaegher, 2013; Gallagher 2013). In light of these advances in cognitive analysis, the Principle of Social Parity (Ludwig, 2015) has been postulated, which is defined as follows:

1) If, when confronting a task T, two agents A and B carry out the cognitive processes P and P* in a way that taken separately would not be functionally sufficient to carry out T.
&
2) If P and P* were performed by a single agent (either A or B), we would have no problem recognizing P or P* as part of A or as part of B, which enables us to consider that P and P* are socially distributed in a singular cognitive system (they are "parts" of an extended cognitive system).

As a cognitive mechanism, the roles/parts of the argumentative process (protagonist, antagonist, direct and indirect audiences, etc.) are distributed among the participants, and each of them can change the role as the process continues, using artefacts that contain beliefs. My point is that every time we argue, we activate a cognitive organization at the group level that is designed, among other things, to form epistemic products and to control their transmission. From this perspective, the cognitive mechanism of arguing to form, transmit and evaluate beliefs manifests, to varying degrees, some of the basic characteristics of distributed and extended cognitive systems (Theiner, 2013). For example, the characteristic of *availability* (the external cognitive tools they contain must be constantly at hand, and can be used by others or by social institutions; or accessibility (easy and constant access to the shared cognitive environment), or direct and mutual transfer between the members of their cognitive capacities (of all or the majority), of the capacities of the collective, or of the intentional point of view in a more or less automatic or habitual way; or control and self-correction (ability and possibility to control the execution of cognitive processes and the distribution of control mechanisms (spontaneous or predetermined); and when it is necessary to correct the actions of all or some of the members in relation to a task; or *regulatory commitment* (explicitly articulated or not).

6. Conclusions

Goldberg (2012) has proposed the generic extended epistemic mind hypothesis to understand testimonial belief, which he describes as an epistemic assessment of the information processing that takes place in the subject's environment. But memory also provides a special case of this hypothesis: when we evaluate a belief through memory, we consider not only the reliability of the process by which the belief was stored and retrieved, but also the reliability of the process by which it was initially produced. Indeed, memory is a temporally extended process. Goldberg points out that we should classify non-human (external) contributions as gross causal, as distinct from and in contrast to cognitive-psychological contributions.

So, let's recall some facts similar to those Goldberg provides to support his hypothesis. First, there are different kinds of epistemic (social) asymmetries (Gilbert, 1989; Fricker, 2007); second, there is a division of cognitive labor every time we argue (Mercier, 2020; Trouche, Johansson, & Mercier, 2018); the acquisition and formation of beliefs in infants depends on the testimony of others; there are cognitive dependencies; there is a default trust for adults; we create physical organizations in the environment as repositories of beliefs; at the group or institutional level, beliefs are monitored in protocols; mutual commitment (voluntary or contractual) with a set of beliefs (expectation): everyone knows them and everyone defends them - to some extent- that the institution believes in and pursues; within groups, members show interdependence in endorsement for completeness; and institutions have memory.

Among the important consequences of these facts are that epistemic vigilance is (automatically) endorsed (because it costs less energy, it is more efficient), and distributed among members of a community (group, institution) and in artefacts. This is a cultural cognitive pattern that has become a stable collective design (or what Tollefsen (2006) calls a stable *collective mind*; or stable *collaborative cognition* (Harnad, 2005; see also Szanto, 2014; Rupert, 2011).

What is the core mechanism by which beliefs and vigilance over those beliefs are transmitted, evaluated and negotiated? Argumentative activity is the process by which these communicative possibilities are enacted, and in which arguers continually change positions for the good of the group. In fact, argumentation is the intelligent superstructure we have created to ensure distributed epistemic vigilance, and in which the best argument emerges, sometimes independently of the arguer's position, but certainly with the help of the others.

References

Bentham, J. van. (2007). Logic in Philosophy. In D. Jacquette (Ed.), *Handbook of the Philosophy of Logic* (pp. 65-99). Amsterdam: Elsevier.

Cash, M. (2013). Cognition without borders: 'Third wave' socially distributed cognition and relational autonomy. *Cognitive Systems Research*, 25–26, 61-71.

Clark, A. (2008). *Supersizing the Mind. Embodiment, Action and Cognitive Extension*. Oxford: Oxford University Press.
Clark, A. & Chalmers, D. (1998). The Extended Mind. *Analysis*, 58(1), 7-19.
Diamond, J. (2012). *The World Until Yesterday. What can we learn from Traditional societies*. New York: Viking.
De Jaegher, H. (2013). Rigid and fluid interactions with institutions. *Cognitive Systems Research*, 25–26, 19-25.
Fricker, M. (2007). *Epistemic Injustice. Power & the Ethics of Knowing*. New York: Oxford.
Gallagher, S. (2013). The Socially Extended Mind. *Cognitive Systems Research*, 25–26, 4-12.
Gallagher, S. & Crisafi, A. (2009). Mental Institutions. *Topoi*, 28(1), 45-51.
Gilbert, M. (1989). *On Social Facts*. Princeton: Princeton University Press.
Golderbg, S. (2012). Epistemic Extendedness, Testimony, and the Epistemology of Instrument-Based Belief. *Philosophical Explorations*, 15(2), 181-197
Harnad, S. (2005). Distributed processes, distributed cognizers, and collaborative cognition. *Pragmatics & Cognition*, 13(3), 501-514.
Kirchhoff, M. (2012). Extended cognition and fixed properties: steps to a third-wave version of extended cognition. *Phenomenology and the Cognitive Sciences*, 11(2), 287-308.
Kissine, M. & Klein, O. (2014). Models of Communication, Epistemic Trust and Epistemic Vigilance. In J. Forgas, O. Vincze & J. László (eds.), *Social Cognition and Communication* (pp. 139-154). New York: Psychology Press.
Kusch, M. (2002). *Knowledge by Agreement*. New York: Oxford.
Ludwig, K. (2015). Is Distributed Cognition Group Level Cognition? *Journal of Social Ontology*, 1(2): 189–224.
Mascaro, O. & Sperber, D. (2009). The moral, epistemic, and mindreading components of children's vigilance towards deception. *Cognition*, 112(3), 367-380.
Mercier, H. (2020). *Not Born Yesterday. The Science of Who We Trust and What We Believe*. New Jersey: Princeton University Press.
Mercier, H., Bernard, S. & Clément, F. (2014). Early sensitivity to arguments: How preschoolers weight circular arguments. *Journal of Experimental Child Psychology*, 125, 102-109.
Michaelian, K. (2010). In defence of gullibility: the epistemology of testimony and the psychology of deception detection. *Synthese*, 176(3), 399–427.
Michaelian, K. (2011). Generative memory. *Philosophical psychology*, 24(3): 323–342.
Michaelian, K. (2012a). Is external memory memory? Biological memory and extended mind. *Consciousness and cognition*, 21(3), 1154–1165.
Michaelian, K. (2012b). Metacognition and endorsement. *Mind & Language*, 27(3): 284–307.
Michaelian, K. (2012c). (Social) metacognition and (self-)trust. *Review of philosophy and psychology*, 3(4), 481–514.
Michaelian, K. (2013a). The information effect: Constructive memory, testimony, and epistemic luck. *Synthese*, 190(12), 2429–2456.
Michaelian, K. (2013b). The evolution of testimony: Receiver vigilance, speaker honesty and the reliability of communication. *Episteme*, 10(1), 37–59.
Michaelian, K. & Sutton, J. (2013). Distributed cognition and memory research: History and current directions. *Review of philosophy and psychology*, 4(1), 1–24.

Origgi, G. (2018). *Reputation. What it is and Why It Matters*. Princeton: Princeton University Press.
Pettit, P. (2003). Groups with minds of their own. In F. Schmitt (Ed.), *Socializing Metaphysics* (pp. 167-193). New York: Rowman & Littlefield.
Rupert, R. (2011). Empirical Arguments for Group Minds. A Critical Appraisal. *Philosophy Compass*, 6(9), 630-639.
Sperber, D. (2013). Speakers are honest because hearers are vigilant. Reply to Kourken Michaelian. *Episteme*, 10(1), 61–71.
Sperber, D. et al. (2010). Epistemic Vigilance. *Mind and Language*, 25(4), 359–93.
Sterelny, K. (2003). *Thought in a Hostile World. The Evolution of Human Cognition*. Oxford: Blackwell Publishing.
Szanto, T. (2014). How to Share a Mind. Reconsidering the Group Mind Thesis. *Phenomenology and the Cognitive Sciences*, 13, 99–120.
Theiner, G. (2013). Transactive memory systems: A mechanistic analysis of emergent group memory. *Review of Philosophy and Psychology*, 4, 65–89
Theiner, G., Allen, C. & Goldstone, R. (2010). Recognizing Group Cognition. *Cognitive Systems Research*, 11(4), 378-395.
Tollefsen, D. (2006). From Extended mind to collective mind. *Cognitive Systems Research*, 7(2/3), 140-150.
Tomasello, M. (2014). *A Natural History of Human Thinking*. Cambridge, MA: Harvard University Press.
Trouche, E., Johansson, P., Hall, L. & Mercier, H. (2018). Vigilant conservatism in evaluating communicated information. PLOS ONE | https://doi.org/10.1371/journal.pone.0188825.

ARGUMENTATION, CRITICAL THINKING, LIBERAL DEMOCRACY, AND NON-WESTERN STUDENTS

MENASHE SCHWED
Ashkelon Academic College, Israel
m.schwed@outlook.com

Abstract

The paper starts with an anthropological fact: The designing and executing courses in argumentation and critical thinking presuppose a prototype of a Western cultured student. The drawback is that these courses lack the ability and sensitivity necessary for teaching them to non-Western cultured students. There is a significant lack of resources for designing and executing such courses. The challenge is threefold: First, designing and executing these courses is how to incorporate Western-liberal key concepts and how to do so without undermining non-Westerner students' cultural identity and values. Second, the challenge is to do so without presupposing only the Western culture and lifestyle but to incorporate the students' local ones into courses. Third, the challenge is to do so while addressing the fact that, in many cases, non-Western communities have low socioeconomic status and suffer various civil and political discriminations. The paper proposes a specific methodology on three levels: The epistemic, the critical and socio-political, and the structural levels.

The working hypothesis of this paper and its main focus is that liberal democracy and argumentation share fundamental values. Thus, the goals of these courses are not only to improve the quality of the student's thinking and academic competence but rather to superimpose liberal values. The thesis is that the courses have an effectual role in educating and maintaining liberal democratic values. Academia has a civic duty to cultivate argumentative faculties among non-Western students.

This talk is a preliminary study of developing a proposal for the EU-ERASMUS funds for a model course in argumentation and critical thinking designed for non-Western students in Israel and the Palestinian National Authority.

1. Introduction

It is almost proverbial to say that the courses in argumentation and critical thinking are essential to citizens' competence to participate in a liberal democratic society. The two share fundamental values, and thus, the goals of the courses are not only to improve the quality of the student's thinking and academic competence but rather to superimpose liberal values. The problem in focus here is that the designing and executing of such courses presuppose a prototype of a Western cultured student. The drawback is that these courses lack the ability and sensitivity necessary for teaching them to non-Western cultured students. Are these courses indeed culturally-laden in such a fundamental way that requires attention?

Turning to the visual sometimes helps to have a sense of the problem and a glimpse of its gravity. Books in argumentation and critical thinking will typically choose as their cover images like _Cicero Denounces Catiline_ by the Italian artist Cesare Maccari (Maccari, 1889) or _The Death of Socrates_ by the French painter Jacques-Louis David (David, 1787). They are such natural choices for 'us' as to be in danger of being pathetic. Compare, for instance, three possible visual icons of arguing: (1) _Pericles Gives the Funeral Speech_ by the German painter Philipp von Foltz (Foltz, 1852); (2) _St Paul Preaching at Athens_ by Raphael (Raphael, 1515); and (3) _The Orator_ by the English engraver Arthur Briscoe (Briscoe, 1926). The first two are conservative, while the first will be more secular in the tone of _Cicero Denounces Catiline_, whereas the second more religious and righteous in the tone of _The Death of Socrates_. On the other hand, the third takes a sharp left turn with a clear political and activist tone. There is little chance that non-Western students can understand this last paragraph.

I, as a Jew, will visualize argumentation differently, similar to the painting _Jewish scene 2_ or _At the Rabbi_ by the Austrian Jewish painter Carl Schleicher (Schleicher, 1860s). The paintings deal with a group of serious Jews discussing the complex meaning of some Talmudic passage, which Jews do six days a week after the morning prayer. However, a more accurate depiction of these debates will be _A Controversy from the Talmud_ or _A Talmudic dispute_ (Schleicher, 1870s; 19th Century). In these two paintings, Schleicher portrays the typical heat of Talmudic controversy along with the protagonists' most common attitudes, mainly opposition, rejection, and astonishment. I cannot even start and explain to those who are not Jews what we really see in these scenes. The Jewish tradition of argumentation deeply diverges from the Greco-Roman one, rooted in Aristotle's _Organon_. It is less schematic and rule-following and more associative and hermeneutical and without any reasonable notions of validity and soundness (Frank, 2004). The Jewish is quite a countermodel to the Greco-Roman. David Frank (2003) adds that Talmudic reasoning and argumentation influenced Perelman's *New Rhetoric Project*. What *cultural capital* will enable one to have a deep and rich understanding of the differences between all these images of argumentation and what it means to choose one over the other?

The problem is twofold. On the one hand, the content of the courses presents themes, contexts, examples, and worldviews which are foreign and unfamiliar to non-Western students. Thus, teaching these courses in non-Western communities can be seen as cultural Colonialism, at least metaphorically. The idea is that the courses would achieve their goals better if they were presented in the context with which the students could identify themselves. However, on the other hand, despite the richness of resources for the courses in Western countries, there is hardly any reference to non-Western students. Especially those from religious and conservative communities respond negatively towards critical thinking and liberal values, to begin with. They are still 'alien' concepts threatening their religious or conservative beliefs and socio-cultural values. It is no wonder these courses are not taught in many non-Western academic institutions precisely because they are considered synonymous with secular and liberal Western culture.

Hence the dilemma: On the one hand, courses in argumentation and critical thinking generate, expands, and strengthen the ideals of liberal democracy and help advance its core values in non-Western communities. They can expand the cognitive and motivational abilities of non-Western students to endorse these ideals and values. However, on the other hand, Western modernity has its darker side that hides the experiences of non-Western people. So, how to avoid the possibility that the textbooks maintain the hegemony of Western culture? Might they be seen as a form of Neo-colonialism?

The goal is to suggest changes in their programs, resulting in a different organization and emphasis. Namely, changes centered on non-Western students rather than an uncritical observance of the Western-rooted tradition. In an important sense, the suggestion is to decolonize the textbooks, bound up with acknowledging their Western origin and essence. It is not a call for non-Western argumentation and critical thinking, which supplies an alternative perspective to the Western one, nor even accommodating "new thinking" from non-Western cultures. Decolonizing here means acknowledging diversity and yet advancing core liberal values needed for critical thinking.

2. Three Interrelated Methodological and Pedagogical Levels

Given these incomplete preliminaries, there is a need for a different methodology for argumentation and critical thinking courses more suitable to non-Western students. The suggested methodology is for obvious reasons only in outlines, and it consists of creating three interrelated pedagogical levels for critical learning: *Structural, Epistemic, and Critical and Socio-Political*.

The first level is *The Structural Level*.

The problem in teaching these courses begins with the fact that the common materials in textbooks are mainly foreign to non-Western students and thus have alienating effects on them, which can produce disengagement with the learning process. The common subjects of textbooks' examples can be classified as scientific, social, technological, fictional, and so on, reflecting Western culture and society. Even seemingly neutral themes are tackled from a Western angle.

For instance, in the concluding chapter of his book *Critical Thinking Skills for Dummies* (2015), Martin Cohen analyses ten arguments that, in his opinion, changed the world. These moral, religious, civilian, political, and scientific arguments are all heritage assets of Western culture. Their concepts and ideas, however, are, to a great extent, unfamiliar and foreign to non-Western students. The life of non-Western students and their education are different from those presented and depicted in the textbooks. The Western's highly modernized lifestyle and its underlying concepts and ideas differ from the students' conservative, tribal, and religious lifestyle of the students. The danger is that they conflict with their worldview, particularly Islam and Orthodox Judaism.

The first type is conflicting or alien concepts. The problem begins with simple Western literary terms that are partially or entirely unfamiliar to the students. Common sources of such simple and 'uncharged' expressions are those from Greek, Roman, and European literature and mythology. The problem gets complicated with more problematic ideas, and one of the more known culprits is Darwin's theory of natural evolution. For example, one of the exercises in *Argumentation: Analysis and Evaluation* uses this concept in an amused and ironic way as if rejecting the theory is, in point of fact, ridiculous (van Eemeren & Henkemans, 2017, pp. 126, c). However, orthodox Judaism and Islam reject this theory in all respects and feel repulsed by the idea that *Homo Sapiens* is just another *Big Ape*. Another such idea is 'religion'. For instance, when Stella Cottrell wants to illustrate the idea of creative comparison in her textbook *Critical Thinking Skills*, she uses the image of the moon Goddess (Cottrell, 2017, p. 112). This passing reference in the textbook shows a lot on the author's part but means a lot, especially for the Muslim and Jewish students. The students would automatically relate argumentation with Western religion. Not to mention that the moon has a special religious significance both in Islam and Judaism, and such use might offend them.

Thus, it will become difficult for the instructors to explain expressions like these without previous knowledge of Western religion and mythology. They would have to explain every expression they use in their Western context, knowing that the non-Western students do not have any previous knowledge of it. Take, for example, the moon goddess. An instructor can use some works of art that represent the goddess, beginning in the Roman era. For example, *Endymion and Selene* (Unknown Artists, Imperial Roman AD), *Luna* by Johan Gregor van der Schardt (Schardt, 1570 c), *Selene* by the Italian painter Pietro della Vecchia (Vecchia, 1600-29) or *Endymion and Selene* by the German painter and architect Januarius Zick (Zick, 1770s). However, since these images involve nudity and paganism, it would probably be the last method one should use, as this will only worsen things.

The second type is 'Ignoring the context'. Textbooks ignore the context of their themes and examples for non-Western readers. A quick look at the exercises in van Eemeren and Henkemans' (2017) textbook shows that although the content is varied, their significant themes reflect exclusively Western society and its culture. Already in the exercises for chapter 1, the examples are about an Italian tenor, the New York city lifestyle, a sucked Roman priest on the charge of homosexuality, gay poets, premenstrual syndrome, or sexual abuse in childhood, and the reliability of traumatic memory. All these topics will be obliviously less motivating and less interesting, not to say highly offensive, for non-Western students with a solid religious background. These students are, in most cases, also from a poor social and economic background, and their lives are less affected by innovative technology, modern lifestyle, or Western art.

Overall, using such examples will make the students disinterested and negatively affect their input level. All the examples tell the same story of Western culture and its social setting, representing its way of thinking and outlook on the physical and social world. These are all odd to Jewish and Islamic ways of thinking and social and cultural values. Thus, we get the problem of alienation and de-motivation all over again.

The third type is alienation toward the pictorial material. The problem of alienation continues with the pictures and images appearing in textbooks. They are subject to the same problems as they are Western and foreign to non-Western students. They are seen as strange, if not intimidating, by non-Western students. A frequent and much-expected problem is the representation of the female figure in a Western manner. The first example is from Peg Tittle's Critical Thinking: An Appeal to Reason (Tittle, 2011, p. 2), in which we see a typical office scene of a group of workers of mixed gender in casual clothes. It is supposedly an innocent and straightforward illustration of women in a Western manner and dressed in a Western style.

However, a strict dress code is essential in defining what is considered a respectful and honorable woman in either Muslim or Orthodox Jewish society. The typical image of a woman in textbooks depicts a bareheaded and tightly dressed woman. Such images will be without the Muslim hijab or the not-so-different head covering for Orthodox Jewish women. Even secular and educated women from the upper class in Muslim countries cannot be seen in public dressed as shown in the image to the left. The other issue is that it depicts a mixed-gender group. Islam and Orthodox Judaism forbid free mixing between genders and enforce gender segregation. Peg Tittle's image depicts a woman inappropriately from a conservative point of view. A similar image with all the above difficulties appears in Eileen Gambrill's *Critical Thinking in Clinical Practice* (Gambrill, 2012, p. 7). It depicts a woman lying on a couch in a simple dress with bare shoulders while a hand pours water on her head. The title says: "The Need for Critical Thinking in Clinical Practice - Treatment of Hysteria". However, the main issue here is not just the female image but the reference to Freud's psychoanalytic theory. This theory is only second to the hostility towards the Evolution theory.

On the whole, the analysis of expressions, concepts, ideas, and images in any textbook in argumentation and critical thinking I reviewed reveals, albeit somewhat predictable, that the authors of these texts have their own cultural, historical, and social roots. They include their worldview in the texts without taking into consideration the reaction from those students who do not share their worldview. This negligence on their part affects the effectiveness of the learning process, to say the least. In most cases, it will cause feelings of alienation and resentment in their non-Western students. The adjustment is straightforward and necessary, and it is the adaptation of all those elements to those more familiar to non-Western students. Among other things, a balance is required between adhering to their cultural sensitivities and the demands for gender equality and non-discrimination in general.

The second level is *The Epistemic Level*.

If structural issues are complicated, the epistemic one is the heart of the dilemma. The first and primary difficulty at the beginning of each course for non-Western students is the clash between the skeptical stance of Western philosophy and its strangeness for non-Western cultures. A profound encounter with the full implications of skepticism can have a mortifying effect on non-Western students. Their relatively dogmatic worldview is threatened once they are made aware of the fallibility, weakness, and fragility of human understanding and knowledge. It is what Wittgenstein mentions in his *On Certainty*, §166: "The difficulty is to realize the groundlessness of our believing" (Wittgenstein, 1969).

However, there is no need to adhere to radical skepticism on the pedagogical level, as more to the point is the skeptical argument that although knowledge is possible, certainty and truth are unattainable, more in line with Wittgenstein or Popper. This more limited approach is all we need to connect argumentation and critical thinking with liberal democracy. This pedagogical focus enables the students to keep a sincere critical stance without dismissing or invalidating other knowledge-cultures, including theirs. In this way, the students maintain an authentic autonomous stand regarding their culture.

We can adopt David Hume's skeptical but optimistic approach to facilitate such a move in the context of non-Western classes. The reason is that his account of skepticism remains valuable even to those disinclined to adopt it at face value. Hume proposes a 'mitigated skepticism' that resolves the tension between philosophical skepticism and ordinary life. The central theme is that there is no conflict between these two outlooks or perspectives. Skeptical doubts do not necessarily context-bound with everyday certainties (Hume, 1748, 1777, pp. 161-2).

On the one hand, teaching argumentation and critical thinking cannot avoid stressing the skeptical implications on the epistemic level. One cannot study them without a deep systematic doubt about the reliability of our knowledge, the reasonableness of our inferences, or the objectivity of our conclusions. We cannot escape the conclusion that our ordinary certainties, judgments, and inferences do not deserve the confidence we attribute to them in everyday practices.

However, on the other hand, this clash of outlooks in the classroom is far from unqualified. We can sustain one outlook in the philosophical level of study and another in everyday life in the class. They can be neither integrated nor reconciled, which can be used to rebuild loyalty and confidence among non-Western students toward their culture and beliefs. We, as instructors, oscillate between them without committing to either of them. This strategy is preferable because it allows instructors and authors of textbooks to remain true to themselves and their beliefs while allowing a safe space for non-Western students.

The third level is *The Critical and Socio-Political Level.*

The limitations of the epistemic level on truth and certainty lead us to the second methodological and pedagogical level, which is adopting a critical stance at the socio-political level. It challenges many non-Western discourses of tribal, religious, or conservative conformity and loyalty to one's religion, community, or tribe. Teaching argumentation and critical thinking emphasizes the critical stance as the right choice for the autonomous and free individual living in a liberal democracy. However, it is a counter-hegemonic approach to non-Western cultural and traditional authorities.

The problem is the inherent negative criticism implied by the critical stance on traditional and religious values. In Islam and Orthodox

Judaism, there is a clear hierarchy of authority in which everyone knows their place. Usually, everyone knows to whom they are supposed to listen and obey, and the penalties for not doing so are known and clear. It is a familiar state of things to non-Western students and, indeed, an unfamiliar one for the authors of most textbooks.

Thus, these authors, probably unintentionally, imply an indirect criticism of the society and culture of their non-Western readers. By stating the critical stance as the *Holy Grail* of rationality, the authors clearly state that the student's culture and politics should be condemned and rejected. It also shows how the West conceives and represents authoritarian and religious societies. Non-Western students will perceive this as an implicit attack on their culture.

In contexts where a critical stance plays an indispensable role in social and political life, the course could hold authorities accountable. The sociopolitical level is concerned with studying how power and oppression operate. This approach enables students to understand the place of critical stance within existing societal and political circumstances while recognizing the limits of such an approach. Here, the pedagogical emphasis is on how the best critical stance could restrict authorities' power. Thus, it draws the student's attention to the complex play of authorities in society.

3. Assimilating liberal-democratic ideas

The idea for this paper draws on my experience teaching these courses at a peripheral college in Israel, where my students are mostly traditional and Orthodox Jews, Muslims, and students of low socioeconomic status. The suggestions stem from my experience bridging and adapting the standard learning material to my students. But above all, the goal is to assimilate liberal-democratic ideas through these courses, ideas lacking in these communities.

The goal is to develop their intrinsic motivation by making the course contents relevant to their past, present, and future lives. Non-Western students in Israel come from orthodox and conservative Islamic and Jewish communities, and their past has been spent chiefly in institutions that emphasize religious education and authorities and the conservative way of life. They are also in the present and, in most cases, will be in their future part of religious and conservative communities.

Adapting the courses to their world can prove a good point for bridging the widening gap between non-Western communities and Western liberal-democratic culture. This situation should provide instrumental in creating more intercultural understanding. Thus, the main methodological change to be made is making course content relevant to non-Western students, which can be a positive step towards making the students less hostile or alienated towards liberal-democratic culture. When confidence is developed and the misunderstanding and unfamiliarity are removed, the ground is set for introducing these values to the students. Learning these courses can be a launching pad for introducing and implementing them.

We must acknowledge that the inclusion of Western cultural and liberal ideas and values in course content is inescapable since the two are inseparable. The history of argumentation and critical thinking reflects the liberal and democratic essence of its origin and essence (Schwed, 2013). However, the advent of liberal democracy values through teaching these courses must bear in mind that numerous changes must be made, not only in teaching methodologies but also in teaching material and course contents. It is our academic duty.

It is 2023, and we are fighting for liberal democracy in Israel.

References

Briscoe, A. (1926). *The Orator [Etching]*. Retrieved from The British Museum: https://www.britishmuseum.org/collection/object/P_1926-0731-3

Cohen, M. (2015). *Critical Thinking Skills For Dummies*. Chichester, West Sussex: John Wiley & Sons.

Cottrell, S. (2017). *Critical Thinking Skills: Effective Analysis, Argument and Reflection* (3rd ed.). London: Macmillan International Higher Education.

David, J.-L. (1787). *The Death of Socrates*. Retrieved from Wikimedia Commons: https://en.wikipedia.org/wiki/File:David_-_The_Death_of_Socrates.jpg

Foltz, P. (1852). *Pericles' Funeral Oration [Painting]*. Retrieved from Private Collection. Wikimedia Commons: https://commons.wikimedia.org/w/index.php?title=File:Discurso_funebre_pericles.PNG&oldid=666985460

Frank, D. A. (2003). The Jewish Countermodel: Talmudic Argumentation, the New Rhetoric Project, and the Classical Tradition of Rhetoric. *Journal of Communication and Religion, 26*, 163-194.

Frank, D. A. (2004). Arguing with God, Talmudic Discourse, and the Jewish Countermodel: Implications for the Study of Argumentation. *Argumentation and Advocacy, 41*(2), 71-86.

Gambrill, E. (2012). *Critical Thinking in Clinical Practice: Improving the Quality of Judgments and Decisions* (3rd ed.). Hoboken, NJ: John Wiley & Sons, Inc.

Hume, D. (1748, 1777). *An Enquiry concerning Human Understanding*. (P. Millican, Ed.) Retrieved 2022, from Hume Texts Online: https://davidhume.org/texts/e/

Maccari, C. (1889). *Cicero Denounces Catiline*. Retrieved from Wikimedia Commons: https://commons.wikimedia.org/wiki/File:Cicer%C3%B3n_denuncia_a_Catilina,_por_Cesare_Maccari.jpg

Raphael. (1515). *St Paul Preaching at Athens [Painting]*. Retrieved from Victoria and Albert Museum, London. Wikimedia Commons: https://commons.wikimedia.org/wiki/File:V%26A_-_Raphael,_St_Paul_Preaching_in_Athens_%281515%29.jpg

Schardt, J. (. (1570 c). *Luna*. Retrieved from Web Gallery of Art: https://www.wga.hu/html_m/s/schardt/luna.html

Schleicher, C. (1860s). *At the Rabbi*. Retrieved from Wikipedia Commons: https://es.wikipedia.org/wiki/Archivo:Carl_Schleicher_Beim_Rabbi.jpg

Schleicher, C. (1860s). *Jewish scene 2*. Retrieved from Wikipedia Commons: https://commons.wikimedia.org/wiki/File:Carl_Schleicher_J%C3%BCdische_Szene_2.jpg

Schleicher, C. (1870s). *Eine Streitfrage aus dem Talmud*. Retrieved from Wikimedia Commons: https://en.wikipedia.org/wiki/File:Carl_Schleicher_Eine_Streitfrage_aus_dem_Talmud.jpg

Schleicher, C. (19th Century). *A Talmudic Dispute*. Retrieved from Wikipedia Commons: https://www.mutualart.com/Artwork/Ein-Talmudstreit/B6B340FF55066EC3

Schwed, M. (2013). Argumentation as an ethical and political choice. In D. Mohammed, & M. Lewiński (Eds.), *Virtues of Argumentation* (pp. 1-14). Windsor, ON: Ontario Society for the Study of Argumentation.

Tittle, P. (2011). *Critical Thinking: An Appeal to Reason*. New York: Routledge.

Unknown Artists. (Imperial Roman AD). *ENDYMION & SELENE*. Retrieved from Theoi Project: https://www.theoi.com/Gallery/Z18.2.html

van Eemeren, F., & Henkemans, F. (2017). *Argumentation: Analysis and Evaluation* (2nd ed.). New York and London: Routledge.

Vecchia, P. d. (1600-29). *Selene*. Retrieved from Wikipedia Commons: https://commons.wikimedia.org/wiki/File:Selene_by_Pietro_della_Vecchia.jpg

Wittgenstein, L. (1969). *On Certainty*. New York: Harper and Row.

Zick, J. (1770s). *Endymion and Selene*. Retrieved from Web Gallery of Art: https://www.wga.hu/html_m/z/zick/son/endymion.html

THE METHODOLOGY OF THE NEW RHETORIC (AND WHY IT STILL MATTERS)

BLAKE D. SCOTT
Institute of Philosophy, KU Leuven
blake.scott@kuleuven.be

Abstract

Despite Perelman and Olbrechts-Tyteca's well-known influence on argumentation studies, it is striking that the theory of argumentation they present in *The New Rhetoric* no longer stands out as a living project in the field. On the one hand, there are those who argue that their theory is inherently relativistic and therefore incapable of offering any normative criteria of argument evaluation. On the other, there are those who argue that, even as a descriptive theory, the new rhetoric fails to sufficiently justify its own systematic ambitions. This paper addresses these dual concerns by returning to one of the most neglected, yet most innovative aspects of Perelman and Olbrechts-Tyteca's project — its methodology. First, I outline the two main lines of criticism mentioned above. Next, I draw on some of Olbrechts-Tyteca's writings to clarify the rhetorical methodology and often-puzzling structure of *The New Rhetoric*. Third, I show how this methodology allows Perelman and Olbrechts-Tyteca to theorize argumentation without making any determinate philosophical commitments of their own. Fourth, I exemplify this last point by examining their discussion of "argument strength". Here it can be seen that Perelman and Olbrechts-Tyteca are not interested in providing practical norms for evaluating arguments, but rather with identifying the conditions under which arguments appear strong or weak to audiences in the first place. Constrained by its rhetorical methodology, the theory of argumentation in *The New Rhetoric* thus aims to provide a neutral framework for describing and interpreting the workings of the *already* norm-laden practice of argumentation. To conclude, I explain why Perelman and Olbrechts-Tyteca's theory of argumentation can withstand the criticisms that have been raised against it. More positively, I argue that that, owing to its rhetorical methodology, its rigorous attempt at philosophical neutrality makes it an ideal framework for building bridges between various branches of argumentation studies and the growing number of disciplines which now intersect it.

1. Introduction

Alongside Whately, Hamblin, Toulmin, and others, the names of Perelman and Olbrechts-Tyteca have become synonymous with the emergence of argumentation as a distinct field of study. In spite of this nearly universal recognition, however, it is striking that their theory of argumentation, presented in their 1958 magnum opus, *The New Rhetoric: A Treatise on Argumentation* [hereafter *Treatise*], no longer stands out as a living project in the field. While many of Perelman and Olbrechts-Tyteca's concepts and distinctions remain influential, their claim of providing the groundwork for a *general* or *systematic* theory of argumentation has, at best, been met with firm skepticism.

Why is this the case? The hypothesis of this paper is that the reception of Perelman and Olbrechts-Tyteca's new rhetoric project has been beset by a misunderstanding, specifically a misunderstanding about the nature of their methodology. My wager, therefore, is that a clarification of their rhetorical method will redress two of the main objections raised by skeptics and make the new rhetoric project a more viable and attractive resource for contemporary researchers. The systematic ambitions of the new rhetoric project are important, I would argue, because many of the most pressing challenges to the field cannot be addressed in a piecemeal way and require systematic coherence.

2. Criticisms of the New Rhetoric

Among the many interpretations of Perelman and Olbrechts-Tyteca's theory of argumentation, two broad strains of criticism can be observed. On the one hand, there are those who argue that their theory is inherently relativistic and therefore incapable of offering any normative criteria of argument evaluation. On the other, there are those who argue that, even as a descriptive theory, the new rhetoric fails to sufficiently justify its own systematic ambitions. Let us look briefly at both criticisms.

(1) The 2014 *Handbook of Argumentation Theory* provides a succinct rendition of the relativist objection. Perelman and Olbrechts-Tyteca, we are told, offer a "descriptive rather than a normative theory"; they "equate sound argumentation with effective argumentation"; and adhere to a "purely anthropological conception of rationality" (van Eemeren et al., 2014, p. 262, 266, 290). Approaches as different as pragma-dialectics, informal logic, and epistemic theories of argument, all share in some version of this objection.

(2) As for the unsystematic objection, many have also taken issue with Perelman and Olbrechts-Tyteca's typology of "argumentation schemes", which is considered to be flawed – or, at the very least, incomplete.

Manfred Keinpointner, for instance, laments the lack of "explicitness" in their taxonomy, and that there are no "clear criteria of demarcation" (Keinpointner, 1993, p. 419), while Douglas Walton regrets that, if they do exist, they are unclearly "woven in with the general themes of the book" (Walton, 1996, p. 2). In a similar vein, the authors of the *Handbook of Argumentation Theory* agree that the *Treatise*'s approach to argumentation schemes "lacks clear definition", and that the schemes are "not mutually exclusive".[1]

In what follows, I will argue that neither criticism stands up to a careful reading of the *Treatise*. To be clear, I say this not because the *Treatise* does actually provide these things and the skeptics have simply failed to take notice of them, but rather because the *Treatise* was intended as a first step towards a systematic theory of argumentation, whose general framework and basic concepts needed to be extended and refined in relation to specific domains (take, for example, Perelman's own attempt to extend the project into the domain of law). In other words, these criticisms do not invalidate the new rhetoric's methodology, but only highlight the incompleteness of its application.

Before explaining the new rhetoric's methodology, a few preliminary remarks are needed on how I propose to interpret the *Treatise*.

3. Interpreting the *Treatise*

The best place to begin is with Perelman and Olbrechts-Tyteca's well-known statement of the aim of their study: "the object of the theory of argumentation is the study of the *discursive techniques* allowing us to induce or to increase the mind's adherence to the theses presented for its assent" (Perelman & Olbrechts-Tyteca, [1958] 1969, p. 4, emphasis modified). This definition, which will appear with little variation throughout many of their writings, provides the most secure point of entry for any interpretation. It specifies that the *Treatise* is first and foremost a study of "discursive techniques" by means of which social actors seek to modify the "adherence" of other people, or even of oneself.

However clear the object of the study may be, the question remains as to how their analysis will proceed. Indeed, as Michel Meyer points out in his Preface to the most recent French edition, the problem is that "nowhere in the *Treatise* is the specificity of the New Rhetoric's approach clearly stated" (Perelman & Olbrechts-Tyteca, [1958] 2008, p. VII, trans. mine). Fortunately, however, there are other places we can look.

---

1 According to the authors of the Handbook: "in drawing up the taxonomy, divergent ordering principles have been used" such that "the taxonomy cannot easily be used as a starting point for carrying out empirical research" (van Eemeren et al., 2014, pp. 292-293).

Of particular help is Lucie Olbrechts-Tyteca's 1974 *Le comique du discours* [*The Comedy of Discourse*]. In this text, Olbrechts-Tyteca draws attention to the method adopted in the *Treatise*, describing it as "regressive" as opposed to "genetic".[2] Where a genetic approach would attempt to derive every kind of argument from a set of simple elements, a regressive approach proceeds in a more interpretative manner, starting from "what is given", and then, in a movement of "successive deepening", proceeding to identify underlying structures.

It is with this regressive method that Perelman and Olbrechts-Tyteca approach the texts and documents which serve as the raw material for their study – namely, "arguments put forward by advertisers in newspapers, politicians in speeches, lawyers in pleadings, judges in decisions, and philosophers in treatises" (Perelman & Olbrechts-Tyteca, 1969, p. 10). What a regressive method looks for are the common argumentative structures in these diverse contexts; it begins only with what is given, the total argumentative situation, and looks for recurring patterns of intelligibility at various analytical levels.

4. The Methodology of the *Treatise*

How then should we understand the new rhetoric's methodology? The answer to this question is synonymous with the answer to the question, "Why rhetoric?". The most fundamental structure revealed by Perelman and Olbrechts-Tyteca's regressive approach is the inherent addressivity of argumentation.[3] In other words, the arguer-audience relation is constitutive of argumentation as a social practice. This structure is not only present in spoken discourse, but also in written discourse. More importantly, this structure is also present across domains. Thus, even Perelman and Olbrechts-Tyteca's own philosophical discourse can be analyzed in terms of its addressivity.

Of course, this raises a clear philosophical problem. The claim that even philosophy addresses itself to an audience has an immediate relativizing effect. If this is true, one might ask, how can philosophical discourse

---

2 "Let us recall that the Treatise adopted a kind of regressive approach... It seems to us today that this regressive approach...responded, perhaps without being fully aware of it at the time, to the need felt by all the sciences at a certain moment to adopt a genetic classification – meaning by this, in the broadest sense, a classification in which the different types of objects are in some way derived from one another. Our chapters correspond to a successive deepening, starting from what is given" (Olbrechts-Tyteca, 1974, pp. 22-23, translation and emphasis mine).
3 "Our rapprochement with [rhetoric] seeks to emphasize that fact that all argumentation develops with a view to an audience" (Perelman & Olbrechts-Tyteca, 1969, 5, translation modified).

establish a philosophically neutral framework of argumentation? Perhaps this is possible to imagine in certain sub-domains where there is sufficient methodological agreement. But, as Perelman points out, philosophers – almost by definition – disagree about the value of different kinds of arguments (Perelman, [1961] 2012, pp. 145-146).[4] This fact seems to rule out the possibility of according any privileged status to philosophy as an ideal model – and *mutatis mutandis* to any kind of "elite" audience. There is thus no philosophically neutral framework of argumentation to which philosophers can appeal; there are always multiple philosophical positions that each construe argumentation in their own way.

This observation brings us to the core philosophical problem raised by Perelman and Olbrechts-Tyteca's work: If every philosophical position prescribes a hierarchy among argument types, how is it possible to study argumentation without presupposing a particular philosophical position from the outset? In other words, how is it possible to study argumentation philosophically without making prior ontological and epistemological commitments that would fix in advance what counts as (good) argumentation?

To overcome this problem, I argue that Perelman and Olbrechts-Tyteca engage in a temporary methodological bracketing or suspension of ontological and epistemological assumptions (borrowing from phenomenology, I call this their *"philosophical epoché"*). If a theory of argumentation aims to be applicable to both everyday arguments and arguments in highly specialized domains such as philosophy, such bracketing is a necessary step in the development of a general theory of argumentation not predetermined by any particular philosophical standpoint.

To see this *epoché* at work, consider, for example, what Perelman and Olbrechts-Tyteca call "arguments based on the structure of reality". Though the name might suggest otherwise, the defining feature of this class of arguments has nothing to do with "an objective description of reality, but the manner in which *opinions concerning [reality] are presented*" (Perelman & Olbrechts-Tyteca, 1969, p. 262). In other words, what "really counts" as objective reality is bracketed. As we will see, Perelman and Olbrechts-Tyteca do not think that it should fall to a theory of argumentation to adjudicate what really counts as objective reality; it

---

4 "Each philosopher is situated in a certain line of thought, which has the effect of giving preference to a certain type of argument. For thinkers who follow Aristotle, the argument that takes essences and essential characteristics into account will be considered strong, whereas for Bentham and the utilitarians, the only argument in favour of a course of action will be the good or bad consequences that result from it... Each philosopher could thus characterize themselves by the kind of arguments to which they deny any value. This is a new way of looking at the history of philosophy, which would correlate a thinker's ontology with their methodological precepts, and shed new light on the history of thought" (Perelman, [1961] 2012, pp. 145-146, translation mine).

is only concerned with the way in which objective reality is presented to audiences by means of argumentative techniques.

In a 1958 letter to French philosopher Raymond Ruyer, Perelman makes this methodological move more explicit, clarifying that his aim in the *Treatise* was not to develop any kind of "axiology" of arguments, but a "phenomenology of our judgments" (Perelman, 1958, translation and emphasis mine).[5] Thus, just as the phenomenologist is not primarily concerned with the existence of an object, but *how* it is given, the new rhetorician is not primarily concerned with the objective qualities of an argument, but *how* it intends to act on the pre-existing adherence of an audience.

If such questions are methodologically bracketed, how is it possible to proceed? Surely, one cannot theorize argumentation without making any philosophical commitments at all. Here we need to introduce the *methodological role of the audience*. Without any pre-argumentative reality to rely on, Perelman and Olbrechts-Tyteca use the "rhetorical audience", an arguer's *construct* of their audience, as a foothold for identifying and classifying argumentative techniques.[6] Thus, if they began by bracketing or suspending philosophical presuppositions, their theory now finds its handhold in argumentation's addressive structure. (Borrowing from phenomenology again, I call this their *"rhetorical reduction"*.)

What change in perspective do these two methodological gestures afford the argumentation theorist? While they lose the ability to speak on

---

5 After congratulating Perelman on the publication of the Treatise, Ruyer questions him on his seemingly "anti-realist" conception of values. This is Perelman's response (note that the Treatise was originally published in two volumes): "I assume that you only wrote your letter after reading the first volume of the Treatise. In the second volume, and especially in the study of dissociations, you will see how close the link between value and reality is. When you speak of my anti-realism of values, for me it is simply a question, not of denying the reality of this or that value, but of insisting on the triple status of our judgements in argumentation; sometimes they are presented as an objective statement, the expression of a fact or a truth, sometimes as a pure subjective attitude, and sometimes, finally, as the basis of an agreement which, without being universal, goes beyond the individual... As to what should be included in each of these three categories depends on a determinate philosophical point of view. [...] My aim was therefore less to develop an axiology than a phenomenology of our judgements, essentially taking into account their dynamic aspect linked to action" (Perelman, Letter to R. Ruyer, 12 July 1958). This and many of Perelman's other letters and correspondences can be accessed online via the PALLAS database "Archives Perelman" of the Université libre de Bruxelles (https://perelman.ulb.be/).
6 For clarity, I use the term "rhetorical audience" to refer to Perelman and Olbrechts-Tyteca's concept "auditoire", i.e. an arguer's construction of their audience, and thus to avoid conflating this concept with the concrete individuals or groups that comprise the concrete, empirical audience.

whether the claims expressed in arguments are objectively true, plausible, valid, and so on, it enables them to describe how or in what way arguments seek to modify audience adherence; it enables them to distinguish between different kinds of argument schemes *in terms of an argument's internal reference to an audience*. Perelman and Olbrechts-Tyteca thus requires no external "first principle" for their typology, their regressive rhetorical method is sufficient to avoid the problem posed by the *de facto* pluralism of philosophy.

5. Theorizing Argumentation

In the chapter of my dissertation on which this paper is based, I examine the structure of the *Treatise* as a whole and show how and where these methodological considerations come into play. To summarize this exegetical discussion, I argue that what I have called Perelman and Olbrechts-Tyteca's methodological bracketing and rhetorical reduction are applied in Parts Two and Three of the *Treatise*. Examples of this method at work in these sections include their treatments of "facts", "values", and their notion of rhetorical "presence" – all of which are introduced as "technical" notions in their theory of argumentation and not as substantive philosophical concepts.

However, the most pertinent example of this is their discussion of "argument strength" in §97, which is typically used as evidence in support of the relativist objection. The first thing to note about this discussion is where it appears within the work as a whole. Note that it is *not* included in the more general philosophical discussions of the Introduction or Part One, but well into Part Three's technical examination of argumentative techniques. Here, argument strength is examined exclusively as a phenomenon of "*argument interaction*" alongside other factors such as "convergence", "amplitude", and "order". What is being discussed here is the fact that the phenomenon of argument strength (and weakness) is the result of a complex interaction of elements within an argumentative situation, including the intensity of the audience's adherence to the premises, the associative links between premises, the relevance of the arguments to the particular discussion, as well as other arguments that might be at play in the argumentative situation (Perelman & Olbrechts-Tyteca, 1969, p. 461).

Just as with the notions of "facts", "values", and "presence", then, this analysis of argument strength is clearly *descriptive*. Perelman and Olbrechts-Tyteca are not claiming to provide their readers with practical norms for evaluating arguments. Rather they are seeking to identify the *conditions* for arguments to appear strong or weak to audiences in the first place.

So where does this leave questions of evaluation? Surely a theory of argumentation must have something to say about the evaluation of arguments. To mitigate what seems to be an otherwise gaping hole in their theory, I think a distinction needs to be made between two "stances" or "perspectives" at work in the *Treatise*. As Perelman and Olbrechts-Tyteca do not name them, I will refer to them as the "theoretical stance" and the "practical stance".

Where the *theoretical stance* refers to the attitude we adopt when we attempt – in our everyday lives or in an academic article – to describe and interpret arguments precisely as they are given, the *practical stance* refers to the attitude we adopt as engaged arguers, when we look to evaluate arguments on the basis of prior interpretation. To evaluate an argument – whether as a philosopher, an argumentation theorist, or a private citizen – involves making a number of difficult judgments (or "inferences", depending on one's preferred vocabulary) involving a degree of personal commitment for which one is ultimately responsible. This is because the assessment of the strengths and weaknesses of arguments inescapably involves the mobilization of our own personal configuration of values, preferences, and sense of reality in relation to what has been argued and by whom the argument was made.

On my interpretation, the *Treatise* primarily theorizes the *theoretical stance*. The practical stance is left open for further elaboration, as the normative question of what strong arguments *ought* to look like will receive difference answers according to the context or domain. The theory of argumentation presented in the *Treatise* is not yet in a position to prescribe what makes arguments strong or weak. At this point, it simply (but admirably) aspires to provide a neutral framework for describing and interpreting the workings of the *already* norm-laden practice of argumentation.

6. Conclusion

To conclude, let us return briefly to the two strands of criticism with which we began.

To the charge of relativism, we can now say that, in a certain sense, Perelman and Olbrechts-Tyteca's theory of argumentation is relativistic. What critics overlook, however, is that this is largely by design: at this stage, an "axiology" of argument, as Perelman calls it, would violate the methodological limits of the *Treatise*; it would involve an unwarranted shift between the theoretical and practical stances we have identified. This worry of relativism should thus not distract us from what it does accomplish: namely, providing a robust vocabulary for describing and interpreting how arguers, audiences – and the arguments between them –

interact at various analytical levels. Far from excluding the normative dimension, this vocabulary equally applies to the process of argument evaluation, which is itself a kind of argumentation – arguments about arguments – an unexploited aspect of the new rhetoric that has yet to receive thorough treatment.

How about the charge of being unsystematic? While it is true that the skeptic will find no foundational "first principles" beneath Perelman & Olbrechts-Tyteca's typology of argument schemes, the systematicity of their project is to be found elsewhere: in its *regressive, rhetorical method*. By focusing on an argument's internal reference to an audience, the *Treatise* proposes a novel, and largely unexplored solution to the philosophical problem posed by taxonomies of argument schemes. In a recent paper that touches on this question, Christopher Tindale, for instance, draws on Perelman and Olbrechts-Tyteca in suggesting that argument schemes might be better classified "in terms of the type of audience they address" – a proposal which dovetails with the interpretation of the new rhetoric I have offered here (Tindale, 2020, p. 266). While such an approach may not provide the same satisfaction promised by a complete and final taxonomy, it has the virtue of being methodologically systematic in the sense that it leaves the door open for extension and revision by researchers of different philosophical persuasions and disciplinary methodologies as Perelman and Olbrechts-Tyteca had intended.[7]

Allow me to conclude with one final remark on the nature of the misunderstanding I detect in the reception of the new rhetoric project. For the Belgians, the purpose of a theory of argumentation is not only, or even primarily, about evaluating arguments and coming to rational agreements. Rather, they regard argumentative exchanges first and foremost as privileged sites of social scientific study, as it is here, on full display, that human beings come to negotiate social meaning. Argumentation, the object of their theory, is thus regarded as a *microcosm of society*, whose study requires careful and patient interpretation.

For this reason, I submit that Perelman and Olbrechts-Tyteca's new rhetoric should not simply be regarded as an exhibit in the museum of argumentation history. While incomplete in its application, its novel rhetorical methodology is capable of withstanding the criticisms that have been raised against it. And, more positively, its rigorous attempt at philosophical neutrality makes it an ideal framework for building bridges between various branches of argumentation studies and the growing number of disciplines which now intersect it.

---

7 Olbrechts-Tyteca even insists that the new rhetoric project should be extended (and revised) by being applied to non-western contexts and other historical periods (Olbrechts-Tyteca, 1979, p. 98).

References

van Eemeren, F.H. et al., eds. (2014). *Handbook of Argumentation Theory.* Dordrecht: Springer.

Keinpointner, M. (1993). The Empirical Relevance of Perelman's New Rhetoric, *Argumentation, 7*, 419-437.

Olbrechts-Tyteca, L. (1974). *Le comique du discours.* Bruxelles : Éditions de l'Université de Bruxelles.

Olbrechts-Tyteca, L. (1979). Les couples philosophiques. Une nouvelle approche, *Revue Internationale de Philosophie, 33(127/8),* 81-98.

Perelman, C. (1958). Letter to R. Ruyer, 12 July 1958. PALLAS database, Archives Perelman, Université libre de Bruxelles. https://perelman.ulb.be/.

Perelman, C. [1961] (2012). L'idéal de rationalité et la règle de justice. In *Éthique et droit*, 2ᵉ éd. (pp. 126-185). Bruxelles : Éditions de l'Université de Bruxelles.

Perelman, C. & Olbrechts-Tyteca, L. [1958] (1969). *The New Rhetoric: A Treatise on Argumentation*, trans. John Wilkinson and Purcell Weaver. Notre Dame: University of Notre Dame Press.

Perelman, C. & Olbrechts-Tyteca, L. [1958] (2008). *Traité de l'argumentation. La nouvelle rhétorique*, 6ᵉ éd. Bruxelles : Éditions de l'Université de Bruxelles.

Tindale, C.W. (2020). Argumentation Schemes and Audiences: What Rhetoric Can Bring to Scheme Theory. In J.A. Blair & C.W. Tindale (Eds.), *Rigour and Reason: Essays in Honour of Hans Vilhelm Hansen* (pp. 252-274). Windsor: Windsor Studies in Argumentation.

Walton, D. (1996). *Argumentation Schemes for Presumptive Reasoning.* New York: Routledge.

A Proof of Concept on Dialogue Games for Explainable Artificial Intelligence

ILIA STEPIN
Centro Singular de Investigación en Tecnoloxías Intelixentes (CiTIUS), Departmento de Electrónica e Computación, Universidade de Santiago de Compostela, Spain
ilia.stepin@usc.es

ALEJANDRO CATALA
Centro Singular de Investigación en Tecnoloxías Intelixentes (CiTIUS), Departmento de Electrónica e Computación, Universidade de Santiago de Compostela, Spain
alejandro.catala@usc.es

JOSE M. ALONSO-MORAL
Centro Singular de Investigación en Tecnoloxías Intelixentes (CiTIUS), Departmento de Electrónica e Computación, Universidade de Santiago de Compostela, Spain
josemaria.alonso.moral@usc.es

Abstract

Recent years have witnessed a groundbreaking number of accurate artificial intelligence-based algorithms. However, their oftentimes obscure nature is known to prevent end users from a safe and responsible use or leads to decreased trustworthiness in their predictions or decisions. In this work, we present a novel dialogue game that serves as an explanatory dialogue model to communicate contrastive, selected, and social explanations from an interpretable rule-based classifier to an end user. In addition, we show how it can address the problem of diversity for such explanations via an empirical human evaluation study.

1. Introduction

Explanations are essential for understanding behaviour of complex Artificial Intelligence (AI)-based decision-making systems and placing trust in them. Ribeiro et al. (2016) state that "if the users do not trust a model or a prediction, they will not use it". It is therefore of crucial importance to enhance their output with automatically generated explanations justifying their reasoning in an efficient and comprehensive manner. Further, eXplainable AI (XAI) is considered to be "essential for users to understand, appropriately trust, and effectively manage this emerging generation of artificially intelligent partners" (Gunning, 2021).

Adadi and Berrada (2018) distinguish four main reasons why state-of-the-art AI-based decision-making systems need explanations. First, explanations justify system's decisions. Second, explanations enable developers to detect errors and debug AI-based systems. Third, explanations make it easier to understand the overall behaviour of the system and therefore facilitate the process of continuous improvement of software in accordance with users' suggestions. Finally, explanations allow the explainee (i.e., the end user) to learn and/or infer new facts about the system's behaviour and therefore gain further knowledge about it.

However, only a limited number of explanation generation algorithms are able to efficiently communicate them to end users, partly due to "the lack of a well-defined protocol for evaluating interactive explanations and the challenging process of assessing their quality and effectiveness" (Sokol et al., 2020). In this work, we propose a transparent argumentative explanatory dialogue model that governs bi-directional interaction between an AI-based agent (hereinafter, the system) and the person responsible for making a decision based on the system's decision guided with the corresponding pieces of explanation (hereinafter, the user). We frame the overall process of explanation communication as a dialogue game between the system and the user. The corresponding dialogue protocol includes turntaking rules, a set of requests and replies for the user and the system, respectively, and a set of transition rules that allow for dialogue state switching to ensure effective communication of automated explanations. In order to assess the utility of the proposed explanatory dialogue model, we carry out a human evaluation study. As a result, we show that explainees find the proposed dialogue model satisfying, specifically, in information-seeking dialogue settings. Further, they particularly appreciate the ability to inquire alternative explanations in the course of iterative explanatory dialogue.

2. Explanation and argumentation

Miller (2019) claims that efficient explanations in the context of XAI are contrastive, selected, and social. The property of contrastiveness implies that the explainee expects to receive a justification for the given decision in terms of hypothetical, non-occurring alternative decisions (e.g., "Why P [instead of Q [and/or $Q_1, Q_2, ..., Q_n$)]]?" where P is the fact (i.e., the actually produced decision to be explained) and Q, Q_1, Q_2, ..., Q_n are foils (i.e., non-made decisions)). In addition, only the most relevant factors leading to the given decision should be included in the explanation, making it selected. Last but not least, explanation is regarded as a social process, i.e., information exchange between the explainer and explainee.

In this section, we exemplify the aforementioned properties of explanation and briefly discuss how argumentation can inherently link them and therefore increase their efficiency.

2.1. Contrastive explanation

Intuitively, an automated explanation for a decision-making system's output justifies the system's decision in terms of the reasons (causes) that led to it. Let us, for example, consider a loan application for a bank client whose monthly income is 1250 euros where the system makes a negative decision upon inspecting the applicant's profile. Example 1 illustrates a piece of explanation justifying the system's decision.

Example 1. *Your loan application is rejected because your monthly income ranges between 1200 and 1500 euros.*

As the explanation from Example 1 indicates only those reasons that motivate the actually made decision (i.e., the fact), we refer to such an explanation as *factual*.

Alternatively, the same decision can be explained contrastively in terms of (one or more) hypothetical non-made decisions (or foils, following the terminology introduced by Lipton (1990)). In the context of XAI, the property of contrastiveness is often captured by means of the so-called *counterfactual* explanations (Stepin et al., 2021). Contrarily to factual explanations, counterfactual explanations oppose two distinct decisions, the fact and a/the foil, while maximising their relevance to the explanandum. Furthermore, Example 2 shows that counterfactual explanations do not only further explain the fact but also offer recommendations on how the system's decision can be changed in favour of the desired alternative.

Example 2. *Your loan application would be approved if your monthly income ranged from 2000 to 3000 euros and if you had less than two active loans.*

2.2. Selected explanation

Automated explanations should offer explainees only a limited number of causes or reasons that led to the decision made (Miller, 2019). Further, all such reasons should be of predominant importance w.r.t. the decision under consideration when it is explained either factually or counterfactually. The number of reasons justifying the automatic decision as well as their contents (e.g., "monthly income is less than 1500 euros") determine the relevance of such an explanation to the decision in question. Indeed, the explanation from Example 1 may sufficiently yet more accurately and concisely motivate the loan rejection decision than, e.g., that from Example 3 for the same client irrespective of how many active loans he or she has got.

Example 3. *Your loan application is rejected because your monthly income is less than 2000 euros and you have more than two active loans.*

Remarkably, the explanations for the system's decision can include imprecise information, i.e., vague linguistic quantifiers (Stepin et al., 2022). Whereas expert users of AI-based systems may require finely-granulated explanations for an automated decision, a wider non-expert audience may be satisfied with those containing (possibly, in part) imprecise information. Example 4 shows how the factual explanation from Example 1 can be approximated with a piece of imprecise information.

Example 4. *Your loan application is rejected because your monthly income is too low.*

Similarly, Example 5 shows how the counterfactual explanation from Example 2 can be simplified using imprecise linguistic quantifiers.

Example 5. *Your loan application would be approved if your monthly income were high and if you had few active loans.*

Examples 1-5 include illustrative rule-based explanations. In such explanations, the consequent represents the system's (possibly, hypothetical) decision that is explained by means of the explanatory features found in the antecedent of a given rule. In this regard, the antecedent-consequent schema of the exemplified explanations may be deemed equivalent to the premise-conclusion schema commonly found in argumentation theories. Further, this representation of rule-based explanations allows us to treat them as arguments following Hempel's theory of explanation, which defines explanation as "an argument to the effect that the phenomenon to be explained, the explanandum phenomenon, was to be expected in virtue of certain explanatory facts" (Hempel, 1965).

In the remainder of this work, we fuse the notions of explanation and argument following Hempel's theory of explanation. In addition, we assume that we have got access to contrastive(-counterfactual) selected explanations generated automatically to explain an arbitrary decision of an interpretable rule-based AI system.

2.3. Social explanation

In general, any phenomenon can be explained in numerous different ways. In the context of XAI, various distinct explanations can be offered for the same (possibly, non-made) decision. Altogether, sets of explanations for all possible (including those non-made) decisions are said to make up an explanation space.

As shown above, distinct explanations may have different degrees of relevance to the phenomenon being explained (see Examples 1 and 3). In his conversational model of explanation, Lipton (1990) stresses the importance of estimating relevance when evaluating explanations. Indeed, (some of) the most relevant explanations (among those generated by the system) may not be found relevant enough by the end user. Instead, the user may be offered the opportunity to examine the explanation space iteratively, i.e., during his or her interaction with the system. To do so, both parties can be engaged in an argumentative explanatory dialogue where the user explicitly or implicitly evaluates explanations presented by the system.

In such an explanatory dialogue, the system's decision can be considered a claim supported by a rule-based explanation whose features from the antecedent serve as premises for that claim. Subsequent system's responses can then be treated as arguments in favour of its own claim while user's requests attack it. Furthermore, every piece of explanation offered to the end user may be regarded as an argument attacked, as the user rejects the previously offered explanations while exploring the explanation space.

3. Explanatory dialogue game

In this work, we address the social aspect of explanation by modelling explanatory dialogue in form of a dialogue game (Prakken, 2005) between an AI-based system and its user. In this section, we outline essential rules that constitute the dialogue protocol that is proposed to govern interaction. The AI-based system is assumed to be capable of generating high-level textual factual (see Example 4) or counterfactual (see Example 5) explanations or their low-level counterparts (see Examples 1 and 2, respectively). On the one hand, low-level explanations are pieces of aggregated information about the decision made in terms of numerical intervals. On the other hand, high-level explanations are produced after applying the so-called "linguistic approximation" to the low-level explanations (Stepin et al., 2022).

3.1. Turntaking

An explanatory dialogue between the system and the user is said to start taking place when the system makes a decision (i.e., a claim) and communicates it to the user. Therefore, the first dialogue move (m_1) is always made by the system. Each dialogue participant is allowed to make only one move at a time. The dialogue presupposes a sequence of request-response pairs by the user and the system. Only the user is allowed to send queries to the system whereas only the system is authorised to respond to them. Therefore, every even dialogue move (m_2, m_4, ...) is made by the user. Complementarily, every subsequent odd move (m_3, m_5, ...) is made by the system. The dialogue terminates when the user makes an informed decision w.r.t. the system's claim – whether it should be accepted or rejected. Further, the user is allowed to end the dialogue at any time. As a result, an explanatory dialogue is said to contain three main building blocks – claim, explanation, and termination – explanation constituting the principal element of the dialogue.

3.2. Requests and responses

Given poor explanatory capacities of many of the state-of-the-art AI algorithms, it becomes essential to find the balance between the information that the system can exploit when explaining its decisions ("supply") and the anticipated user's requests ("demand"). Provided that the system is equipped with an explanation generation module that is able to offer textual rule-based (factual and counterfactual) explanations upon request, it has to be able to explain not only its own decision but also all the components that constitute its explanations (i.e., the features and the corresponding values found in the antecedent of the related rule).

Following the taxonomy of dialogue moves for earnings conference calls that was introduced by Budzynska et al. (2014, p. 22), we address this challenging problem by proposing the following explanation-related user's requests:

1) *The requests of explanation*: these include general requests for factual (e.g., "Why is my loan application rejected?", see Example 1) and counterfactual explanations (e.g., "What can I do to have my loan application approved?", see Example 2);

2) *The request of clarification*: this request is sent when a definition of a feature is needed (e.g., "What do you mean by *monthly income*?" (see Example 4));

3) *The request of detailisation*: this request aims to specify a feature value if it contains an imprecise quantifier, allowing to switch from a high-level explanation to its low-level equivalent (e.g., "Could you specify how low is *too low*?" (see Example 4));

4) *The request of alternative explanation*: this request allows the user to further explore the explanation space with the aim of finding the piece of explanation that is the most relevant to his or her needs (e.g., "Could you offer me another (counter-)factual explanation?") by rejecting or attacking the pieces of explanation that are found unsatisfactory.

In summary, the set of user's requests includes the request of factual explanation ("why-explain"), the request of counterfactual explanation ("why-not-explain"), the request of clarification ("what-is"), the request of detailisation ("what-details"), the request of an alternative factual explanation ("why-alternative"), and the request of an alternative counterfactual explanation ("why-not-alternative").

Possible system's responses mirror all the user's requests. For each type of requests, the system provides either of the two types of responses: *positive* if the system is able to adequately respond to user's request (i.e., to generate a piece of (alternative) explanation, retrieve the corresponding numerical intervals for the given feature, or provide a definition for it) or *negative* – otherwise. The set of system's responses is therefore said to include the following items: the positive factual explanation response ("explain-f"), the negative factual explanation response ("no-explain-f"), the positive counterfactual explanation response ("explain-cf"), the negative counterfactual explanation response ("no-explain-cf"), the positive clarification response ("clarify"), the negative clarification response ("no-clarify"), the positive detailisation response ("elaborate"), the negative detailisation response ("no-elaborate"), the positive alternative factual explanation response ("alter-f"), the negative alternative factual explanation response ("no-alter-f"), the positive alternative counterfactual explanation response ("alter-cf"), and the negative alternative counterfactual explanation response ("no-alter-cf").

All in all, the proposed requests and responses are considered to sufficiently communicate all the components of rule-based explanations. Further, the alternative explanation requests enable the user to interactively evaluate explanations offered to him or her during iterative exploration of the available explanation space.

3.3. Dialogue state transitions

While the user may desire to have explanations for a (non-)made automatic decision, he or she is by no means obliged to request and subsequently receive any. Furthermore, recent worldwide legal regulations concerning AI (e.g., the European Union's General Data Protection Regulation – GDPR – or the AI Act) aim to ensure the user to have the "right to explanation" (Wachter, 2018, p. 860). Therefore, the explanation-related dialogue block is made optional in all cases.

Given the turntaking rules defined in Section 3.1, the initial dialogue move is made by the system. Recall that the system formulates the claim in favour of its decision at this stage. The user is subsequently allowed to accept the claim, reject it, or express his or her doubts over it by requesting a factual explanation. Upon receiving a factual explanation, the user is allowed to demonstrate disagreement by asking for (an) alternative factual explanation(-s) or inspecting counterfactual explanations for other (non-made) decisions. Importantly, processing the given piece of explanation is considered finalised when a request for explanation of a different (possibly, non-made) decision is made.

Table I presents all possible dialogue state transitions. The dialogue proceeds from user's request to system's response (a move from the left-most to the central column). Subsequently, the dialogue is processed in a loop. The user can choose a follow-up request (a move from the central to the right-most column). Once selected (a move from the right-most to the left-most column), the request is processed by the system following the same scenario (a move from the left-most to the central column). The dialogue continues until the termination state (i.e., accept or reject) is reached.

Table I. User's requests, possible system's responses, and user's follow-up requests.

User's request	Possible system's response	Possible user's follow-up request
why-explain	explain-f	why-not-explain, what-details, what-is, why-alternative, accept, reject
	no-explain-f	why-not-explain, accept, reject
why-not-explain	explain-cf	why-not-explain, what-details, what-is, why-not-alternative, accept, reject
	no-explain-cf	why-not-explain, why-alternative, accept, reject
what-details	elaborate	what-details, what-is, why-not-explain, why-alternative, why-not-alternative, accept, reject
	no-elaborate	
what-is	clarify	
	no-clarify	
why-alternative	alter-f	what-details, why-not-explain, what-is, why-alternative, accept, reject
	no-alter-f	what-details, why-not-explain, what-is, accept, reject
why-not-alternative	alter-cf	what-details, why-not-explain, what-is, why-not-alternative, accept, reject
	no-alter-cf	what-details, why-not-explain, what-is, accept, reject

Table II provides the reader with a generic template-based example of an explanatory dialogue modelled following the principles described above.

Table II. An illustrative example of the modelled explanatory dialogue.

Move	Agent	Utterance
m_1	System	The test instance is of class y_1.
m_2	User	Could you explain why you think so?
m_3	System	It is of class y_1 because <feature$_1$> is <value$_1$>.
m_4	User	What do you mean by <feature$_1$>?
m_5	System	<feature$_1$> is <definition for feature$_1$>.
m_6	User	But why is the test instance not of class y_2?
m_7	System	It would be of class y_2 if <feature$_1$> were <value$_2$> and <feature$_2$> were <value$_3$>.
m_8	User	Could you specify how <feature$_1$> is defined?
m_9	System	<feature$_1$> is defined to be <value$_2$> because it is found in the interval <[value$_{2min}$, value$_{2max}$]>.
m_{10}	User	But why is the test instance not of class y_3?
m_{11}	System	It would be of class y_3 if <feature$_1$> were <value$_3$> and <feature$_3$> were <value$_4$>.
m_{12}	User	I am not quite satisfied with your explanation. Could you offer me another one?
m_{13}	System	Sure! It would be of class y_3 if <feature$_1$> were <value$_5$> and <feature$_4$> were <value$_2$>.
m_{14}	User	Ok, now I trust your decision.

4. Insights from the human evaluation study

4.1. Experimental settings

In order to validate the utility of the proposed dialogue model, we carried out a human evaluation study. The participants were first to choose one of the three given use cases: basketball player position classification[1], beer style classification[2], or thyroid disease diagnosis[3]. In all the cases, the study participants were given numerical characteristics of an object following the selected use case (i.e., those of a basketball player, an instance of beer, or hormone levels of a hypothyroid patient, respectively). The details of the objects are outlined in Table III. The system's decision determined the position of the basketball player, the style of the beer instance, or the type of the thyroid disease that the patient had (if any).

---

[1] https://gitlab.citius.usc.es/jose.alonso/basketballplayers-dataset
[2] https://dx.doi.org/10.13140/RG.2.2.20313.67680
[3] https://doi.org/10.24432/C5D010

Table III. The characteristics of the objects upon which the system made a decision (i.e., classification).

Use case	Characteristics of the object	System's decision	Alternative decisions
Basketball player position	Height = 1.85; minutes played = 21.19; points scored = 9.2; two-points field goals percentage = 43.1; three-points field goals percentage = 40.0; free throws percentage = 81.9; rebounds = 1.9; assists = 3.8; blocks = 0.0; turnovers = 0.7; global assessment = 8.8	Point-guard	Shooting-guard, Small-forward, Power-forward, Center
Beer style	Colour = 2; bitterness = 18; strength = 0.049	Blanche	Lager, Pilsner, IPA, Barleywine, Stout, Porter, Belgian Strong Ale
Thyroid disease	Thyroid-stimulating hormone (TSH) = 4.6; triiodothyronine (T3) = 1.2; total thyroxine (TT4) = 48; thyroxine utilisation rate (T4U) = 0.89; free thyroxine index (FTI) = 54	Secondary hypothyroid	No hypothyroid, Primary hypothyroid, Compensated hypothyroid

The study participants were placed in the information-seeking dialogue settings (following the dialogue type classification by Walton and Krabbe (1995)). They interacted with the system without having any prior knowledge about the system or the dataset used. Instead, they were only given the object's characteristics outlined in Table III and the system's decision. The participants interacted with the system until they could make an informed decision on whether the system's decision was credible enough. Notably, the system's decision was correct in all the cases. However, the study participants had not been informed about it prior to their interaction with the implemented dialogue system.

Decision trees were used as classifiers representing the AI-based system, with branches of trees translated into IF-THEN rules for the purpose of explanation generation. Multiple counterfactual explanations were generated using the XOR algorithm for counterfactual explanation generation for rule-based classifiers (ibid). Only continuous features were used for training the system in all the use cases.

Overall, we collected 60 explanatory dialogue transcripts: 14 participants selected the basketball player position use case, 37 – the beer style classification scenario, 9 – the thyroid disease diagnosis scenario. All the collected data were anonymous and obtained strictly after receiving an explicit informed consent from the study participants.

4.2. Request utility

All the proposed requests were extensively used by the study participants in all the use cases. Table IV shows the absolute numbers of all the types of requests made by the users as well the relative numbers to the overall number of explanation-related requests in all the use cases.

Table IV. Numbers of explanation-related requests submitted to the system by the study participants.

User's request	Use case		
	Basketball player position	Beer style	Thyroid disease diagnosis
Factual explanation	12 (17.91%)	36 (15.32%)	9 (31.04%)
Counterfactual explanation	18 (26.87%)	50 (21.28%)	8 (27.59%)
Detailisation	15 (22.39%)	78 (33.19%)	6 (20.69%)
Clarification	13 (19.40%)	46 (19.57%)	3 (10.34%)
Alternative counterfactual explanation	9 (13.43%)	25 (10.64%)	3 (10.34%)
In total	**67 (100%)**	**235 (100%)**	**29 (100%)**

As it can be observed from Table IV, a majority of the study participants requested (at least, factual) explanations to the system's decision. Furthermore, most of such study participants accepted the system's decision at the end of their interaction with the system (91.67% in the basketball player and the beer style scenarios, 55.56% in the thyroid disease scenario).

Remarkably, responses to all the types of requests are found to trigger users' final decisions. In the case of the basketball player position scenario, 38.46% of the participants accepted the system's decision once an alternative counterfactual explanation was offered to them. In the case of the beer style scenario, 29.41% of the participants accepted the decision when high-level counterfactual explanations as well as low-level explanations of either kind were presented to them. Finally, 40.00% of the study participants accepted the system's decision once their clarification requests were responded to in the thyroid disease scenario.

4.3. Free-form user feedback

Upon completion of the experimental task, the study participants were asked to optionally leave free-text responses to the following questions and/or suggestions: (Q1) "If you could add other types of requests to the system, what would those be?"; (Q2) "Did the interaction with the system change your initial (dis-)belief in the system's prediction? Why (not)?"; (Q3) "If you have any other comments for us, please leave them in the textbox below". Below we summarise the most informative comments.

As for Q1, the study participants would like to extend the actual dialogue model so that it could further inform them about the domain knowledge available to the system as well as the technical details on how the classification was obtained. In addition, some users wished to have access to the previously processed explanations.

As for Q2, a number of commentators expressed their satisfaction with the offered explanations. Further, the automated explanations were largely deemed convincing w.r.t. the system's claim. In addition, some study participants positively assessed the ability to query the system for counterfactual explanations and further details and clarifications. Finally, the study participants positively commented on the ability to request counterfactual explanations for hypothetical system's decisions.

As for Q3, some study participants pointed to the following limitations of the proposed explanatory dialogue model. First, the system's responses appeared unnaturally fast. Second, a use of visualisation tools was desired to support the textual explanations. Finally, the difference between the detailisation and clarification requests seemed rather unclear to one study participant.

5. Conclusion

In this work, we designed a dialogue game that serves the task of communicating automatically generated rule-based explanations for an AI-based decision-maker. We showed that the proposed rule-based dialogue protocol represents a transparent mechanism of factual and counterfactual explanation communication.

The designed dialogue game has multiple potential applications. Due to the modularity of the protocol, it is flexible enough to be adapted for running experiments on the trustworthiness, satisfaction, and/or persuasive capability of automatically generated explanations. In addition, it can be used as a benchmark for evaluating the effectiveness of automatic rule-based explanation generation algorithms, as it operates on the full explanation space.

Both quantitative and qualitative results of the human evaluation study carried out to validate the dialogue model confirm the necessity in all the proposed requests and responses for explanatory information-seeking human-machine dialogue. In addition, user free-form feedback shows that the proposed dialogue model is, in principle, found to be an efficient explanation communication tool to support collaborative human-machine decision-making.

The present dialogue game-based framework opens the door for several prospective lines of research. First, the dialogue protocol needs to be enhanced with the capability to handle explanations making use of non-numerical (e.g., categorical) features that cannot be quantified to ensure the protocol's operability for any decision-making scenario. This implies necessary changes in the set of dialogue state transitions, as detailisation requests may need to be made unavailable for categorical features whose meaningful numerical interpretations are unavailable (e.g., gender). In addition, further experiments are necessary with classifiers whose feature space is poorly interpretable (if at all). Second, the explanatory dialogue settings as well as user's expectations may impose the requirement of tackling visual or multi-modal explanations. It therefore appears important to integrate mechanisms of communicating textual explanations with those of non-textual nature. Finally, further human evaluation experiments need to be designed to validate the aforementioned dialogue protocol improvements.

Acknowledgements

Ilia Stepin is an *FPI* researcher (grant PRE2019-090153). This work was supported by the Spanish Ministry of Science and Innovation (grants PID2021-123152OB-C21, and TED2021-130295B-C33) and the Galician Ministry of Culture, Education, Professional Training and University (grants ED431G2019/04 and ED431C2022/19). All the grants were co-funded by the European Regional Development Fund (ERDF/FEDER program).

References

Adadi, A., & Berrada M. (2018). Peeking Inside the Black-Box: A Survey on Explainable Artificial Intelligence (XAI). *IEEE Access*, 6, 52138-52160. https://doi.org/10.1109/ACCESS.2018.2870052

Budzynska, K., Rossi, A., & Yaskorska, O. (2014). Financial dialogue games: A protocol for earnings conference calls. *Proceedings of the Conference on Computational Models of Argument*. IOS Press, 19-30. https://doi.org/10.3233/978-1-61499-436-7-19

Gunning, D., Vorm, E., Wang, J. Y., & Turek, M. (2021). DARPA's explainable AI (XAI) program: A retrospective. *Applied AI Letters*, 2(4):e61. https://doi.org/10.1002/ail2.61

Hempel, C. G. (1965). *Aspects of Scientific Explanation and other Essays in the Philosophy of Science*. New York: Free Press.

Lipton, P. (1990). *Contrastive explanation*. Royal Institute of Philosophy Supplement, 27, 247-366. https://doi.org/10.1017/S1358246100005130

Miller, T. (2019). Explanation in Artificial Intelligence: Insights from the Social Sciences. *Artificial Intelligence*, 267, 1-38. https://doi.org/10.1016/j.artint.2018.07.007

Prakken, H. (2005). Coherence and Flexibility in Dialogue Games for Argumentation. Journal of Logic and Computation, 15(6), 1009-1040. https://doi.org/10.1093/logcom/exi046

Ribeiro, M. T., Singh, S., & Guestrin C. (2016). "Why Should I Trust you?": Explaining the Predictions of Any Classifier. *Proceedings of the International Conference on Knowledge Discovery and Data Mining*, 1135-1144. https://doi.org/10.1145/2939672.2939778

Sokol, K. & Flash, P. (2020). One Explanation Does Not Fit All: The Promise of Interactive Explanations for Machine Learning Transparency. *KI – Künstliche Intelligenz, 34,* 235-250. https://dx.doi.org/10.1007/s13218-020-00637-y

Stepin, I., Alonso, J.M., Catala, A., & Pereira-Fariña, M. (2021). A Survey of Contrastive and Counterfactual Explanation Generation Methods for Explainable Artificial Intelligence. *IEEE Access, 9,* 11974-12001. https://doi.org/10.1109/ACCESS.2021.3051315

Stepin, I., Alonso, J.M., Catala, A., & Pereira-Fariña, M. (2022). An empirical study on how humans appreciate automated counterfactual explanations which embrace imprecise information. *Information Sciences, 618,* 379-399. https://doi.org/10.1016/j.ins.2022.10.098

Wachter, S., Mittelstadt, B., & Russell, C. (2018). Counterfactual Explanations without Opening the Black Box: Automated Decisions and the GDPR. *Harvard Journal of Law & Technology, 31(2),* 841-887. https://dx.doi.org/10.2139/ssrn.3063289

Walton, D., & Krabbe, E. C. (1995). *Commitment in dialogue: Basic concepts of interpersonal reasoning*. New York: SUNY Press.

FORENSIC ARGUMENTATION IN EDUCATION

SERENA TOMASI
University of Trento
serena.tomasi_1@unitn.it

Abstract

In this paper, argumentation, education and law are recomposed in a complementary way: the aim is to show not only that argumentation is a component of the practice of teaching / learning in legal education, but that legal argumentation can serve as a model for democratic education of the citizen. In particular, I will first show the intersection between the judicial dimension and society, through an analysis of the judicial procedure in its primitive forms that appeared in Archaic Greece (*neikos*). The examples of judgments drawn from classical Greek literature are re-examined in the argumentative framework and are functional to delineate the archetype of forensic rhetoric. In the following sections, I question the widespread image of trial as an arena, valuing forensic rhetoric as a model, capable of enhancing the ethical demands of cooperative argumentation, as opposed to a competitive dialectic. Not even the trial, in fact, ends in a confrontation between two opponents, but develops in a relationship between several parties (a polylogue).

1. Introduction

This contribution intends to be part of the European and International debate on the relationship between education and argumentation, focusing on the practical forms of exercise and application of argumentative strategies in educational contexts (Khun, 1992; Muller Mirza & Perret-Clermont, 2009; Rapanta & Macagno, 2016); it also offers some important starting points in the discussion concerning the teaching of law (Tomasi, 2022).

The purpose of this research is to enhance the relationship between argumentation and education in the legal context by showing the double meaning of this interdisciplinary connection: a sense of the relationship consists in pointing out the role of argumentation theory as an essential component in jurist training (Atienza, 2019; Canale & Tuzet, 2020; Manzin, 2014); but the link is two-way since, as we will argue, the particular model of forensic argumentation can be used for the development of skills in other learning contexts.

I argue that forensic argumentation can give rise to training practices and provide tools and methodological approaches in all contexts of social life, in which citizens must deliberate in the face of multiple choices, validating common practices and values.

To explain how argumentation, and in particular forensic argumentation, can be a privileged model for citizenship education, it is necessary to define the forensic argumentation model and clarify its ability to settle conflicts and, therefore, its cognitive and ethical aspects.

2. Argumentation and Judgement in Archaic and Classical Greece

The value of judicial experience emerges from its primitive forms in Western legal culture. Law in classical Greece is a way to reflect on law before law (in the sense of positive law): how was the life of regulated communities before law was born?

From the literature of Homeric poems and Attic tragedies it emerges that legal experience has a value and its value is shown in the capacity of law, as a judgement, to give meaning and stability to concrete situations that occur in the experience of life community (Jellamo, 2015).

The procedural dimension contains the germ of a substantial principle: the sense of justice. The archaic Greek language, the cradle of the legal and cultural tradition of the West, as is well known, does not have a word corresponding to the Latin *ius* and uses the word *dike*, which means justice and right at the same time (Jellamo, 2005).

Justice in ancient Greece was conceived as a procedure (Gagarin, 2020) and is proper of language and the judicial dimension. Already in the Homeric poems, justice is linked to contexts that have a controversy as their object. The activity of judging is expressed with the word *dikazein* and attests to the constant, if not exclusive, use of the term justice in contexts related to the dispute.

The most typical and most famous case is that of the scene represented on Achilles' shield (*Iliad* xvm. 497–508): it is significant that the scene of the quarrel appears in the city of peace. The dispute takes place between two contenders on the square of the city, in front of a noisy crowd, which discusses and divides taking the defenses of both parties, and to the elders of the community, appointed to judge the quarrel. Each man was seeking a limit [*peirar*], in the presence of an arbitrator [*histôr*], and the people took side; the elders sat on their seats of stone in a solemn circle and each in his turn gave judgment [*dike*].

These verses contain the description of the archaic form of trial, which confirms that the appearance of law in western culture exhibits itself in social experience through the administration of a dispute in the non-violent form of public discussion.

The Homeric representation of judgement conveys the original need for a peaceful resolution of the dispute, which distances itself from the mechanisms of private revenge, through a discussion open to the public, by referring the judgment to a third party impartial and authoritative.

The literary and historical fact attributes to the law in the judicial dimension the function of recomposing the disagreement. The sentence will emerge from the confrontation between the parties and will depend on the ability of each party to state their reasons.

The presence of the *histor* confirms this meaning: the *histor* is a particular figure of the Homeric legal world, both witness and third-party actor of the facts in relation to which the dispute arises. In the scene depicted on the shield, the end of the quarrel depends on the *histor*, in the sense that his words can confirm or deny the version of each side. But the jurisdictional function, in this case, belongs to the *gerontes* who in turn, standing up, pronounce sentences.

In the Homeric pomes, *neikos* is the greek term that designates the judicial dispute (Marks, 2005): the same term is used in Hesiod's work as opposed to *eris*. *Eris* is the boundless, violent discord that explodes in *hybris* and can only be resolved with physical force (*bia*). Hesiod's use of the terms follows the Homeric use: *eris* and *neikos* express two forms of disagreement; *neikos* is a form of discord but marked by a measure, the one given by procedural justice (Jellamo,2005). It is a barely sketched image of the law, but it is also the sign of a new need which bases the common measure of action on the judicial dimension of reasonableness.

Attic tragedy brings judicial justice to the stage: in the tragedy of the *Eumenides*, Aeschylus represents the transition from a violent vengeful justice to the justice of judgment. Athena realizes that she cannot resolve the conflict with the tools at her disposal and gives life to the Areopagus. The new judgment ends with the acquittal of Orestes and with the reintegration of all the parts of the dispute in a new order, in a new equilibrium. The justice of the judgment depends not only on the formal structure and composition of the new court, but on the words that are used and the arguments that turn to be persuasive:

And there, with judges of your case and speeches
of persuasive charm, we shall find means to release
you completely from your labors. For I persuaded
you to take your mother's life.
(*Eum.*, 81-83)

Let me persuade you.
The lethal spell of your voice, never cast it

down on the land and blight its harvest home.
(*Eum.* 839–41)

Your magic is working... I can feel the hate, the fury slip away.
(*Eum.* 909)

Spirit of Athens, hear my words, my prayer
like a prophet's warm and kind.
(*Eum.* 933–4)

Athena invokes Persuasion to convince the Erinyes not to pour a grudge on the city and their renunciation of violence takes on a fundamental meaning: persuasion is the link between divine and earthly justice, since the split is averted by the persuasive use of words. Persuasion is invoked as the means of dissolving the conflict and, at the same time, the guarantee of social ties, since tragedy is the narrative construction with which Aeschylus re-establishes a continuity between the old social order and the new social order (Hewitt, 2019).

The concept of trial in Homeric poems and Greek tragedy play a paradigmatic role to reflect on the model of forensic argument as a place of resolution of disputes, in the realization of the need for social reconciliation.

3. The ongoing judicial argument

In our common sense the idea of judgment as a duel is strongly rooted, that is, the image of the court as an arena for competition between the contending parties; likewise, the image of the lawyer as a conflictual and aggressive party.

Yet if we look more closely at the types of argumentation in early Greek thought, a different view of judgment is delivered. Namely, the conflict situation does not involve the radicalization of positions and this emerges from the following elements, which constitute the pillars of the model of forensic argumentation.

First, judgment is never a two-sided duel. The parties exchange, face to face, is an integral part of a more complex relationship (polylogue: Lewiński & Aakhus 2014). The structure of the judgment involves the third party (public, *histor*, judge). The third, as demonstrated by the scene of the judgment represented in the shield of Achilles, is a necessary element: the presence of the third, spectator-witness-judge, establishes a triangular relationship. The third party is an unavoidable protagonist of the confrontation and his presence modifies the opposition, expanding it. According to Francesca Piazza (Piazza, 2019, p. 86), the Homeric duels are always triangular duels and this would be the reflection of the peculiar sociality of the human animal and is linked to language. The need for the third party in the duel is, in other words, a further indication of the fact that we are dealing with a specifically human conflict, in which the confrontation between two people has an impact on the society in which they live. The relationship between a duel and a third party can be explained not only in the sense that the judgment of the third party depends on what the contending parties say, but also in the reverse sense: what the parties say is influenced by the evaluation of the third party. The triangular structure reveals an unavoidable negotiation component that reflects the interweaving between the antagonists and the other parties (judge/public/society). Specifically, the Homeric duels show us that not even aggressive verbal exchanges can be reduced to a two-way question and answer, since combat is enriched with procedural rituals: the words spoken before and after the physical confrontation integrating the duel in the strict sense are acts, conventional linguistics, that contribute to carrying out the practice (Piazza, 2019, p. 89). The ritual that accompanies even violent verbal exchanges has a negotiating value: it recalls a connection to values/practices that go beyond the duelists and that concern the society in which the duelists are inserted. Rituality, which in law takes the form of procedures, guaranteed by the third party, has the function of re-balancing the fate of the clash, sanctioning the substantial equality of the dueled and creating that space thanks to which the clash is never destructive but compositional.

Another aspect concerns the strength of language: faced with hostility, resolution is possible only with the use of language. Athena, in the judgment of the Eumenides, invokes *Persuasion*: Athena's speech to the Erinyes can be considered an exemplary case of rhetoric based on traditional values. Athena tries to bring the ancient deities right and uses a persuasive strategy: Athena's words are yet another confirmation of the fact that what is at stake is a social order.

The first act, the opening of the play, sees Oreste acknowledging that he has committed the matricide, but also explaining the reasons that led him to this extreme act. It recalls the murder of his father and his betrayal; recalls Apollo's suggestion that led him to revenge. Such a complex situation requires an equally complex judge: a court composed of sworn judges, chosen from among the best citizens and presided over by the goddess Athena.

At trial, words, reasoning, persuasion, proof dominate. Reasoning takes the place of the instinct for revenge; arguing and motivating take the place of mystery. To better understand this passage of the tragedy, we need to take a step back: the first scene of the tragedy portrays the Erinyes; the very sight of the Furies reduces to silence; in the same scene, when the Pythia enters the temple of Apollo and sees the Erinyes, she explodes into an exclamation of terror and impotence and is unable to describe them. The contrast with Athena stands out from the beginning of the trial: Athena, at their sight, asks her to have everything explained with a "perspicuous speech". Athena bases all her action on persuasion, *Peithó*, goddess explicitly evoked also at the end of the trial (vv. 970-975): Athena's "new" justice is all based on the voice, on arguing, on dialogue. The heart of the judgment is the word, persuasion: *Audiatur et altera pars* is the fundamental procedural rule of due process, based on the adversarial process and can be perfectly traced back to this first, mythological, process of Western civilization (Manzin, 2014). Here, Aeschylus stages the attitude of listening, the first of the virtues required of a judge: listening to the accusation, listening to the defense, listening to interested third parties. Listen, first of all. What is at stake is not only the rationality of the arguments, but the ability to involve the listener (Piazza, 2017).

4. The role of forensic rhetoric

Contemporary argumentative theories basically assume rhetoric as a tool for a "reasonable persuasion". Rhetoric, that is, when contemplated in the argumentative study, is reduced to an accessory component to reasonableness (Rocci, 2015).

Rhetoric is therefore contrasted with logic and, in comparison with it, disqualified as a study of the forms of embellishment of the discourse or stratagem for the search for consensus (Puppo, 2019).

The account of the ancient archetype of judgment, as a form of social justice, reveals the essence of a verbal practice in which rhetoric is not ancillary, but constitutive. The legal context is paradigmatic: the purpose of the judgment is not finding an agreement, but to resolve the dispute in a particular case. After the judgment, the parties may still disagree: so much that an appeal or procedural means of reviewing the judicial opinion are envisaged. Why does judicial disagreement survive the sentence? Since, argumentation is inherently subjective, it depends on people, on the discourses that are introduced by the parties in that judgment, on variable values, on the relationship of trust between the speaker and the listener, on the asymmetries of power, on the emotional dimension. The practical judicial argument is, in a nutshell, characterized not only by logos, but also by pathos and ethos.

The theories of argumentation that assume as a goal the consensus, privileging the *logos*, underestimate the personal dimension of the argument (Ferry&Zagarella, 2015).

By considering the difference (and dialectics) as the indisputable engine of democratic life, most argumentative theories, also applied in educational process, shift their attention to the functions of disagreement, as well as on the constructive nature of the conflict (*eris*) and to the competitive dimension of the argument.

Debate (in the Greek sense of *eris*) is, in fact, the widespread model of argumentation in education.

By referring to the ancient model of *neikos*, our purpose is to reverse the agreement and disagreement pair and thinking that the starting point of argumentation is a form of preliminary agreement between speakers, based on beliefs, preferences, emotions, feelings, relationships of trust, relationships of power and, in general, about one shared background.

The starting point is not the conflict, but an agreement: this agreement is, in forensic practice, symbolized by the procedure, which constitutes the constant point of reference between the litigants, what they agree on.

This reversal of perspective coincides with a decentralization of the debate and open to a more complex dimension, a personal and social dimension.

5. From courts to schools

Why study forensic rhetoric at school? A first, obvious answer to this question can be referred to the field of social education: since in today's life there are many opportunities for comparing positions and for discussion, education in democratic coexistence requires a citizen capable of understanding different points of view, supporting an opinion, presenting it in a effective way for the type of context and interlocutor, as well as knowing how to evaluate the speeches of others.

But there is a deeper sense that legal argument, especially forensic rhetoric, plays a decisive role in society.

I argue that it is important that school does not set aside or neglect the legal dimension of argument, as if it were a remote or marginal aspect to teaching a skill: legal argumentation can become, in fact, a strong point of teaching argumentation *tout court*, allowing to experience, even in the context of class, how to evaluate different opinions and solve a controversial situation.

There are two fundamental ways of understanding the act of arguing:

– in a dialectical key, argumentation is a competitive procedure between two arguers aimed at reaching a rational agreement; the rules of exchange normatively exclude the use of fallacious techniques, which involve the appeal to emotional or purely agonistic factors, aimed at devaluing the proponent;

– in a rhetorical key, the argument is a practical knowledge which develops in the production of speeches. Unlike normative theories, rhetorical argumentation is characterized by its openness to the personal dimension of the discourse relationship. The purpose of the rhetorical argumentative act is not to eliminate disagreement, but to create the conditions of harmony, that is, that continuum that ranges from reaching an agreement to ensuring coexistence in disagreement (Amossy, 2011).

The dialectical argument highlights the conflict; in this frame, the argumentative action is oriented towards the elimination of the disagreement with prevalence of one over the other or with eliminating disagreement by means of strictly rational arguments and with normative commitment of the contending parties to accept the resolution of the conflict.

The rhetorical perspective, on the other hand, appears to be more suitable for the change of perspective invoked for the trial and inspired by the classical concept of *neikos*: rhetoric is an art of the person, since speaking to persuade (others, but also ourselves) is a natural and spontaneous activity (Piazza, 2017). From this point of view, one of the lessons of rhetoric is that to seek preliminary agreement on the premises, bring out shared values and encourage the construction of a reasoning starting from common principles.

This perspective leads us to renounce the ideal of normativism, shifting attention to the rhetorical technique. The rhetorical technique requires not only knowledge, but also the ability to know how to recognize the 'right' moment to advance a certain topic or present a certain argument (*kairos*) (Tindale, 2004).

In the forensic context, rhetoric used by lawyers is an example: it is not obvious that the availability of a certain authoritative legal opinion in favor of one's position or of a certain piece of evidence guarantee the success of the case. The work of the lawyer consists in the construction of an argument: the set of interpretative hypotheses and allegations are not something abstract, but shall be communicated to the counterparties and to the judge. The good argument depends not only on the ability to find good arguments, but from the ability to "feel" the right moment and the right word to present it. This verb "to feel" echoes Aristotelian *synesthesia* (Ferry&Zagarella, 2015), that common feeling.

6. How to educate to forensic argumentation?

It is clear that the teaching guidelines for the analysis and evaluation of the arguments change according to the way of understanding the argument: in the framework of competitive argumentation, the best argument is the most effective one to affirm one's point of view; in the perspective of rhetorical argument, the best argument is that more suited to the specific situation.

In the spirit of promoting a pedagogy of attitudes and relational moves, I argue that it is preferable to teach rhetoric, to learn not to impose one's position, but to relativize it and to seek a common horizon from which to face the conflict positively, not by eliminating but by valuing the differences.

In school practice, exactly as in courtrooms, the competitive argument is not rejected as *per se* harmful; rhetoric has the advantage of highlighting not the competition, but the space shared by the speaker and the listener.

Setting up in class a triadic argumentative relationship, based on common values and rules (such as judicial rules and principles) enables students to carry out the practice of giving space to conflicting phenomena and finding room for a compromise or a resolution that guarantees coexistence.

7. Conclusion

The rhetorical argument in courts is never a mere confrontation, necessarily implying a negotiating component: the argument is built on a series of intersubjective negotiation processes.

The concrete search for a preliminary agreement involves the identification of support points, on which the argument is built. The development of the argument presupposes the recognition of a set of conditions that binds the parties: the existence of a common language, a series of social rules that establish how a conversation should take place, the attribution of value to the consent of the interlocutor.

In the legal context, the argument takes place in rhetorical form and can be a privileged model for the recovery of rhetoric and education for democracy: the controversy is never a two-way duel, but is always triangular; the third party is the one who guarantees the existence and use, in the argumentative act, of shared speech acts. The participation of the judge helps to create that universe of reference, which the parties share and which delimits the argumentative space. Competitive argumentation, the one that is experimented with the question-and-answer technique and with refuting exercises, has the advantage of equip students to face the argument of others. School practice cannot, however, neglect the characteristic cooperative situation of argumentation, which is more frequently performed in trial.

References

Atienza, M (2019). *Diritto come argomentazione. Concezioni dell'argomentazione*. Napoli: Editoriale scientifica.

Di Piazza, S. and Piazza, F. (2016). Building Consensus. An introduction to a rhetorical approach. *RIFL Rivista Italiana di Filosofia del Linguaggio, 1-3.*

Ferry, V., Zagarella, M.R. (2015). Sentir en commun. Une approche rhétorique de la sociabilité. *RIFL Rivista Italiana di Filosofia del Linguaggio, 95-108.*

Gagarin, M. (2020). *Democratic Law in Classical Athens*. Austin: University of Texas Press.

Hewitt, A. (2019). Aeschylous'Eumenides and Political Impass. *ISLL Papers, 12, 1-24.*

Jellamo, A. (2005). *Il cammino di Dike. L'idea della giustizia da Omero a Eschilo.* Roma: Donzelli.

Jellamo, A. (2015). La giustizia sulla scena/ Justice on stage. *Parolechiave, 1, 209-230.*

Kuhn, D. (1992). Thinking as argument. *Harvard Educational Review*, 62, 155–178.

Lewiński, M., Aakhus, M. (2014). Argumentative polylogues in a dialectical framework: a methodological inquiry. *Argumentation, 28 (2)*, 161-185.

Manzin, M. (2014). *Argomentazione giuridica e retorica forense: dieci riletture sul ragionamento processuale*. Torino: Giappichelli.

Marks, J. (2005). The Ongoing Neikos: Thersites, Odysseus, and Achilleus. *The American Journal of Philology, 126, 1, 1-31.*

Muller Mirza, N. & Perret-Clermont, A.-N. (Eds). (2009). *Argumentation and education: theoretical foundations and practices*. Dordrecht, Heidelberg, London, New York: Springer.

Piazza, F. (2017). Rhetoric as Philosophy of Language. An Aristotelian Perspective. *Res Rhetorica, 1, 3-16.*

Piazza, F. (2019). *La parola e la spada. Violenza e linguaggio attraverso l'Iliade*. Bologna: Il Mulino.

Puppo, F. (2019). Retorica. Il diritto a servizio della verità. In A: Andronico, T. Greco & F. Macioce (Eds.), *Dimensioni del diritto* (pp. 239-318). Torino: Giappichelli.

Rapanta, C., Macagno, F. (2016). Argumentation Methods in Educational Contexts: Introduction to the Special Issue. *International Journal of Educational Research*, 79, 142-149.

Rocci, A. (2016). Ragionevolezza dell'impegno persuasivo. In P. Nanni, E. Rigotti & C. Wolfsgruber (Eds.), *Argomentare: per un rapporto ragionevole con la realtà* (pp. 88-120). Milano: Fondazione per la Sussidiarietà.

Tindale, C.W. (2019). Informal Logic and the Nature of Argument. In F. Puppo (Ed.), *Informal Logic: A 'Canadian' Approach to Argument* (pp. 375-401). Windsor: Windsor Studies in Argumentation.

Tindale, Christopher W. (2004). *Rhetorical Argumentation: Principles of Theory and Practice*. Thousand Oaks (CA): Sage.

Tomasi, S. (2021). *L' argomentazione giuridica dopo Perelman. Teorie, tecniche e casi pratici*. Roma: Carocci.

Tomasi, S. (2022*). Argomentazione, educazione, diritto. La retorica forense come strumento di formazion*e. Bari: Cacucci.

EXPANDING DEEP DISAGREEMENTS: INCOMPATIBLE NARRATIVES AS INTERPRETATIVE REPERTOIRES IN POLITICAL DISAGREEMENTS

MEHMET ALİ ÜZELGÜN
IFILNOVA, Universidade Nova de Lisboa
CIES-ISCTE, Instituto Universitário de Lisboa
uzelgun@fcsh.unl.pt

Abstract

Deep disagreements, defined as proceeding from a clash in framework propositions, are not about some isolated and exceptional propositions (Fogelin, 1985). What impedes rational interaction and common understanding in deep disagreements instead is a "whole system of mutually supporting propositions" (Fogelin, 1985, p. 5). This paper interrogates the relations of (mutual) support and contradiction that framework propositions are theorized to have with each other. To explore those relations of support and contradiction in public political discourse, I draw on the concepts of interpretative repertoires (Wetherell and Potter, 1988) and cognitive environment (Sperber and Wilson, 1986) as well as narrative rationality (Fisher, 1987). Accordingly, I consider framework propositions as constituting the interpretative layer of sense-making, lying beneath the argumentative layer of claims-making. To argue that the interpretative layer of political communication is organized in a narrative format, I use examples of public disagreement that originate in contexts of violent ethnic conflict. The interactional trouble in political deep disagreements then becomes evident as resulting from incompatible narratives that cannot be called into question and negotiated in a given context. I discuss the notion of narratives underlying disagreements - or sustaining arguments - in connection to the recent scholarship on narrative arguments (Olmos, 2017), as well as the political uses of deep disagreements from a narrative argumentation perspective.

1. Introduction

In his article drawing attention to deep disagreements, Fogelin (1985) defines deep disagreement as proceeding from a clash of framework propositions. If these problematic sets of deeply held beliefs and preferences could be called out and discussed, Fogelin maintains, it would be possible to manage such disagreements. However, the trouble of a deep disagreement concerns not just isolated beliefs or preferences, but "a whole system of mutually supporting propositions" (Fogelin, 1985, p. 6). It is the intractability of this whole "system" that seems to constitute the problem of deep disagreement, but the literature is largely silent about this system of mutually supporting propositions. In other words, very few contributions focus on the "mutual support" with which background propositions are connected to one another and form the intractable frameworks that characterize deep disagreements.

This study aims to address the composition of the "system of mutually supporting propositions" that prevents the parties of a deep disagreement from arguing reasonably with each other. Before moving forward to hypothesize that the socio-cognitive "framework" underpinning public political disagreements is organized in a narrative structure, two questions need be addressed. The first concerns the usefulness of the notion of deep disagreement and whether they exist: against the restricted definitions that focus on the conditions of possibility of deep disagreements, I employ a broad understanding of deep disagreement, which I think was adopted in Fogelin's thesis and examples. The dispute over "affirmative action" or "reverse discrimination" quotas, for instance, is presented as "a group phenomenon", and "in fact, one concerning moral standing" (Fogelin, 1985, p. 7). This means that Fogelin clearly suggested the relevance of values together with definitions in the clash of underlying framework propositions. And this brings us to the second question of deep disagreements of the public political kind, and whether examining bloody public controversies as deep disagreements is apt to the task of providing a critical perspective. In both regards, I'd like to call on the idea that "enacting the posture of deep disagreement - whether or not sincerely felt - is politically useful", and therefore may apply to many present-day polemics and controversies (Zarefsky, 2016, p. 10).

2. Deep disagreement in public political controversies

Several social scientific concepts can be considered in accounting for the composition of "a system of mutually supporting propositions" (Fogelin,

1985, p. 6), and the relations of mutual support such propositions have. Social representations, belief systems, interpretative repertoires, and master narratives are just some of the candidates. The crucial point to note is that framework propositions are not just abstract linguistic or semantic systems, they are essentially socio-cognitive and cultural resources, meaning they are both public and personal. Following such insight and drawing on Wittgenstein's *On Certainty*, Godden and Brenner (2010) offer the notion of *weltbild* (world-picture) to capture the deep-seated and relatively stable character of the socio-cognitive systems underpinning deep disagreements: *weltbild*, like language, is acquired through imitation and instruction, and not through rational processes of reasoning and argument (ibid.). Further, world-pictures are not upheld or justified by reasons, notwithstanding that they constitute the framework in which reasons are generated and evaluated. This means that the systems of propositions mentioned by Fogelin are not as rational as reasoning and argument would have it.

Before moving on to consider - as it were - "lesser" degrees of rationality, let us explore what a system of propositions could look like. Goodwin's (2020) "system-level" investigation of argumentative activity in climate science and policy controversies is an invitation to account for the complex argumentative relations unfolding in societal debates. Accordingly, at the system level, arguments are "abstract entities which cannot be equated with the specific makings, 'products' or 'speech acts' that instantiate them" (Goodwin, 2020, p. 169). The "hypocrisy argument" directed at climate scientists, for example, consists of several propositions and instantiations, but denotes a broader perspective established by the skeptic/contrarian camp. Goodwin's analysis shows that, within a given controversy, particular arguments will take place primarily in relation to such established perspectives, which constitute the topical knowledge or "argumentative content knowledge" (Goodwin, 2019) of that controversy. This topical background of a debate or controversy is not merely a methodological concern for a system-level perspective, it is above all the content that is available, many times mutually manifest (Sperber and Wilson, 1986), to the contenders in a debate. It is simply about what participants do - or have done - in a debate, i.e., how that content knowledge is deployed in a controversy (See e.g., Üzelgün et al., 2016).

Although topically limited to particular debates and controversies, Goodwin (2020) entertains the idea that "argumentative content knowledge" can be regarded in more generalized terms - take the examples of parliamentary democracy or nationalism for such grand controversies. When moving from particular controversies to such generalized - or even historical - debates, the apparatus of discourse theories becomes relevant. And there we can find concepts such as "interpretative repertoires" (Potter and Wetherell, 1987) to reconsider the nature of the system(s) of propositions pointed out by Fogelin. Interpretative repertoire is offered as a "summary unit" that describes "the explanatory resources to which speakers have access" in their discursive engagement with a topic (Wetherell and Potter, 1988, p. 172). Unlike social representations (Wagner and Hayes, 2005), they are not bound to particular social groups and are closer to being topical in the construction and warranting of explanations. In that service of providing warrants, they are recurrently used "systems of terms" for characterizing and evaluating actions and events (Potter and Wetherell, 1987, p. 149). The empiricist and racist repertoires, for instance, enable and obstruct particular ways of seeing and acting the world.

The notions of world-picture(s) and repertoire(s) clearly refer to layers or system(s) of sense-making that make disagreements meaningful when shared to some extent. They also enable the understanding of these socio-cognitive systems as the less called on, the more stable. Yet, they give us just too little insight into the composition or structure of such socio-cognitive systems. It would be a futile effort too if those systems of terms and propositions were solely cognitive artefacts - subsisting only in the minds of the contenders. The proposal of the next section is that what structures whatever cognitive systems is the way they are shared and distributed - their communicative structure.

3. Narrative paradigm: a lesser rationality?

A word of caution before proposing (master) narratives as a tool to better understand some argumentative quandaries: with the notion of narrative being used indiscriminately for a variety of purposes, it is crucial to recognize the distinct patterns and functions of different enunciative modes such as explication, argumentation, and narration (Reisigl, 2020). The distinction between narration and argumentation, specifically, is indeed well-established. Let us take the example of the paradigmatic and the narrative "cognitive modes" through which audiences acquire information (Bruner, 1991). The so-called paradigmatic mode concerns facts, weighing evidence, and evaluating arguments, whereas the narrative mode concerns understanding the causes and the succession of events as well as the actors (motives) connecting them. In other words, whereas nonfictional narratives are concerned with veracity only, as sources of information on what happened in past time (Plumer, 2017), arguments enjoy a well-delineated verbal format and propositional content ordered in inferential structure (van den Hoven, 2017).

That said, the boundaries between the narrative and the argumentative are blurrier than what definitions may suggest (Olmos, 2017). Rather than focusing on the intersections, here I propose to employ the two perspectives in a joint analytical framework. For this, it is crucial to distinguish between the conception of narrative (a) as an enunciative mode, which recapitulates past experience, and (b) as an epistemological mode, a way of understanding the world (de Fina, 2018): While the first conception refers to everyday discourse practices with specific characteristics, the latter is understood, especially in Critical Discourse Analysis, as "overarching structures that underlie and organize discourse and interpretation" (p. 233). It is important to recognize the specific use of the latter in studies of public political discourse as "master" narratives, to examine "the imposition of frames of understanding by the powerful" (de Fina, 2018, p. 237).

Notice that the notion of "master" narratives as frames of understanding social issues is in line with Bruner's "cognitive mode" of narrative appraisal. In such a framework, narrative can be understood as a human cultural effort to transform the feelings and responses associated with certain events and memories into a coherent sequence in a way to explain and learn from them (van den Hoven, 2017). The transformative act of narration connects two casualties: some precedent event - known as complication - changes the circumstances of an actor, requiring her to respond, creating a succeeding causality. This action is central for narration in that it connects the two in one causal sequence that helps to construct experience and drive lessons. "If a human culture develops standard strategies to respond to complications that disturb an existing equilibrium", this heuristic strategy constitutes an abstract rule or regularity that can "account for the choice of a response to a situation" (van den Hoven, 2017, p. 111). This transition from the encoded sequence of events and emotions to abstract and normative rules represents the shift from narration to practical argumentation (van den Hoven, 2017).

Building on van den Hoven's anthropological account that links the two enunciative modes in an evolutionary explanation, one can consider the narrative and the argumentative as continuous capacities ordered in a more-or-less hierarchical manner. Here, we can resort to a basic insight from Fisher (1987) who conceived the reasons that justify certain acts and feelings as founded upon narratives that pertain to particular societies. Fisher offers narrative rationality as the basis or "paradigm" of human communication, based on the idea that narrative is the paradigmatic mode for encoding and understanding human experience (see also de Fina, 2018; van den Hoven, 2017). His narrative rationality "does not deny that discourse often contains structures of reason that can be identified as specific forms of argument..." but claims that "reason occurs in human communication in other than traditional argumentative forms" (Fisher 1987, p. 48). These less institutionalized and structured forms of Fisher's "expansive understanding of logic" (Tindale, 2017) will involve values, character, and experience, and are evaluated with regard to two broad principles, coherence and fidelity to values (Fisher, 1987).

The conception of reasons as founded upon narratives is an intriguing claim when looked at from the trouble of deep disagreement. This does not mean to suggest a shift to narrative mode and analysis when rational argumentation does not hold, while it also can be a worthy track for dealing with deep disagreements. The more limited proposal here is to regard and examine the public political controversies with a polarized disagreement space as ensuing from incompatible narrative patterns. In the remainder of this paper, I try to show that the incompatibility of such (master) narratives is due to their capacity in encoding the mutually manifest features of controversy into divergent political projects and imaginaries. To echo Wittgenstein and replace the concept of "principles", my attempt is to show that: where two political narratives or imaginaries really do meet which cannot be reconciled with one another, "then each man declares the other a fool and heretic" (quoted in Fogelin, 1985, p. 6).

4. The Teacher Ayşe Case

What is called here the Teacher Ayşe Case is part or a small episode of Turkey's long-standing Kurdish conflict. After the peace talks between the PKK and the Erdoğan government halted in 2015, a new period of military action started, using heavy weaponry and a state of emergency in Turkey's Kurdish cities. In January 2016, with media bans and bombardments devastating several cities, Ayşe Çelik remotely connected to a live TV entertainment show and made an intrusive appeal. Instead of asking personal questions to the Beyaz Show guests, she addressed both the guests and the wider public. Her appeal had some characteristics of witness testimony:

Example 1 Teacher Ayşe's intrusive message

Are you aware of what's going on in the East, Southeast of the country? Here, children, mothers, people are killed. As an artist, as a human being, you shouldn't be silent and say stop somehow. ...I want to say one more thing, there are pathetic people who get happy as children die. I cannot, I mean, we cannot say anything to those people, shame on you. I want to say one more thing, excuse me, I am a teacher, and I want to appeal to teachers who have abandoned their students... How will they get back, how will they look at the faces of those beautiful, innocent children? I can't talk really, things experienced here are represented very differently on screens. I mean, I can't talk really, don't remain silent. As human beings, show more sensitivity. Hear us, help us. Don't let people die. Don't let children die. Don't let mothers cry. This is what I want to say, thank you.

[Anchorman Beyaz: Miss Ayşe, first an applause for Miss Ayşe!] [applause]

Actually I want to say so much, I can not say anything because of emotional intensity. You notice, my voice breaks. Bomb sounds, bullet sounds, people are struggling with hunger, thirst... Especially babies, children, please be sensitive, do not remain silent, I beg you, please.

Her telephone interruption from Diyarbakır, one of the cities then under curfew, received applause from the audience and support from the Show's anchorman Beyaz. After the Show's smooth handling of the interruption, a long-lasting mediated conflict took over, involving the cancellation of Beyaz Show, Teacher Ayşe's trial, and her eventual prison sentence due to terror propaganda. Having followed how this controversy unfolded in Turkey's mainstream mediascape - across 14 national newspapers and over several months - I identified the critical moments of the Teacher Ayşe controversy. Because the critical moments that featured in national print media are too many to cover here, I list the first ten of them, together with the standpoint reconstructed for each episode (Table I).

In the identification and reconstruction of each critical moment, I paid attention both to the micro-context of the news article and the macro-context of the controversy. Once an outline of the key players, positions, and places through which the controversy unfolded (Aakhus and Lewinski, 2017) was achieved, it became evident that (a) the disagreement space was clearly polarized, (b) in a way that splits the media organizations into two camps also as active players of the controversy rather than mediating agents, (c) argument analysis as a normative endeavor had little or no traction, that is, beyond serving as a fallacy hunt, (d) without the argumentative content knowledge (Goodwin, 2019) or background information (van Eemeren, 2011) necessary to make sense of the referential and intertextual relations, it was not possible to even follow the logic of the controversy. While a full-length study is required to adequately contextualize the arguments (re)presented by the two opposing camps of the Turkish press, what follows is restricted to discussing how reasons are grounded in conflicting narratives.

5. The useful posture of deep disagreement

Teacher Ayşe's disruptive telephone message in the live Beyaz Show - which opened another chapter of the broader controversy on the rights of the Kurdish minority in Turkey - can be reconstructed as: "You should be more sensitive about what is happening in the East of the country and do something about it" (Table I). The substandpoints and complex argumentation culminating in this position cannot be examined in this short outline. Yet, it is clear from the verbatim quotation in the preceding section (my translation) that one of the main arguments upholding this position is "people are killed", another one "people are striving for hunger". The ambiguity of the subject of the killing here indicates a whole baggage of argumentative content knowledge (Goodwin, 2019) as well as the power relations therein.

Table I. Critical moments of the Teacher Ayşe Controversy (first ten)
1. Intervention of Teacher Ayşe: "You should be more sensitive about what is happening in the east of the country and do something about it"
2. The first reactions on media: "Live terrorist propaganda on Beyaz Show"

3. The Ministry of Education leaks: "Ayşe Çelik is not a teacher"
4. Accusations of deception: "Beyaz and D-Media participated in terror propaganda"
5. Defense of Channel D: "Channel D is facing provocations and tarnishing efforts"
6. Commentary of opposition leader Demirtaş: "A peace massage is being used to create fear over the media"
7. Apology of Beyaz: "I am loyal to my nation"
8. Prosecutor opens lawsuit: "There is the crime of terror propaganda in this case"
9. Involvement of supporters in the lawsuit: "If that is a crime, we commit the same"
10. Admonition by a special operations policeman: "Beyaz's betrayal will not be forgotten"

One of the first headlines to fire the controversy became available online a few hours after the closing of the Show, and is repeated exactly by four newspapers in the first day: "PKK propaganda live broadcasted on Beyaz Show" (*Yeni Şafak*, January 9, 2016). The variants of this headline are reconstructed as the second position in the Teacher Ayşe Controversy (Table I). Upholding this position, for instance, the news piece from Yeni Şafak, features "a spectator, allegedly a teacher, who connected to the Show through telephone and who spoke with the PKK mouth" and concludes with "[she] was permitted to talk without disruption and was applauded by the studio guests with the invitation of Beyazıt Öztürk [the anchorman of the Show]". Notice that such news presentations, the "alleged" teacher is stripped of virtually all agential qualities, including her name, intention, authenticity/voice, and profession. Furthermore, the *ad hominem* attack constituting a complete evasion (Üzelgün et al., 2022) of Teacher Ayşe's argumentation is expanded towards the anchorman - and eventually also to the commercial TV channel and its owner - who is portrayed as permitting and supporting terror propaganda.

For a linguistic argument analysis, it may be startling to render and summarize how an argument that bears some characteristics of *ad misericordiam* argument is framed and reported as terror propaganda. Beyond the political polarization in Turkey, and the disruptive presentation of the message, I maintain that the answer resides in the divergent frameworks of interpretation. The key here is the locus of responsibility, left ambiguous in Teacher Ayşe's testimony, namely the source of the threat that forced the teachers to leave their schools - or "abandon their students" - and flee the region. This locus, presented and experienced as taboo (rule 1, van Eemeren & Grootendorst, 2004), represents one of the key missing premises in arguments by both sides of the controversy - yet it appears to account for considerable part of the relations of support in the divergent frameworks of interpretation. Instead, the discussion pivots on Teacher Ayşe's person, who in due process transforms into yet another character interwoven with many others in the divergent narratives over Turkey's Kurdish question.

To be sure, showing or exemplifying "the imposition of frames of understanding by the powerful" (de Fina, 2018, p. 237) in various settings does not constitute sufficient proof to establish the link hypothesized between the argumentative encounter and the narrative framework(s) of interpretation. Let me invite you to another critical moment of the Teacher Ayşe Case - and to many other public political controversies - to indicate the embeddedness of argumentative reasons in divergent narrative frameworks.

On the second day, as Teacher Ayşe was still missing, and attacks by the government-sided media organizations focused on the Show's anchorman Beyaz and the owner of Channel D participating in a plot, Beyaz announced his first message, which had the form of an apology (Table I, Move 7). Beyaz's apology appeared in the *Hürriyet* newspaper which belonged to the same conglomerate with Channel D, and had the headlines "Beyazıt Öztürk: Don't exploit me for political purposes". In this newspiece, Beyaz is quoted saying:

Example 2 Anchorman Beyaz's defense (*Hürriyet*, January 10, 2016)

My position is obvious for many years, my attitude is obvious. My loyalty to the nation, to the motherland, to the flag is certain. When I first encountered "Are you aware that there are children dying there?", I say with all my sincerity, it was the point of breaking for me. My brain stopped there. My great mistake is that I lost my concentration. (...)

  I'm a policeman's son [Title of Subsection]
A lot of people in the studio also couldn't understand really that there was such a conversation, such an intention behind it.

  I am sorry [Title of Subsection]
I think what the whole Turkish people thinks. Of course, I am sincerely wishful that the terrorist organization will drop weapons and that these issues are resolved as soon as possible. God help all our security forces there. We are on the side of our nation.

Beyond its many intricate details, this defense of the TV Show's anchorman has the characteristics of the classical *eikos* argument, that is, an appeal to one's customary behavior, in the form of "I did not do it, because I would not do it" (Jansen, 2023). While on one layer Beyaz presents his apology for failing to abide by the standards he would customarily do, he also assures that it was neither his intention nor a plot he was aware of. While he presses all the necessary buttons to that end, in a fascinating way, he calls yet another figure into the controversy: his father once being a policeman appears as an argument in his defense - I'd like to add, perhaps his argument with the top traction in the context of the controversy. While logical-dialectical analyses and rational resolution of this ever-expanding disagreement would probably exclude the father's profession as a relevant reason for Beyaz's innocence, this work on one's ethotic rating has an important role in providing coherence and fidelity to his speech acts and positions before and after the personal attacks absorbing him into the controversy.

How come the irrelevant facts about his father's profession turn to play a key role in Beyaz's transformation from a suspect into a witness in Teacher Ayşe's court case? Following Goodwin (2020), I'd say what seems to be an irrelevant reason at the micro-context of the news article, turns out to be not just relevant but a critical part of the macro-context or at the system level of the controversy. What appears, without access to Beyaz's character and family story, as just another of the many fallacious moves in a terrible conflict is indeed interwoven with the values Beyaz evokes (loyalty, nation, security) and his speech acts. It is precisely the alignment of the values, ethos, and the speech acts that has some appeal and traction in the controversy that is otherwise just a show of name calling and crude power.

This brief demonstration of the meta-level awareness and divergent "manifestness" of the topical knowledge and values may add up to the claim that at the system level - here understood as significant issues for a society - arguments sometimes have a narrative form or dimension. While that may well be the case, I'd like to appeal to the relevance of character and motives, as well as their coherence and continuity, as essential constituents of persistent disagreements. As such constituents appear in argumentative discourse only at times and as brief appeals, they subsist as at the system level codified in narratives, to be called on when necessary or useful.

6. Conclusion

To conclude with some specificities of the Teacher Ayşe Case, even though superficially, let us draw on three perspectives of argument. From the informal logical perspective, the controversy is almost completely made up of *ad hominem* attacks, and their defenses, besides several other fallacies. As such, it cannot be regarded as a rational or critical discussion. From the perspective of pragma-dialectics, the discussion has remained in the opening phase, unable to proceed towards a critical test of opinions; and this is due to violations of various critical discussion rules, above all the first rule. From the perspective of disagreement management (Aakhus & Lewinski, 2017), the disagreement has expanded faster than the ability of the participants to question the moves, submit reasonable arguments, and negotiate their status on a shared ground. This was never the intention, we could say, as the mere posture of deep disagreement seems sufficient to achieve certain ends, especially for the power-holding party (Zarefsky, 2016).

To add one more word to the foregoing, the expansion of the disagreement - also absorbing public figures not initially associated with the conflict - was largely a function of the *ad hominem* attacks characterizing the case, which were rarely reasonable. This - declaring the other a fool and heretic - feature of the Teacher Ayşe Case clearly exhibits the meta-level argumentative engagement characteristic of deep disagreements. And the constant attempts at re-framing issue and re-writing the conflict essentially draws on a polarized disagreement space with two incompatible narratives.

Much of the recent work linking the argumentative and the narrative (e.g. Olmos, 2017) focuses on the intersection of the two discourse modes. In this paper, instead, I tried to focus on their relationship as capacities of speakers in reading and writing controversies, and as congruent analytic frameworks that can be used together in the exploration of public political disagreements. The proposal draws on the idea that argumentative reasons are grounded in narrative practices and conventions (Fisher, 1987; van den Hoven, 2017). Accordingly, beneath the argumentative layer(s) of claims-making, we can assume a narrative layer of sense-making and interpretation. Fogelin's, or rather, Wittgenstein's framework propositions then emerge as instants or maxims of narrative patterns, still constituting the framework in which reasons are generated and evaluated. But this move explains their stable character without being upheld or justified by reasons: it is the relations of "mutual support" that keep them in place. Mutual support in this narrative framework means values, character, and experience, which at times wired so tightly by power and politics that become self-sealing.

Drawing on an argumentative layer(s) of claims-making together with a narrative layer of interpretation or sense-making can be particularly useful in dealing with deep disagreements in the public political domain. When a disagreement is polarized by design - of one or both parties - it is not subject to rational resolution, in the sense used by Fogelin. The only way towards some resolution, in such non-normal circumstances, may be to slowly interweave different narrative patterns and a new interpretative framework that is both coherent and just to both poles. And that involves dealing with values, characters, and deeply moralized experiences, a direction the studies of persistent disagreement and argumentation may take fruitfully. Such studies then would reveal new insights on the extent and possibilities of rational resolution of deep disagreements.

References

Aakhus, M., & Lewiński, M. (2017). Advancing polylogical analysis of large-scale argumentation: Disagreement management in the fracking controversy. *Argumentation*, 31, 179-207.

Aikin, S. F. (2019). Deep disagreement, the dark enlightenment, and the rhetoric of the red pill. *Journal of Applied Philosophy*, 36(3), 420-435.

Bruner, J. (1991). The narrative construction of reality. *Critical Inquiry*, 18(1), 1-21.

De Fina, A. (2017). Narrative analysis. In R. Wodak, & B. Forchtner (Eds.), *The Routledge handbook of language and politics* (pp. 233-246). Routledge.

Fisher, W. R. (1984). Narration as a human communication paradigm: The case of public moral argument. Communications Monographs, 51(1), 1-22.

Fisher, W. R. (1987). *Human communication as narration: Toward a philosophy of reason, value, and action.* University of South Carolina Press.

Fogelin, R. (1985/2005). The logic of deep disagreements. *Informal Logic*, 25(1), 1-7.

Godden, D., & Brenner, W.H. (2010). Wittgenstein and the logic of deep disagreement. *Cogency: Journal of Reasoning and Argumentation*, 2(2), 41.

Goodwin, J. (2019). Sophistical refutations in the climate change debates. *Journal of Argumentation in Context*, 8(1), 40-64.

Goodwin, J. (2020). Should climate scientists fly? A case study of arguments at the system level. *Informal Logic*, 40(2), 157-203.

Jansen, H. (2023). High costs and low benefits: Analysis and evaluation of the "I'm Not Stupid" argument. *Argumentation*, 37, 529-551..

Olmos, P. (Ed.). (2017). *Narration as Argument*. Springer International.

Potter, J., & Wetherell, M. (1987). *Discourse and social psychology: Beyond attitudes and behaviour*. London: Sage.

Plumer, G. (2017). Analogy, supposition, and transcendentality in narrative argument. In P. Olmos (Ed.), *Narration as Argument*, 63-81.

Reisigl, M. (2021). "'Narrative!' I can't hear that anymore'. A linguistic critique of an overstretched umbrella term in cultural and social science studies, discussed with the example of the discourse on climate change. *Critical Discourse Studies*, 18(3), 368-386.

Sperber, D., & Wilson, D. (1986). *Relevance: Communication and Cognition*. Cambridge, MA: Harvard University Press.

Tindale, C. (2017). Narratives and the Concept of Argument. In P. Olmos (Ed.), *Narration as Argument*, 11-30.

Üzelgün, M. A., Lewiński, M., & Castro, P. (2016). Favorite battlegrounds of climate action: Arguing about scientific consensus, representing science-society relations. *Science Communication*, 38, 699-723.

Üzelgün, M. A., Fernandes-Jesus, M. & Küçükural, Ö. (2022). Reception of climate activist messages by low-carbon transition actors: Argument evasion in the carbon offsetting debate. *Argumentation and Advocacy*, 58(2), 102-122.

van den Hoven, P. (2017). Narratives and pragmatic arguments: Ivens' The 400 million. In P. Olmos (Ed.), *Narration as Argument*, 103-121.

van Eemeren, F. H. (2011). In context: Giving contextualization its rightful place in the study of argumentation. *Argumentation*, 25(2), 141-161.

van Eemeren, F. H., & Grootendorst, R. (2004). *A Systematic Theory of Argumentation: The Pragma-Dialectical Approach*. Cambridge University Press.
Wagner, W., & Hayes, N. (2005). *Everyday Discourse and Common Sense: The Theory of Social Representations*. Bloomsbury Publishing.
Wetherell, M., & Potter, J. (1988). Discourse analysis and the identification of interpretative repertoires. In C. Antaki (Ed.), *Analysing Everyday Explanation: A Casebook of Methods*, pp. 168-183. London: Sage.
Zarefsky, D. (2016). On deep disagreement. In R. von Burg (Ed.), *Dialogues in Argumentation*. Accessed at:
www.windsor.scholarsportal.info/omp/index.php/wsia/catalog/book/12

THE FALLACY OF POPULARITY

JAN ALBERT VAN LAAR
University of Groningen
j.a.van.laar@rug.nl

Abstract
How to understand and assess arguments in which the popularity of an opinion is put forward as a reason to accept that opinion? There exist widely diverging views on how to analyse and evaluate such arguments from popularity. First, I define the concept of an argument from popularity, and show that typical appeals to the popularity of a policy are not genuine arguments from popularity. Second, I acknowledge the importance of some recent probability-based accounts according to which some arguments from popularity are epistemically strong arguments, but also contend that despite these strengths such arguments have at most limited value in argumentative discussions. Finally, I show that there are at least five different ways that arguments from popularity can be fallacious, and examine what this means for an account of the Fallacy of Popularity.

1. Introduction

How to understand and assess arguments in which the popularity of an opinion is put forward as a reason to accept that opinion? There exist widely diverging views on how to analyse and evaluate such arguments from popularity. According to strong critics, such as David Godden (2008), Trudy Govier (2005), and Henrike Jansen (2020), arguments from popularity are always or typically unconvincing and fallacious. According to enthusiasts, like Don Dedrick (2019), and Ulrike Hahn and Jos Hornikx (2016), there are important settings in which arguments from popularity are legitimate and rationally persuasive arguments. Mild critics, such as James Freeman (1995), Douglas Walton (1999), Ralph Johnson and Anthony Blair (1994), and Frans van Eemeren and Rob Grootendorst (1992, p.166), hold that arguments from popularity, as I will conceptualise them in this paper, can provide reasonable and useful contributions to an argumentative discussion although they are, on their own, weak arguments that often are used in a faulty and fallacious way.

This paper has three aims. It deals with the conceptualisation of arguments from popularity, and aims at clarifying a possible confusion in the concept of argument from popularity. It deals with evaluation of arguments from popularity, and aims at showing that it has a modest epistemic role to play in argumentative dialogue. It deals with settings where it makes sense to charge the proponents of such arguments with committing the Fallacy of Popularity, and aims at finding a solution to the problem that on the one hand there does not seem to exist one Fallacy of Popularity and on the other that the concept of the Fallacy of Popularity often is useful and important to keep an argumentative exchange on a good track.[1]

2. Argument from popularity

First, I will delineate the concept of an argument from popularity. I will do so by elaborating on the theory of argumentation schemes (Walton, Reed and Macagno, 2008). Being somewhat precise about what counts as an argument from popularity will allow us to distinguish arguments from popularity from a related but different kind of argument, in which the popularity of a policy is advanced as a reason for the policy's democratic legitimacy. This is a point worth making, as it implies that the virtues of the latter cannot without further ado be attributed to the former.

I start with two simple examples.
1. "This vaccine is unsafe. Why? Almost everyone thinks so."
2. "We should stop administering this vaccine. Why? Almost everyone thinks we should!"

I am interested in how such arguments fare in critical examination dialogues. A critical examination dialogue is modelled as a conversational exchange in which: (a) a proponent advances a thesis; (b) an opponent critically challenges the thesis; (c) the proponent justifies the thesis by responding to the challenges; (d) the opponent critically challenges the justification; (d) and where these dialectical moves - back and forth - is part of a cooperative effort to arrive at a shared view on what is a correct judgment regarding the thesis (cf. the concept of a critical discussion, van Eemeren and Grootendorst 2004; cf. the concept of persuasion dialogue, Walton and Krabbe 1995).

I follow a tradition of regarding arguments from popularity as instances of a specific argumentation scheme (Walton, Reed and Macagno 2008; Wagemans 2016; Hinton and Wagemans 2022), and shall in this paper define "argument from popularity" by means of the following pattern of reasoning, that I label "argumentation scheme from popularity", where T may stand for any proposition (i.e.: of any kind, and of any complexity): "T, because most people in group G accept T." I define "argument from

---

[1] The sections in this paper are included in (van Laar 2023).

popularity" as any instance from the argumentation scheme from popularity.

A natural distinction between two kinds of argument from popularity is between those that concern practical issues, such as issues of policy choice, and those that are not action-oriented, such as those dealing with the truth or correctness of some descriptive or evaluative claim. The idea that arguments from popularity are more easily to be evaluated as reasonable when they concern policy choice in a democracy has been supported (Walton 1999) as well as criticised (cf. Jansen 2020). But how about the presupposition shared by both positions: are such arguments really arguments from popularity?

If we specify T so that it concerns policy, we obtain a *candidate* scheme such as: "Policy P should be implemented, because most people in group G accept that policy P should be implemented."

Jansen treats this as a subtype of argument from popularity, but criticizes it as a conversation-stopper. I think that they are, despite appearances, not really arguments from popularity. Because the phrase "should be" is ambiguous. Because, sometimes one can say that one both endorses and rejects that some policy *should* be implemented. *Suppose*, my government implements a policy that I disagree with, but results from a democratically impeccable procedure. *Then* I could, without contradiction, say: "well, it's a shame, but this policy should be implemented" or "this policy should not be implemented as far as its substance is concerned, but then it should be when seen from a procedural stance." The first reading is a *substantive reading*, and the second a *procedural* one.

The first disambiguation of the scheme, I label the *substantive argument from popular policy*: "Policy P should be implemented, when evaluated from a substantive perspective, because most people accept that policy P should be implemented, when evaluated from a substantive perspective."

Suppose the issue is: "should we – in the Netherlands – abandon Dutch as the national language, and change to Frisian?" Then an example would be: "I wholeheartedly believe that changing to Frisian is the correct choice. After all, most people wholeheartedly believe that changing to Frisian is the correct change." The conclusion does not concern legitimacy, but the policy's correctness, or epistemic justifiability. This is a real argument from popularity.

By the way, the reasoning can be quite convincing. If everyone in the Netherlands welcomes this change, and people abroad do not object, then the popularity of the policy *makes* it a change for the better. When there is no truth 'out there', then the collective of people is not only politically but also epistemically sovereign, and popularity produces epistemic value.

But, most scholars have a different disambiguation in mind, which I label *from substantive to procedural acceptability*: "Policy P should be implemented, when evaluated from a *procedural* point of view, because most people accept that policy *P* should be implemented, when evaluated from a *substantive* point of view."

For example: "Banning this vaccine is the democratically legitimate choice to make. Because, most people wholeheartedly believe that banning the vaccine is the correct and prudent choice to make." This, indeed, has plausibility. But this is not an argument from popularity - in our sense. *Yes*, there is an appeal to popularity. But *No*, it is not the popularity-of-the-thesis that is advanced as support for the thesis. Instead, the popularity of the substantive correctness of the policy is support for the legitimacy of the policy.

3. Are arguments from popularity rationally persuasive?

I now turn to assessment. Can arguments from popularity provide a relevant, or even a sufficiently convincing, reason to accept its conclusion? This depends on the reliability of the reference group.

If the proponent lacks such information, we have a blatant version of the argument, and the proponent cannot answer critical questions regarding reliability. In the proponent has such information, he could argue that these people are, on average, *outright reliable*, that is: more likely to be right than wrong. Or, he could argue that these people are *relatively reliable*, that is: at least equally reliable as the addressed opponent, so that the opponent should see the reference group as her epistemic peers or epistemic superiors. The enthusiasts focus on cases where there is information available about the outright reliability. With probability theory they show that popularity can have epistemic weight.

In their words, Hahn and Hornikx provide a Bayesian foundation for the evaluation of arguments. They identify the strength of an argument thesis with the probability that the thesis is true, provided the reasons are true. The bottom line is that the strength of an argument from popularity depends on three items of information:

- First the prior probability of thesis T, independent of the reason. The higher the prior probability of the thesis, the more quickly the reason provides sufficient support.
- Second, the probability of "T is popular", provided T is true.
- Third, the probability of "T is popular", provided T is false.

The theorem specifies how argument strengths depends of this information: $\text{Prob}(T/\text{"}T\text{ is popular"}) = (\text{Prob}(T) * \text{Prob}(\text{"}T\text{ is popular"}/T))$, divided by $(\text{Prob}(T) * \text{Prob}(\text{"}T\text{ is popular"}/T) + \text{Prob}(\sim T) * \text{Prob}(\text{"}T\text{ is popular"}/\sim T)$.

This explains how the strength of such an argument depends on contextual details. If T is very improbable then plausibly the argument has insufficient strength. And if the people are likely to be collectively misled, then the argument also falters. But in some situations it provides a strong argument.

For example, if the prior probability of T is 0,4, and the probability that T is popular if T is true is 0,66, whereas the probability that T is popular if T is false is 0,33, then when we apply Bayes theorem, the strength of the argument is 0,57.

Does this make Bayes' theorem the foundation for the evaluation of these arguments? *Yes*, if they mean to say that when the requisite information is available this provides a way to support the strength of an argument from popularity. But *No*, if they mean to say that an appeal to Bayes theorem is necessary or sufficient for justifying its strength. Before getting there, I reconstruct the Bayesian justification as a two-part argumentation.

Argument A:
Conclusion A1: T, because
Premise A2: T is popular (in group G), &
Premise A3: T's popularity is a serious reason for T, and more specifically: Prob(T/Popular) = 0,57,

Argument B:
Conclusion B1: T's popularity is a serious reason for T, and more specifically: Prob(T/Popular) = 0,57, because
Premise B2: Prob(T) = 0,4 &
Premise B3: Prob(Popular/T) = 0,66 &
Premise B4: Prob(Popular/~T) = 0,33 &
Premise B5: Bayes' Theorem.

Argument B is the subordinate argument that bolsters Argument A, the argument from popularity. Argument B is an interesting kind of argument, also because it suggests similar justifications for other argument types, and further because it connects to the rich field of Bayesian epistemology. But then, the proponent does not need to advance this very argument to convince an opponent who challenges a connection premise (i.e. A3), so that a Bayesian justification is not indispensable for the participants of a dialogue in which this foundational issue has arisen. Alternatively, the proponent can justify the connection by appealing to some relevant expert ("that what my teacher says"), or to an analogy ("the majority was right regarding that other issue"), or by way of some well-chosen examples ("look at her, her opinion is right"), or by means of statistical information ("this percentage in that group answered question such-and-so correctly").

One special way of providing a justification of premise B3 can be provided by yet another result in the intellectual history of probability, namely Condorcet's jury theorem – also discussed by Hahn and Hornikx. This theorem states that groups may outperform individuals as regards the correctness of their judgments. If under specific conditions a majority of people (the "jurors") judges a proposition to be true rather than false, then the probability of the majority being right is higher than the probability of each of the individuals being right, and the larger the group, the more the group outperforms each of the individuals. In such settings, as Dedrick says, "Popularity is a truth tracker" and "It is not a fallacy to appeal to the popularity of a belief as evidence for its truth" (2019, pp. 158-159).

In short, if information about the outright reliability about the reference group is available then a Bayesian justification becomes a serious option for any user of an argument from popularity. In those cases, can a Bayesian justification be expected to provide a sufficient justification? I conclude this section by discussing two related considerations that suggest that it can only be of limited use in critical examination dialogue.

First, the point of a discussion is to examine an issue on the basis of considerations that participants accept and understand. Nothing but an improved understanding of the issue, and of the perspectives on the discussants, should prompt an opponent to concede the thesis. But, an argument from popularity at most yields that the majority verdict is correct, without shedding light on why the outcome is correct. For all we know, the group could be reliable by sheer luck. Arguments from popularity suffer from a lack of transparency problem.

Second, the addressee typically has considerations that motivate her to resist the thesis. The proponent should tailor his arguments at his opponent by responding to whatever makes her doubtful. Typically, an argument from popularity does not speak to the addressee's concerns. Arguments from popularity suffer from a lack of responsiveness problem. Any argument that suffers from a lack of transparency and a lack of responsiveness, whatever its other merits, flops in discussion.

Third, the epistemic support of an argument from popularity only bears on the issue indirectly, rather than directly. The higher order evidence provided by popularity can have a bearing on the opponent's attitude to her own belief (commitment) without necessarily changing anything to the belief (commitment) and its strength itself, except that it affects her confidence with respect to that credence.

For a discussion on settings where the proponent appeals to the relative reliability of people, I refer to (van Laar 2023).

4. The Fallacy of Popularity

It is possible to advance a variety of fallacy charges against a proponent who, allegedly, uses an argument from popularity fallaciously. What to conclude from this? One thing is that there's not much unity in the way arguments from popularity can decay to fallacies. This then prompts the question whether it makes sense to include anything like "the Fallacy of Popularity" in our normative theory.

I will make my case with a speech from Geert Wilders, a right-wing, populist member of the Dutch parliament's second chamber. The speech dealt with the high number of asylum seekers from war-ridden Syria. Wilders party, the Party for Freedom (PVV), strongly opposed the reception of these people, mainly for the reason that many of them were Islamic, which was framed as a threat for the country's national identity: "I said so yesterday: if anything, it has become clear during this general budget debate that this government and this chamber [the second chamber] no longer represent the Dutch people. An enormous catastrophe is coming towards us. The whole of the Netherlands feels it. The whole of the Netherlands sees it. The whole of the Netherlands is crying out for action. But no action is coming from the chamber or the government." (Handelingen der Staten Generaal 2015)

One element of this fragment is the standpoint that it is really the case that a catastrophe consisting of the arrival of asylum seekers from Syria is imminent, and that what justifies the correctness of this thesis is that the whole of the Netherlands feels and sees that this catastrophe is heading their way. I focus on this specific argument from popularity. I use the pragma-dialectical theory of fallacies as specific kinds of violations of rules for critical discussion (van Eemeren and Grootendorst 2004, pp. 190-196).

1. The opponent could point out that the proponent employs an argumentation scheme (the argumentation scheme from popularity) that she does not accept, and that she regards as unacceptable for the conversation at hand. What is more, she could add, the proponent presents the scheme as unproblematic, thereby falsely suggesting that the reasoning has more credit than the dialectical situation warrants.

In our example, Wilders could be told that it is up to the representatives to weigh all the substantive considerations and how they bear of the various policy options themselves, and that any such argument from popularity, and thereby the use of the underlying scheme, is inappropriate in this conversational context. In the pragma-dialectical theory, this charge amounts to the claim that the proponent has violated the *Argument Scheme Rule* by using an – in that context – inappropriate argumentation scheme.

2. The opponent may point out that, even though there may not be anything intrinsically wrong with the argumentation scheme in the conversation at hand, the scheme has been applied incorrectly by having a premise that is false or doubtful.

Characteristic of contemporary populism as a political movement, is that it claims that there exists something like "what the people want," and that their spokesmen have privileged access to this, and thus can act as the mouthpiece of an allegedly unified will of the people. Arguments from popularity that include such a claim as a premise are fallacious when the highly questionable, if not false, premise gets presented as a matter of course. Wilders could be charged with falsely representing his view as being supported by almost all Dutch citizens, and that is what actually happened in the follow-up of the debate, where it was pointed out that a majority of citizens actually voted for representatives who favoured a much more lenient policy regarding asylum seekers. We could label this mistaken usage of the argument from popularity a populist fallacy.

The proponent may (implicitly) appeal to the idea that the people who support his thesis have independently arrived at that thesis, so as to boost the epistemic value of that majority view (in line with the reasoning behind the Condorcet's jury theorem), whereas the persons involved lack the required independent judgment, and merely parrot some propagandistic opinion leader (see van Laar 2023).

When in such cases a problematic premise is presented as unproblematic, the opponent can charge the argument from popularity as fallacious. The opponent could provide a two-fold diagnosis. First, she could explain that the proponent applies the argumentation scheme from popularity, which in itself need not be bad thing, and that he does so in an incorrect way, by including a premise that is unacceptable. Thereby, the proponent violates the *Argument Scheme Rule*, but in a different manner than before. Second, she could point out that this unacceptable premise is presented as a matter of course, which violates the *Starting Point Rule*.

3. Freeman stresses that the limited degree of support that popularity can provide, can easily be inflated (1995, pp. 266-267), in which case the proponent of the argument from popularity can be charged with committing the Fallacy of Hasty Conclusion (Johnson and Blair 1994, pp. 70-75). The proponent may assign a higher probability to his thesis than is warranted by the level of popularity and the (average) reliability of the reference group. Or, the proponent may act as if the critical consideration that motivates the opponent to challenge the thesis no longer requires a response. In such cases the argumentative strength of the argument gets inflated, or so the opponent may explain when charging the proponent with the Fallacy of Popularity. Wilders could be told that even when all these people do believe that the said catastrophe is imminent, the conclusion cannot be drawn that that we cannot but accept that the catastrophe is, indeed, imminent. After all, even when on average these

citizens have some reliability, the degree of certainty with which the conclusion gets drawn may not be warranted.

Again, the opponent could provide the same two-fold diagnosis, but now with a focus on the justifying force of the combined reasons in support of the conclusion, rather than on the one popularity premise. Thus, she could explain that the proponent applies the argumentation scheme from popularity incorrectly by relying on an argumentative connection that is too weak. Thereby, the proponent violates the *Argument Scheme Rule*. Because the sufficiency of the argumentative connection has been presented as a matter of course, the proponent also violates the *Starting Point Rule*.

4. The opponent could claim that the proponent's argument functions to supress an in-depth discussion about the merits of the thesis. The indirect evidence provided by the beliefs of people then is said to divert from the examination of substantial evidence and considerations directly bearing on the issue at hand (cf. on the Fallacy of Diversion: Johnson and Blair 1994, p. 93). In Jansen's terminology, the popularity of a policy gets misapplied when used to beat a discussion to death and to bypass genuine deliberation on the policy (2019, p. 362).

Wilders could be told that he is diverting attention from considerations such as the basic needs of people, human rights or the willingness of Dutch citizens to support refugees. Such a contribution can be regarded as a violation of the *Obligation-to-Defend Rule*, that requires a proponent not to evade his burden of proof: even if the argument from popularity provides some epistemic support, in these cases they also turn the attention away from considerations that (also) merit attention.

5. The proponent can commit a Fallacy of Intimidation by advancing an argument from popularity (Johnson and Blair 1994, pp. 167-190). The argument may exert pressure on the opponent to yield to the thesis, and to defer to the proponent's point of view - motivated by a fear for being considered a social, or an epistemic, or a moral aberration, the odd one out, difficult and uncooperative, impertinent and disrespectful, abnormal, and so forth. What does the trick is the veiled threat of leaving the addressed opponent alone and abandoned, outside of the community of people who are in their right mind and have a proper (moral or epistemic) sense - if she does not yield to the thesis. In such cases, the argument from popularity exploits the human need for recognition, and damages the critical self-steered thinking and decision making of the addressed opponent.

Wilders could be criticized on account of his trying to blacklist people who tend to empathise with war refugees, and would be willing to accept them as asylum seekers; after all, they apparently do not befit the Dutch community.

When the opponent wishes to point out that the proponent's argument from popularity functions as an intimidation tactics, she can admit that on the one hand the proponent offers an argument, so that he invites the opponent to inspect and assess his reason in support of his thesis. But then she can explain that on the other hand, due to the way the proponent frames and words his argument, the argument from popularity functions as a non-argumentative, manipulative, and primarily emotion-based device to causally affect the opponent, rather than to rationally persuade her.

Thus, the proponent can be criticized as violating the *Freedom Rule*, according to which it is impermissible to prevent others from advancing their positions by exerting pressure on them. And given the causal role of the emotions of wishing to belong to a community, of being appreciated by peers, and of the fear of being ostracised, the opponent may equally well point out that the proponent violates the *Relevance Rule*, according to which (among other things) it is not allowed to merely exploit pathos when argumentation is required.

Even though arguments from popularity are epistemically interesting and potentially even strong arguments, it is a type of argumentation that easily derails. What is more, there are at least five quite different ways in which such arguments can turn out to be fallacious. What to conclude from this?

We can take a practical approach towards any classification of fallacies, even when taking a more principled stance on the importance of fallacies and on the definition of "fallacy." The concept of a popular opinion is a highly salient one in public discourse, and it is easy to find both cases where real-life arguers use it to support their own opinion, and where they use it to diagnose what they regard as fallacies of popularity. In public discourse, people seem to have an ambivalent stance towards the argumentative merits of popularity, but the term is clearly a part of the meta-language used by real-life arguers. The concept of the Fallacy of Popularity is useful, because it allows the opponent to say (quoting Johnson and Blair): "Hold on, are you saying that because everyone in your group believes it, therefore it is true?" (Johnson and Blair 1994, p. 177). Having this fallacy charge at one's disposal is useful, for it provides an entry point for a more fine-grained evaluation, and if the need arises, the charge "Fallacy of Popularity!" can be specified along the way.

5. Conclusion

This paper dealt with the conceptualisation of arguments form popularity, with the epistemic assessment of such arguments, and with the question in what circumstances the proponent of any such argument could reasonably be charged with having committed the Fallacy of Popularity.

After having presented an argument scheme based definition of "argument from popularity", I showed that not all argumentative appeals to popularity are really arguments from popularity, and that some quite plausible democratic appeals to popularity fall off the wagon.

Then I contended that arguments from popularity can have real epistemic merit, because in some circumstances people can be regarded as reliable, so that an appeal to the popularity of a thesis may become a serious reason for that thesis. But in the context of a critical examination dialogue, arguments from popularity have a number of serious drawbacks, and as a result, there is neither cause for enthusiasm about such arguments, not for being dismissive about them.

Finally, I distinguished five different ways in which an argument from popularity may be fallacious. At the same time, for practical reasons I stopped short of concluding that we better drop the concept of the Fallacy of Popularity.

References

Aikin, Scott F., and Robert B. Talisse (2019). *Why We Argue: A Guide to Political Disagreement in an Age of Unreason* (second edition). New York, et al: Routledge.

Dedrick, Don (2019). Is an appeal to popularity a fallacy of popularity? *Informal Logic*, 39, pp. 147–167. Doi: 10.22329/il.v39i2.5101.

van Eemeren, Frans H., and Rob Grootendorst (1992). *Argumentation, Communication, and Fallacies: A Pragma-Dialectical Perspective*. Hillsdale, NJ: Lawrence Erlbaum.

van Eemeren, Frans H., and Rob Grootendorst (2004). *A Systematic Theory of Argumentation: The Pragma-dialectical Approach*. Cambridge: Cambridge University Press.

Freeman, James B. (1995). The appeal to popularity and presumption by common knowledge. In: H. V. Hansen and R. C. Pinto (Eds.), *Fallacies: Classical and Contemporary Readings* (pp. 265–73). University Park: Pennsylvania State University Press.

Godden, David M. (2008). On common knowledge and Ad Populum: Acceptance as grounds for acceptability. *Philosophy & Rhetoric*, 41, pp. 101-129. Doi: 10.1353/par.0.0000.

Govier, Trudy (2005). *A Practical Study of Argument* (6th edition). Toronto: Thomson, Wadsworth.

Hahn, Ulrike, and Jos Hornikx (2016). A normative framework for argument quality: Argumentation schemes with a Bayesian foundation. *Synthese*, 193, pp. 1833–1873. Doi: 10.1007/s11229-015-0815-0.

Handelingen der Staten Generaal (2015). De Algemene politieke beschouwingen naar aanleiding van de Miljoenennota voor het jaar 2016; Tweede Kamer, 3e vergadering, donderdag 17 september 2015. *Handelingen der Staten Generaal*, Tweede Kamer, zittingsjaar 2015-2016.

Hinton, Martin and Jean H. M. Wagemans (2022). Evaluating reasoning in natural arguments: A procedural approach. *Argumentation*, 36, 61-84. Doi: 10.1007/s10503-021-09555-1

Jansen, Henrike (2020). "The people want it": Analysis and evaluation of the populist argument in the context of deliberation. *Journal of Argumentation in Context*, 9, pp. 342–367. Doi: 10.1075/jaic.17028.jan

Johnson, Ralph H., and J. Anthony Blair (1994). *Logical Self-Defense* (3rd edition). Toronto: McGraw Hill Ryerson.

van Laar (2023). Arguments from popularity: Their merits and defects in argumentative discussion. *Topoi*, 42, pp. 609–623. https://doi.org/10.1007/s11245-022-09872-4.

Wagemans, Jean H. M. (2016). Constructing a Periodic Table of Arguments. In P. Bondy & L. Benacquista (Eds.), *Argumentation, Objectivity, and Bias: Proceedings of the 11th International Conference of the Ontario Society for the Study of Argumentation* (OSSA), 18-21 May 2016 (pp. 1-12). Windsor, ON: OSSA.

Walton, Douglas N. (1999). *Appeal to Popular Opinion*. University Park: Pennsylvania State University Press.

Walton, Douglas N., Reed, Chris, & Macagno, Fabrizio (2008). *Argumentation Schemes*. Cambridge: Cambridge University Press.

COLLECTIVE ETHOS, AD HOMINEM, AND AUDIENCE JUDGMENT: THE CREDIBILITY OF THE INSTITUTIONAL "FACES" OF FACEBOOK

JIANFENG WANG
Center for Studies in Rhetoric & Argumentation
College of Foreign Languages
Fujian Normal University
8 Shangsan Rd., Fuzhou 350007, Fujian Province
China
414856370@qq.com

CHRISTOPHER W. TINDALE
Department of Philosophy
University of Windsor
Ontario, Canada

Abstract

By engaging the contemporary standard readings of the Aristotelian ethos, we focus on the credibility of the arguer or the argument itself that can be established through the projection of a credible persona in the individual and collectie senses. Argumentative credibility is an important dimension in social argumentation simply because it involves relations between arguers at a particular moment and in a particular context. Credibility should thus be an important dimension in the reconstruction of the Facebook crisis as a case study of social argumentation, involving the dual dimensions of individual and collective ethos. In this paper, we address the issue of credibility by looking at it from the perspectives of individual and collective ethos. The central argument of this paper is that the collective image of an institution (e.g. The Facebook company) is an act of individualization, while individual ethos is collectivized in such a way that it becomes the very idealized or much-anticipated collective ethos.

1. Introduction

Not everyone fully agrees that the Facebook company has encountered or is encountering what is called "a historic crisis" (Allyn, 2021). But hopefully, we all agree to a certain degree that we are facing an unprecedented privacy crisis vis-à-vis the Facebook crisis, which is both brought about by the shared pseudo-reality created by the Internet and by social media platforms including Facebook, Twitter, Instagram, etc. Therefore, the crisis we're describing here is rendered in a symbiotic relationship. And it is meant less in an economic or financial sense, but more in a discursive or rhetorical sense, i.e., a crisis over the institutional ethos or credibility of the social media company in question. In particular, it seems that in recent years, the Facebook company has been facing a credibility crisis, culminating in a situation involving the quartet of Frances Haugen as the whistleblower, the mass media, the Facebook company, and the U. S. Senate.[1] At its center is the credibility of the Facebook company's public image.

One of the earliest conceptions of discursive credibility dates back to Aristotle in his *On Rhetoric* (Aristotle, 2007), and this is done in his discussion of ethos as one of the three means of persuasion. Aristotle's comments on the relationship between ethos and credibility in his *On Rhetoric* still reverberate in today's social argumentation. According to Aristotle, the credibility of the arguer or the argument itself could be established through the projection of a credible persona, as is shown in the Aristotelian defintion of ethos: "[There is persuasion] through character whenever the speech is spoken in such a way as to make the speaker worthy of credence; for we believe fair-minded people to a greater extent and more quickly [than we do others], on all subjects in general and completely so in cases where there is not exact knowledge but room for doubt" (1356a). Argumentative credibility is an important dimension in social argumentation simply because it involves relations between arguers at a particular moment and in a particular "rhetorical situation" (Bitzer, 1968). The crucial importance of credibility is also exemplified in the fact that argumentation is social in nature involving not only social norms but

---

[1] But clearly, this is the result of a series of discursive incidents over the years. It is estimated that Mark Zuckerberg has testified 7 times before the U.S. Congress. He went into a Congress hearing organized by the Senate Commerce and Judiciary Committees for the first time on April 11, 2018, the themes of which covered privacy, data mining, regulations and Cambridge Analytica. Here is the transcript of Mark Zuckerberg's Senate hearing: https://www.washingtonpost.com/news/the-switch/wp/2018/04/10/transcript-of-mark-zuckerbergs-senate-hearing/. On another occasion, there was the Congress hearing before the U.S. Senate Committee on the Judiciary on November 17, 2020. Here is a transcript available at: https://www.judiciary.senate.gov/download/zuckerberg-testimony.

also arguers, audiences, and cultural traditions, etc (Tindale, 2021). Credibility should thus be significant consideration in the reconstruction of social argumentation. But where does credibility come from or what is the source of the credibility of argumentative discourse? How should credibility be defined? How are personal and institutional credulities important in social argumentation? All these questions are yet to be answered in a crystal clear way.

Here in this paper, by drawing upon the Aristotelian conception of rhetorical ethos and its contemporary developments such as discursive persona (Arendt, 2003), pragmatic identity (Tindale, 2015), argumentative argument ad hominem (Walton, 1998), and collective corporate image (Wei, 2002), we treat all of these as variants of ethotic argument (Brinton, 1986). We argue that in the present-day information age, rhetoric as the most authoritative form of persuasion still dominates the activities of social argumentation on social media platforms such as Facebook, Twitter. In this sense, the Haugen vs. Zuckerberg case serves our purpose well. In responding to a series of accusations initiated by Frances Haugen, now widely acclaimed as the brave "Whistleblower," Mark Zuckerberg's reaction to Facebook's institutional crisis is seriously flawed in several ways, as will be shown in the case study.

That said, the Facebook crisis could certainly be interpreted in many other legitimate ways. We do not suggest that the Facebook crisis be treated as only rhetorical in nature, but that the controversy can be more fully unpacked from a rhetorical perspective. In this sense, we propose addressing it as a war of words centered over the institutional ethos or image, which becomes a site of rhetorically inventing and reinventing an image to influence the audience's judgment over a certain issue. However, what also interests us in this project is not just how different parties concerned "attacked" each other's "ethos," but also how different parties manage to modify the audience's cognitive environment (e.g. in the form of collective consciousness) in the overall arena of social argumentation and how the audience adjusts his or her judgment on the issue accordingly. In so doing, we will first provide a brief introduction to the Aristotelian concept of ethos and the contemporary advances in its reconceptualization (Charland, 2001; Amossy, 2001), especially the relationship between individual and collective ethos, and the source of an authorized ethos (Perelman & Olbrechts-Tyteca, 1969; Reynolds, 1993). Three dimensions of ethos will be discussed in terms of audience judgment. Then follows a critical discussion featuring the abovementioned theoretical tools of the Zuckerberg Note as counterarguments against the public image crisis, especially its strategy of ad hominem. Finally, we will point out, as the case of Frances Haugen shows, that there is always a tension between an individual and collective (group or institution) image (Wang, 2020).

2. Ethos in the Aristotelian Standard and Its Problems

The projection of an individual or collective image matters in social argumentation. The earliest systemic account of image construction dates back to at least the Aristotelian concept of ethos in his *On Rhetoric*. Ethos in Aristotle's *On Rhetoric* is understood as the "most authoritative form of persuasion" which can be made available in each particular case through the projection of an approapriate "character" of the speaker (1356a), that could influence the audience's judgment on an issue in a political, legal or epideictic context. On the Aristotelian standard, ethos as one of the three means of persuasion is the discursive projection of the rhetor or arguer's own character, and the rhetorical speech or text per se is the only source of ethos. In Aristotle's terms, ethos is actually the persona or image of the individual rhetor or arguer. On the standard reading of this, ethos is only authorized by the speech or discourse per se, which has nothing to do with the previous reputation and social status of the rhetor or speaker. This is the standardized reading of Aristotelian ethos: there is only the discussion of an individual ethos of the speaker without any discussion of a collective ethos of a group or institution or entity (Charland, 2001), and there is only a linguistic dimension of ethos (Amossy, 2001).

These limits are not without their problems. Prior to his definition of ethos as one of the three means of persuasion, Aristotle differentiates between non-artistic (atechnic) and artistic (entechnic) means of persuasion. The former set of things are preexisting, not provided or rhetorically maneuverable by the potential speaker. On the list of non-artistic means are things such as "witnesses," "testimony from torture," "contracts," etc. Collected in the domain of artistic means are pathos, logos, and ethos, all of which are "prepared by method" or rhetorically maneuverable by the potential speaker (1355b). Therefore, as Aristotle points out himself, "one must use the former and invent the latter" (1355b). To paraphrase what Aristotle means in these words, the non-artistic means of persuasion are not subject to invention, while the artistic ones are. The potential speaker can "use" the non-artistic means of persuasion and "invent" the artistic means to win over the audience. Apparently, this has been the discursive reality of the Aristotelian concept of ethos, at least in the contemporary sense. There seems to be no room for doubt.

However, when we consider how the particularized "presentation of the self" by the potential speaker can impact the judgment of the audience on a certain issue (Goffman, 1957), and how the audience judges the crediblity of the argument presented for their assent, we probably need to rethink the Aristotelian distinction between artistic and non-artistic means. As Aristotle points out, the potential speaker can legitimately "use the [non-artistic means]" and "invent the [artistic means]" in each particular case of the persuading effort. It is in the minds of the audience that an image of the potential speaker appears. It is the audience who will judge what kind of person the potential speaker is. And it is the audience who receives the argument presented to them in ways entrenched in the spatial and temporal matrix. In a nutshell, artistic and non-artistic means of persuasion could be legitimately employed in impacting the way the audience receives an argument. Therefore today, we can probably reject the standardized Aristotelian distinction between artistic and non-artistic means. For example, witnesses, testimonies, and contracts of the Aristotelian sense are all usable, interpretable, and inventable, as we will see in the testimony Frances Haugen provided in her opening remarks on the Capitol Hill on October 4, 2021.

Also problematic is the standardized reading of the Aristotelian ethos as being individual in nature. Aristotle, on this account, stops short of talking about a collective ethos (Charland, 2001). This claim is corroborated when Aristotle refers to the potential speaker in the singular third person pronoun, as is the case here: "[There is persuasion] through character whenever the speech is spoken in such a way as to make *the speaker* worthy of credence" (1356a; emphasis added). But here, it is unclear whether "the speaker" refers to the individual speaker him-/herself or whether the speaker represents an institution (for example, it could be a judge speaking on behalf of a court in ancient Athens). More importantly, essentially speaking, an individual ethos could be understood as a personalized representative of shared values within a community, while a collective ethos could be a collectivized representative who serves to activate, emphasize, and simply make present the shared values deeply entrenched in a community. There could be an intense relationship between the individual ethos and the collective ethos (Frances Haugen vs. Facebook). As we will see, a conflict of values is inherent in the different "faces" in the case of the institutional images of Facebook (the diversified faces of Facebook, for example).

3. The Individualized "Faces" of Facebook

At the heart of the controversy over the institutional image of the Facebook company is a defense of "its carefully crafted, decade-old image as a benevolent company just wanting to connect the world" (Ortutay, 2021). Therefore, the discourse production of "ethos" here becomes a battleground of rhetorical contention, which deserves our attention. In Mark Zuckerberg's first Senate Hearing focused on "data privacy and Russian disinformation" jointly organized by the Commerce and Judiciary committees on April 10, 2018,[2] which was ignited by the then Cambridge Analytica incident, there is a set of conflicting images of the Facebook company. The positive group of images includes "the successful storyteller of the American Dream," "the social media giant," and "a platform for all ideas," etc. The negative group includes "a platform for all ideas that is not operating impartially," a "technological monopoly," and a "digital monster," so on and so forth. Both groups of images are in stark contrast to each other, occupying the two extremes of the linear line in ethotic reasoning. Seen this way, ethotic reasoning is an act of "operational essentialism" or "strategic essentialism," i.e., "the process of making an identity ingredient in the core part of one's persona, which legitimizes the right to speak"(Palczewski at al., 2022, pp. 197-198). The credibility issue is best captured here in the pair of questions raised by Senator John Thune in the hearing:

> **[Case #1]** After more than a decade of promises to do better, how is today's apology different? And why should we trust Facebook to make the necessary changes to ensure user privacy and give people a clearer picture of your privacy policies?

Here in this quote, the keyword is "credibility" or the credible ethos of the Facebook company. In Senator Thune's view, Facebook has made one promise after another to do better in protecting users' privacy data in the past decade, but none of the promises has come true. The apologies also turn out to be lip services. Under such circumstances, the credibility of the company is suspicious. Noteworthy is the fact that here since Senator Thune is addressing Mark Zuckerberg in the form of Q&A, the individual company and its CEO are combined ("your privacy policies") to suggest the persona of Zuckerberg represents the institutional image of the company. Therefore, the collective image of the company is an act of individualization.

On another occasion, the same strategy of individualization is adopted. In the Senate's Committee on Commerce, Science, and Transportation Hearing that dwelled on "Children & Social Media Use" on October 4, 2021, Senator Richard Blumenthal, also the chairman of the Committee, directly challenged the moral integrity of Mark Zuckerberg by calling on him to look at himself in the mirror. The Mark Zuckerberg in the mirror is a person of irresponsibility and arrogance, who adopts the personal style of "no apologies, no admission, no action, nothing to see" as his "new modus operandi." Blumenthal hints at whether the "Mark Zuckerberg" in the

---

[2] A transcript is available here provided by The Washington Post: https://www.washingtonpost.com/news/the-switch/wp/2018/04/10/transcript-of-mark-zuckerbergs-senate-hearing/. Unless otherwise noted, the discussion in this section makes references to the transcript.

mirror is recognizable to himself, i.e., the "real" or "familiar" Zuckerberg. Contemporary German philosopher Hannah Arendt rejects the idea that there could be an "inner self" from which all the other versions of the self are its radiations (Arendt, 2003, p. 13). By making present the inconsistency of his persona before his colleagues, the mass media, Frances Haugen, and the American audience who are watching the live broadcasting of the hearing, Senator Blumenthal actually paves the way for the projection of an image of the Facebook company that is "putting astronomic profits before people."

An individualized collective ethos of the Facebook company must be activated in the personal experiences of the audience. For this to happen, Senator Blumenthal appeals to American parents in two ways. First, he accuses the social media company of "targeting children pushing products on pre-teens not just teens, but pre-teens that it knows are harmful to our kids' mental health and wellbeing," concealing these facts and seeking to "to stonewall and block this information from becoming public" to parents and the committee as well. Before concluding his opening remarks, he reminds the target audience of the "heartbreaking and spine chilling stories about children" he had heard in the past weeks and days from parents, in which children were "pushed into eating disorders, bullying online, self injury of the most disturbing kind, and sometimes even taking their lives because of social media." Here, the extremely negative image of the social media as the de facto killer of victimized children is now reactivated in the minds of the parents.

Reminiscent of the moral bankruptcy of the institutional image of Facebook is what is called "Big Tobacco" moment faced by the tobacco industry in the 1990s in the U. S.[3] Prior to the Senate hearing on October 4, 2021, the Facebook company was compared to the tobacco companies in the series of reports entitled "the Facebook files" by the *Wall Street Journal*. The central finding of this series, based on a review of internal Facebook documents provided by Frances Haugen, is that "Facebook Inc. knows, in acute detail, that its platforms are riddled with flaws that cause harm, often in ways only the company fully understands."[4] The intertextuality of consciousness is established not only between the *Wall Street Journal* series and the Senate hearing, but also

---

3 According to a news report by the Los Angeles Times, in 1994, the top five U.S. tobacco executives stood before Congress and lied about the addictiveness of their products. The hearings had a ripple effect. As the tobacco industry's history of deceit and harmful products was laid bare, its public credibility crumbled. Retrieved from: https://www.latimes.com/opinion/story/2021-10-27/op-ed-big-oil-congress-hearings.
4 For a brief introduction to the Facebook Files, here is the link: https://www.wsj.com/articles/the-facebook-files-11631713039?mod=series_facebookfiles

between the "Big Tobacco" moment and the credibility crisis of Facebook, in the opening remarks of Senator Blumenthal, noted here:

> **[Case #2]** Facebook and big tech are facing a big tobacco moment, a moment of reckoning, the parallel is striking. ...big tobacco knew that its product caused cancer but that they had done the research, they concealed the files, and now we knew and the world knew. And big tech now faces that big tobacco jaw dropping moment of truth. It is documented proof that Facebook knows its products can be addictive and toxic to children. And it's not just that they made money, again, it's that they valued their profit more than the pain that they caused to children and their families.

Here, the effectiveness of analogical reasoning is determined by two factors: the comparisons are invited between the social media platform and the tobacco companies; the "relevant similarities" are so striking that they could not be outweighed by "the way in which they are dissimilar" or relevant dissimilarities (Tindale, 2023, pp. 127-130). Here in this case of argument from analogy, the two analogues are respectively the tobacco companies (the first one) and the Facebook company (the second or primary one). The relevant similarities here could include the following: both companies had done the research but had both chosen to conceal the truth that their products could be harmful to health or even cause diseases; the products of both companies are addicted and toxic to children; both companies valued profit more than people; both sought to control people and play a dominant role in the relevant fields (Frenkel & Kang, 2021). But of course, there are several ways in which they are dissimilar: in the case of the tobacco companies, people are addictive to nicotine, a poisonous alkaloid which causes lung cancer, while in the case of the social media platform, people get addicted to nonphysical things such as computer games, ads, misinformation, and exposure to pornographic products; smoking cigarettes is expensive, while using social media does not cost much. However, the relevant similarities in this case are stronger than these dissimilarities and suffice to strengthen the conclusion that Facebook knows that "its platforms are riddled with flaws that cause harm, often in ways only the company fully understands." If the audience makes comparisons between the two analogues and accepts the striking relevant similarities, then this can serve the epistemic purpose of drawing inferences to a conclusion that the institutional image of the Facebook company, by putting profit before people and concealing the research findings, is unethical and immoral.

4. The Collectivized Persona of Frances Haugen

The unethical and immoral nature of the institutional image of Facebook runs counter to the idealized image Frances Haugen had envisaged for Facebook. As Geller and O'Brien put it in the Facebook Papers series collaborated on by 17 American news organizations, "The idealism she and countless others had invested in promises by the world's biggest social network to fix itself had been woefully misplaced" (2021). This could account for her leaving the company in May 2021 after serving in it for about two years.[5] Haugen herself explains it this way, "I joined Facebook because I think Facebook has the potential to bring out the best in us" (2021). "The best" of what "in us" remains unclear here, but she could be read as being identified with the idealized group ethos. "The best" also suggests something of the common values that both sides could stand on. Almost immediately after saying this, she explains the reason for the breakup: "But I am here today [testifying before the Senate committee] because I believe Facebook's products harm children, stoke division, and weaken our democracy" (2021). What she means here is not only that the idealism of the individualized collective ethos is not warranted, but also that the individual ethos existing in the person and the idealized collective ethos expected of the social media platform are heading in the opposite directions. That said, Haugen also insists she does not intend to damage Facebook but wants to change it. That is why she decides to speak out. Therefore, this will be a process of negotiation. If that anticipated change happens, she may be back. If not, she just walks away. This also reminds us of the unobserved complex interactive relationship between an individual ethos and a collective ethos. In many cases, a collective ethos could be just the individualized ethos of a specific person, and an individual ethos could be a collectivized identity of a group or institution.

In the case of Frances Haugen, the different versions of her persona as the brave whistleblower, risk-taker, truth-teller, and "21st century American hero" go far beyond their literal senses. "Frances Haugen" suddenly becomes a symbolic name and is solidified into a collective consciousness or memory. Canadian rhetorician Maurice Charland proposes the theory of "constitutive rhetoric." According to Charland, constitutive rhetoric "presumes and asserts a fundamental collective identity for its audience, offers a narrative that demonstrates that identity, and issues a call to act to affirm that identity" (1987, p. 638). This is exactly what is on the agenda of Senator Richard Blumenthal. His point

---

[5] But of course, her leaving the company and speaking out as a product manager of the company could be motivated by other factors. There could be explicit reasons (made available to us by Haugen in her opening remarks, interviews, etc.) and implicit reasons that remain unknown. In this respect, more explanations could be expected from her forthcoming book The Power of One (2023).

is that Haugen has set a leading example for "other whistleblowers out there" and "other truth-tellers in the tech world who want to come forward." She has shown them "a path to make this industry more responsible and more caring about kids and about the nature of our public discourse generally, and about the strength of our democracy." Clearly, an individual ethos is collectivized in such a way that it becomes the very idealized or much-anticipated *collective ethos*.

But of course, from the perspective of the social media platform, they do not like to see this individual ethos collectivized in the very way assumed by Senator Blumenthal. Rather than addressing the lost idealism seen in the promises the Facebook company has made for its countless users, the Facebook team chooses to attack or play down Frances Haugen's personal character. This is best captured in the official statement issued by Lena Pietsch, Director of Policy Communications of the company shortly after the Senate hearing:

> [Case #3] Today, a Senate Commerce subcommittee held a hearing with a former product manager at Facebook who worked for the company for less than two years, had no direct reports, never attended a decision-point meeting with C-level executives -- and testified more than six times to not working on the subject matter in question. We don't agree with her characterization of the many issues she testified about. Despite all this, we agree on one thing; it's time to begin to create standard rules for the internet. It's been 25 years since the rules for the internet have been updated, and instead of expecting the industry to make societal decisions that belong to legislators, it is time for Congress to act.

The gist of the statement here is to play down Haugen's credentials. Working for the company for less than two years, having no direct reports, never attending a decision-point meeting with C-level executives, and testifying more than six times to not working on the subject matter in question are all inferences leading to the implicit conclusion that Haugen is not at all in the position where she is qualified to talk about the subject matter. There is also the implicit argument ad hominem here, because Frances Haugen as a lower-level product manager is manifestly between the lines. The team is not addressing whether the accusations raised by Haugen are worthy of being considered, but taking issue with the hierarchical status of the accuser or utterer. This is also the strategy taken by Mark Zuckerberg himself in his "Note," which is addressed firstly at an inner audience and seemingly designed to boost the collective low morale within the company resulting from the perlocutionary force of Haugen's speech act of revelations. Expressions or phrases such as "mischaracterization," worrying "incentives," a "disheartening...false narrative," and the "moral, business, and product incentives" which all point in the opposite direction to Haugen's remarks, all suggest that the audience should reject the messages provided by the whistleblower, without mentioning the whistleblower's name. This last point reflects the collective strategy of playing down the moral character of the whistleblower.[6]

5. Some Concluding Remarks

As could be seen in the collective responses made by the Facebook team in general, and in particular, their responses to a series of accusations initiated by Frances Haugen, the Whistleblower, Mark Zuckerberg's and his team's responses to Facebook's institutional crisis are seriously flawed in several ways. This can be viewed as a typical case of social argumentation seen from the perspective of rhetorical ethos that includes the individual and collective dimensions. First, Zuckerberg's "rhetoric of silence" could be a rhetorical miscalculation which allows more discursive mutations to emerge to his disadvantage. This silence could also create more space for people in the mass media and Frances Haugen (the whistleblower) to challenge Facebook's institutional image and move forward their plea for more outside regulations to be imposed (by US Congress, for example) on social media platforms such as Facebook. Second, rather than addressing all those accusations and quesitons raised by different parties concerned, Zuckerberg in his "Note" projects himself as an irritative and emotional persona, easily recognizable between the lines. Third, Mark Zuckerberg and some senior officials of the Facebook company have collectively underestimated the pragmatic force of Frances Haugen's arguments by undercutting the ethos of Frances Haugen or worse by resorting to the fallacious argumentum ad hominem.

References

Allyn, B. (2021). Here are 4 key points from the Facebook whistleblower's testimony on Capitol Hill. *NPR*, October 5, 2021. https://www.npr.org/2021/10/05/1043377310/facebook-whistleblower-frances-haugen-congress

Amossy, R. (2001). Ethos at the crossroads of disciplines: rhetoric, pragmatics, sociology. *Poetics Today* 22.1, 1-23.

---

6 Facebook Whistleblower Frances Haugen Testifies on Children & Social Media Use: Full Senate Hearing Transcript. Retrieved from: https://www.rev.com/blog/transcripts/facebook-whistleblower-frances-haugen-testifies-on-children-social-media-use-full-senate-hearing-transcript.

Arendt, Hannah. (2003). *Responsibility and judgment.* New York: Schocken Books.

Aristotle. (2007). *On rhetoric: A civic theory of discourse.* Translated with Introduction, Notes, and Appendices by George A. Kennedy. 2nd edition. Oxford and New York: Oxford UP.

Bitzer, L. (1968). The rhetorical situation. *Philosophy and Rhetoric* 1, 1-14.

Brinton, A. (1986). Ethotic argument. *History of Philosophy Quarterly* 3.3, 245-258.

Charland, Maurice. (2007). Constitutive rhetoric. In Sloane, T. O. (Ed.), *Encyclopedia of Rhetoric* (pp. 638-641). New York and Oxford: Oxford UP.

Geller, A. & O'Brien, M. (2021). How one Facebook worker unfriended the giant social network. *The Associated Press*, October 10. Retrieved from: https://apnews.com/article/facebook-science-technology-business-congress-frances-haugen-80e92043b7211590b6be84dcc7a05b4a

Goffman, I. (1956). *The presentation of self in everyday life.* Edinburgh: University of Edinburgh Social Sciences Research Center.

Frenkel, S. & Kang, C. (2021). An ugly truth: Inside Facebook's battle for domination. New York: HarperCollins Publisher Inc.

Haugen, F. (2021). Statement of Frances Haugen. Released by the United States Senate Committee on Commerce, Science, and Transportation, October 4. Retrieved from: https://www.commerce.senate.gov/services/files/FC8A558E-824E-4914-BEDB-3A7B1190BD49#.

Haugen, F. (2023). *The power of one: how I found the strength to tell the truth and why I blew the whistle on Facebook.* Forthcoming by Little Brown company.

Ortutay, Barbara. (2021). People or profit? Facebook papers show deep conflict within. *The Associated Press*, Oct.25. https://apnews.com/article/the-facebook-papers-whistleblower-misinfo-trafficking-64f11ccae637cdfb7a89e049c5095dca.

Palczewski, C. H., Ice, R., Fritch, J., & McGeough, R. (2022). *Rhetoric in civic life.* 3rd ed. State College: Strata Publishing.

Perelman, C., & Olbrechts-Tyteca, L. (1969). *The new rhetoric: A treatise on argumentation.* Notre Dame: Notre Dame University Press.

Reynolds, N. (1993). Ethos as location: new sites for understanding discursive authority. *Rhetoric Review* 11.2, 325-338.

Tindale, C. W. (2015). *The philosophy of argument and audience reception.* Cambridge: Cambridge University Press.

Tindale, C. W. (2021). *The anthropology of argument: cultural foundations of rhetoric and reason.* New York and London: Routledge.

Tindale, C. W. (2023). *How we argue: 30 lessons in persuasive communication.* New York and London: Routledge.

Walton, Douglas. (1998). *Ad hominem arguments*. Tuscaloosa and London: The University of Alabama Press.
Wang, J. F. (2020). Place, image and argument: The physical and nonphysical dimensions of a collective ethos. *Argumentation* 34, 88-99.
Wei, Y. K. (2002). Corporate image as collective ethos: a poststructural approach. *Corporate Communications: An International Journal* 7.4, 269-276.

COGNITIVE BIAS IN SOCIAL ARGUMENTATION

MARK WEINSTEIN
Montclair State University
weinsteinm@mail.montclair.edu

Abstract
This paper looks at recent research indicating the deep connection between cognitive processes and information storage, retrieval and salience. Models of cognitive functioning challenge a simple resolution of the problem of belief entrenchment with particular relevance to social argumentation.

1. Introduction

Arguments about social and political issues are notorious for being unpersuasive. Entrenchment of social and political postures is increasingly obvious as divisions on such issues are the basis for the new tribalism and other deep divisions within democratic societies, where political and social argumentation is both common and freely exercised (see Edsall, 2022a, for a comprehensive analysis). The unwillingness of people to alter their views on issues of social and political concern in the face of counter argument and contrary evidence has been generally construed as bias within the psychological literature. Early research was focused on the persistence of racial bias, and its resistance to evidence in support of prior beliefs (Ehrlich, 1973). More recent work has focused on social issues and economic issues (Lewandowsky et. al. 2012) offering possible insights into what Nobel Prize winning economist Paul Krugman called 'zombie ideas,' ideas that 'should have been killed by evidence but keep on lurching along' (Krugman, 2020).

Research in cognitive science offers insight into these phenomena, building upon a long-standing concern with problematic reasoning. The early literature focused on performance errors, reasoning that fails to meet normative standards from both formal and inductive logic (Tversky & Kahneman, 1973). Research identified psychological mechanisms supporting such intergroup behaviors as stereotyping (Hamilton, 1981), with particular relevance to racial attitudes (Dovidio & Gaertner, 1986). More recent efforts have offered processing accounts that offer structural analyses that offer insights not available from the earlier psychological literature (Bargh & Ferguson, 2000; Evans, 2008). Speculative accounts look further, indicating the deep connections between emotion and memory, information retrieval and resistance to refutation. We will eventually focus on the work of Damasio (2012) and Thagard & Aubie (2008), who drawing upon a rich basis in recent experimental work on the physiology of the brain and nervous system, offer comprehensive models of cognitive functioning that offer an account of cognitive processes that questions the possibility of a simple resolution of problems such as belief entrenchment. It is the thesis of this paper that such an understanding of cognitive processes offers both a challenge and an opportunity for argumentation theorists, pointing to the need for a deep analysis of underlying commitments, implicit and explicit warrants that determine how we construct and evaluate arguments.

2. Why Cognitive science

Cognitive scientists, rather than looking at behavior alone, build functional models that account for the behavior using theoretic constructs (Gardner, 1987). I see this to have a clear analogy with early physical chemistry. In the history of physical chemistry, the increasing degree of articulation in the details that chemical theories explained, what I call 'consilience,' combined with breadth, that is, increase in the scope of a theory, were all predicated on a concern with deep theory. Consilience, breadth and especially depth is arguably the source of physical chemistry's enormous epistemic power (Weinstein, 2011). A parallel analysis of cognitive science can be seen as plausible evidence that cognitive science is in a position to sustain indefinite empirical growth and increasing theoretic strength (Weinstein, 2015). The promise of increasingly sophisticated computer simulations of mind offers possibilities for the description of the complex theoretic structures put forward. Complex descriptions that require computer modeling for their articulation offers a test of consilience unlike anything in the prior history of psychology. Computer simulations of interactions employ theoretic constructs based on a vastly increased knowledge of the structure of the brain, available through powerful advances in instrumentation, brain scans of various sorts. This enables the analysis of the range of cognitive behaviors.

We do not know which theories in cognitive science are correct, but if they can be developed consistent with the available evidence, they have the potential to grow in scope and detail as the theoretic predictions of ever-finer models of complex systems can be ascertained through computer simulations corresponding to the increasingly detailed experimental knowledge of the brain. That is, cognitive science shows potential for consilience. Like early physical chemistry, we don't know which theories in cognitive science will be sustained, but if a theory continues to yield important explanations, the potential for a growing and all-encompassing theoretic structure of psychology becomes plausible.

Cognitive science is, if nothing else, exceptionally broad in the scope of its concerns. The *Cambridge Handbook of Cognitive Science* (Frankish & Ramsey, 2012) lists eight related research areas that reflect different aspects of cognition, including perception, action, learning and memory, reasoning and decision making, concepts, language, emotion and consciousness. In addition, they list four broad area that extend the reach of cognitive science from human cognition standardly construed to include animal cognition, evolutionary psychology, the relation of cognition to social entities and artifacts and most essential, the bridge between cognitive science and the rest of physical science: cognitive neuroscience. Each of these is a going concern, and none of them is free of difficulties. Yet in all cases there is a sense of advance, of wider and more thoughtful articulation of theoretical perspectives that address a growing range of cognitive concerns. But as compelling as these characteristics are, it is depth that cognitive science shares with physical science, as both structures enable micro-explanations that can be seen to yield an overarching ontology (Weinstein, 2002).

The key to the epistemological power of cognitive science is its foundation in neuro-science. Speculations of instantiated neural mechanisms have systemic power much greater than their evidentiary weights. Such speculations offer an image of enormous potential warrant. For their enterprise, bridging between fundamental pre-cognitive processes such as physiological control and emotions to build the functional potential for memory and cognition offers deep structural warrants supported by reliable evidence and accepted theories. Moreover, their materialist assumptions permit a deep reduction to physiology, neurobiology, biochemistry and electrochemistry. A materialist foundation that any adequate theory of brain function must ultimately depend on. The question for us is what cognitive science has to offer to argumentation theorists, whose concern is understanding the role of evidence and underlying belief commitments in explaining the strength of arguments and especially their resistance to, what seems to be the heart of rationality, change in the face of counter-evidence.

3. Cognition and Bias

Research over decades indicates that our past associations affect our ability to alter our beliefs (Jacoby, et. al, 1989). A study of political beliefs showed resistance to argument that challenge our memories and commitments: "the persistence of misinformation might better be understood as characteristic of human thinking" (Lewandowsky et al., 2012, p. 114). Much of the available research relevant to the role of emotions in cognition focuses on bias and stereotyping. For example, the studies of unacknowledged bias indicate "influence of implicit stereotypes on judgment and behavior" (Blair, Ma, & Lenton 2001, p. 828). Unacknowledged, such attitudes may remain disconnected from a person's avowed beliefs: "Dissociations [between implicit and explicit attitudes] are commonly observed in attitudes toward stigmatized groups, including groups defined by race, age, ethnicity, disability, and sexual orientation" (Greenwald & Krieger 2006, p. 949). Such implicit biases create emotional disturbance when in the face of social pressure such views are put into question. "When one denies a personal prejudice (explicit bias) that co-exists with underlying unconscious negative feelings and beliefs (implicit bias] leading to diffuse negative feelings of anxiety and uneasiness" (Dovidio and Gaertner 2005, p. 42).

There are neural mechanisms that account for such phenomena. The prefrontal cortex which processes conscious thought and the so-called "executive functions," planning, goal setting, evaluation, and cognitive control is connected to other parts of the brain organizing input together into a coherent whole. Under the prefrontal cortex is the orbitofrontal cortex, which broadly supports self-regulation: physical, cognitive, emotional and social. These regions combine inputs to create the image of our physical body as well as perceptions of the external world and mental constructs (Dehaene, 2014). An interesting detail relevant for social cognition are so called "mirror neurons," neurons that fire both when you act and when you perceive another performing the same action and which allow us to infer or predict others' intentions (Iacoboni, et. al. 2005). Research indicates that mirroring of emotions, the degree of empathy we show others, is modifiable by real or perceived social relationships supporting ethnic or gender stereotypes (Amodio & Devine, 2006). There is evidence that biasing emotions reach deep into our biographies and are expressed in implicit biases. Evidence indicates that, despite expressed personal opinions "early and affective experiences may influence automatic evaluations more than explicit attitudes. In addition, there is growing evidence that systemic, culturally held beliefs can bias people's automatic evaluations" (Rudman, 2004, p. 81). Childhood based biases cause strong reaction such as fear of unfamiliar others, which has been correlated with activation in the amygdala (Dunham, Baron, & Banaji 2008). Biases interfere, on a neural level, with the ability to experience others. When "European-American subjects looked at the face of another European-American, there was a larger neural response than when they looked at African-American faces" (Lebrecht, et. al., 2009, p. 3). The result: "people do not mentally simulate the actions of [members of] outgroups.

Their mirror-neuronsystems are less responsive to outgroup members than to ingroup members" (Gutsell and Inzlicht 2010, p. 844).

Such results have been generalized in a theory of the "automaticity" of higher mental functions, which sees ordinary cognition as dependent on environmental and social factors (Bargh & Ferguson, 2000). Evans (2008) in response to the then prevailing dual-processing model that distinguished between System 1 (unconscious/automatic/low effort) and System 2 (conscious/explicit/high effort) offers a complex image of the interaction between what he terms unconscious and conscious cognition, seeing a variety of distinct and possibly incompatible systems. The work continues with the development of neural models that indicate the integration of cognition and emotion through abstract structures based on the known physiology of the brain.

4. Models of the knowing brain

Speculations as to the neural mechanisms have systemic power much greater than their evidentiary weights. Although speculative and very likely inadequate, they offer an image of enormous potential warrant. For their enterprise, bridging between fundamental pre-cognitive processes such as physiological control and emotions to build the functional potential for memory and cognition, offers deep structural warrants supported by reliable evidence and accepted theories. As indicated, materialist assumptions point to the deep reduction to physiology, neurobiology, biochemistry and electrochemistry that an adequate theory of brain function would depend on. This seems to me to parallel my account of the structure of scientific reasoning as exemplified by physical chemistry (Weinstein, 2016). As the models, indicated below, show, the brain coordinates functions across an array of inputs permitting an integrated response that enables perception, memory and purposes to bring together information necessary for coordinated action in the world. I see this as a clear parallel with consilience, the increasing systematic effectiveness across areas on concern as the sciences develop and new problems are confronted. Second the brain integrates the broad array of disparate information, proprioceptive, hormonal, electrical, and chemical, integrating new input with stored impute and modifying content in relation to newly acquired stimuli of many kinds. This seems to me parallel to breadth, the range of concern typified by physical chemistry. Most importantly, all of these functions are accounted for on increasingly defined more abstract levels, moving from gross physiological function to the operation at the cellular level, and if we accept materialism, to the molecular level, as we understand the functions of the neurological array on the deepest physiological levels. This has a clear parallel with the depth characteristic of physical chemistry, the reinterpretation of a theory in terms of a higher order, more abstract and more deeply ontological sense of the ultimate realties behind the phenomena. And this is despite the enormous gap between the simple models of neurological activity proffered and the brute facts of the living brain: 30 billion neurons making countless trillions of connections and sensitive to a wide array of known biochemical agents, with more perhaps to come. We turn to two such accounts, the ambitious attempts of Thagard and Aubie, (2008) and Damasio (2012) to bridge the gap between abstract structure and available physiological knowledge.

Thagard and Aubie draw upon both neurophysiology and computer modeling. This enables both theoretic depth and the possibility of increasing adequacy, even if the latter is no more that computer simulations of simplified cognitive tasks. They cite ANDREA, a model which "involves the interaction of at least seven major brain areas that contribute to evaluation of potential actions: the amygdala, orbitofrontal cortex, anterior cingulate cortex, dorsolateral pre-frontal cortex, the ventral striatum, midbrain dopaminergic neurons, and serotonergic neurons centered in the dorsal raphe nucleus of the brainstem" (Thagard and Aubie, 2008, p. 815). With ANDREA as the empirical basis, they construct EMOCON, which models emotional appraisals, based on a model of explanatory coherence, in terms of 5 key dimensions that determine responses: valance, intensity, change, integration and differentiation (pp. 816ff). EMOCON employs parallel constraint satisfaction based on a program, NECO, which provide elements needed to construct systems of artificial neural populations that can perform complex functions (fig. 11, p. 827; pp. 831 ff. for the mathematical details). The upshot is to show "how interactions between cognition and emotion can be understood in terms of parallel constraint satisfaction, if mental representations are assumed to

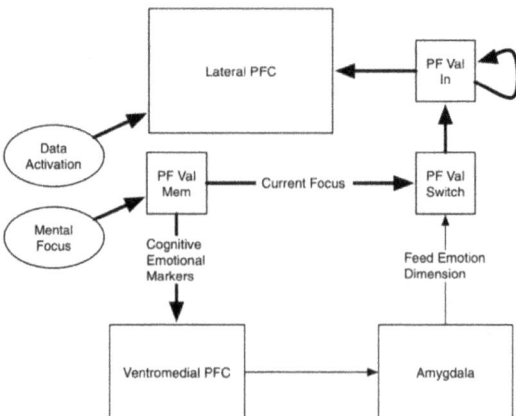

Fig. 11. NECO model of emotional coherence. Lateral PFC is lateral prefrontal cortex. Ventromedial PFC is ventromedial prefrontal cortex. Mental focus is an external signal corresponding to working memory.

have an emotional value, called a valence, as well as a degree of acceptability" (p. 827). The construction of EMOCON points to the potential power of their approach. Computer models, even if gross simplifications, permit of ramping up. A logical basis with a clear mathematical articulation has enormous potential descriptive power as evidenced by the history of physical science.

Damasio (2012) has a similarly ambitious program. He begins with the brain's ability to monitor primordial states of the body, for example, the presence of chemical molecules (interoceptive), physiological awareness, such as the position of the limbs (proprioceptive), and the external world based on perceptual input (extroceptive). He construes this as the ability to construct maps and connects these functions with areas of the brain based on current research (pp. 74ff.). This becomes the basis for his association of maps with images defined in neural terms, which will ground his theory of the conscious brain.

Given that much, Damasio gives an account of emotions elaborating on his earlier work, but now connecting emotions with perceived feelings. As with the association of maps and images, he associates emotions with feeling and offers the following account: "Feeling of emotions are composite perceptions of (1) a particular state of the body, during actual or simulated emotion, and (2) a state of altered cognitive resources and the deployment of certain mental scripts" (p. 124). As before he draws upon available knowledge of the physiology of emotional states but the purpose of the discussion is not an account of emotions *per se*, but rather to ground the discussion of memory, which becomes the core of his attempt at a cognitive architecture (p. 139ff.). The main task is to construct a system of information transfer within the brain and from the body to the brain. The model is, again, mediated by available physiological fact and theory about brain function and structure. The main theoretic construct in his discussion of memory is the postulation of 'convergence-divergence zones' (CDZs, fig. 6.1, p. 152), which store 'mental scripts' (pp. 151ff.). Mental scripts are the basis of the core notion of stored 'dispositions,' which he construes as 'know-how' that enables the 'reconstruction of explicit representation when they are needed" (p. 150). Like maps (images) and emotions (feelings) memory requires the ability of parts of the brain to store procedures that reactivate prior internal states when triggered by other parts of the brain or states of the body. Dispositions, unlike images and feelings are unconscious, 'abstract records of potentialities' (p. 154) that enable retrieval of prior images, feelings and words through a process of reconstruction based in CDZs, what he calls 'time-locked retroactivation' (p. 155). CDZs form feedforward loops with, e.g., sensory information and feedback to the place of origination in accordance with coordinated input from other CDZs via convergence-divergence regions (CDRegions) by analogy with airport hubs, optimizing traffic between regions and zones (pp. 154ff.). Damasio indicates empirical evidence in primate brains for such regions and zones (p. 155) and offers examples of how the architecture works in understanding visual imagery and recall (pp. 158ff.).

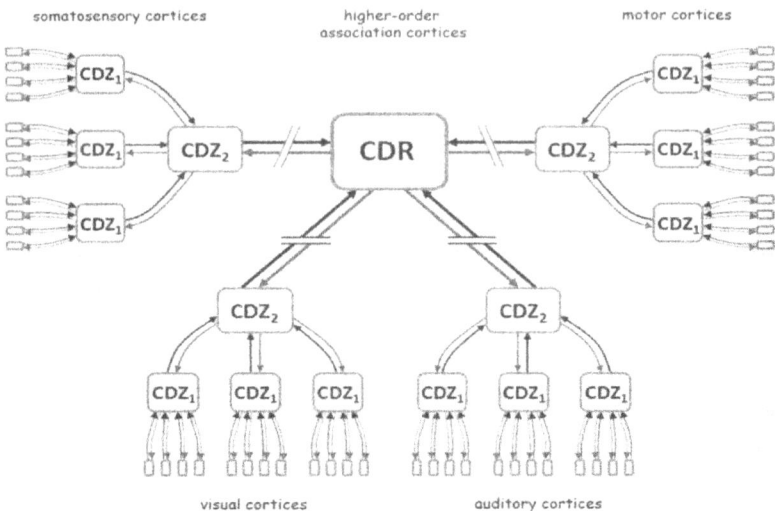

Figure 6.1: Schematics of the convergence-divergence architecture. Four hierarchical levels are depicted. The primary cortical level is shown in small rectangular boxes, and three levels of convergence-divergence (larger boxes) are marked CDZ1, CDZ2, and CDR. Between CDZ levels and CDR levels (interrupted arrows), numerous intermediate CDZs are possible. Note that, throughout the network, every forward projection is reciprocated by a return projection (arrows).

Damasio concludes: "the CDZ framework posits two somewhat separate brain spaces. One space constructs explicit maps of objects and events during perception and reconstructs them during recall. In both percept and recall, there is a manifest correspondence between the properties of the object and the map" (p. 163). The question is how far does this model translate into verbal understanding? How far can we extrapolate from perceptual memory to conceptual understanding? His answer is both fundamental and far-reaching. "Images of the internal and external world can be organized in a cohesive way around the protoself and become oriented by the homeostatic requirements of the organism...The eventual presence of the core [the CDZ framework] is followed by the expansion of the mental processing space, of conventional memory and recall, of working memory and of reasoning" (p. 304).

In Damasio's model the demand for coherence, and the optimization of competing concerns is built right into the neural array, and is thus more profoundly grounded than the mere need for conceptual coherence. If we accept the basic premise as relevant to argumentation, it is no wonder that reasoned arguments are less persuasive than demanded by normative models of claim and challenge. Relinquishing a core belief in response to counter-evidence may demand complex structural reorganization, if a core element representing that belief is deeply embedded in persistent brain architecture. If this image is compelling, there is no simple structural account that isolates individual elements of the brain's storage, thus making them available for piecemeal replacement. Quite the opposite, the brain architecture strives to optimize and so holding on to a core commitment is more likely preferable than a wholesale reorganization of commitments in the face of counter-evidence. This comports with Thagard and Augie's account in terms of 'parallel constraint satisfaction,' the balancing of information in terms of the valence of previously held beliefs, all tied to an emotive complex, built into the processing of information.

Damasio like Thagard and Aubie offer speculative models that reference current physiological knowledge, rely on concepts from computer science and information theory and bypass the deep philosophical issues that are seen by many to create an unbridgeable gap between the mental and the physical short of deep metaphysical reorientation (Chalmers, 1996). Yet, whatever the ultimate verdict on these two authors, the rich program in cognitive science persists and has a strong appeal. The reason is the potential strength of the warrants, that is to say, if such models prove to be correct, the epistemic force of the warrants that support them will be enormous, for they are presumptively warrants with increasing consilience and breadth, and most importantly warrants that have great ontologically depth. And thus, they are warrants that swamp the alternative approaches that rely on, for example, psychological generalizations alone.

5. Conclusion

If my analysis is at all correct, the consequences for argument theory are significant. For, if as I maintain, arguments both in successful areas of inquiry like physical chemistry and. cognitive science have analogous structural to those found on a neural level, then argument theorists should consider expanding their analysis of argument structure in significant ways. As a start, the model of argument as a structure whose form offers an indication of its adequacy must be expanded. Argument must move from description of the surface structure to the underlying functions the structures exemplify. In particular, argument theorists should consider strategies to identify the substructure of warrants that reflect the relative strength of underlying networks of commitments. These deep structural warrants, identified by their role in organizing diverse claims, afford additional support by virtue of the variety of the claims they support and by their increasing effectiveness in directing and sustaining argument. Generally, more abstract than the variety of claims they support, and I think here of, for example, broadly political or even moral commitments, such underlying warrants explain the tenacity of positions in social argument that are resistant to counter-argument, and even contrary evidence. This is describable in terms of a model of warrant strength based upon the structure of scientific discourse in such theoretically advanced inquiry such as physical chemistry. I have developed such a model, grafting my analysis of warrant strength onto a formal model of adaptive logic developed by Strasser & Seselja (2011), (Weinstein, 2013, chapter 4). Although, perhaps not the last word in understanding how a theory of warrant strength may be articulated and combined with an account of evolving argumentation, it is at least a step in the direction that argument theorists might find worth considering.

A position such as mine requires a refocusing of recent efforts in argumentation theory. For example, moving critical questions as pioneered by Walton (1996) beyond the framework of fallacies to a concern with belief structures and underlying commitments. It requires more than a complication of argument diagramming (Freeman, 2011) and moves argument analysis to models of belief stores that reflect dimensions of commitment (Paglieri & Castelfranchi, 2005). Such a perspective challenges the idea of a topic neutral theory of argument, requiring movement into the epistemic and even emotive details of support networks that underlie claims, and the justification, tacit or explicit, that warrant their support. The most critical question of all becomes: How does the level of commitment to warrants and the networks of beliefs that they represent alter the evaluation of evidence, both evidence sought and evidence already available? (Weinstein, 2006).

As important, the hard and fast distinction between fact and value needs to be overcome, especially in areas of social significance, for values affect the way we look at facts. The gloss of value as emotions, is not the main concern, it is rather the affect-laden nature of our commitments that must be taken into account, for the force of value-laden commitments in making determinations of facts used as evidence are all too often more powerful than the force of facts alone (See Edsall, 2022b for an interesting account of voting preferences). How does the value of individual freedom as compared to the value of lives possibly at risk, determine the gun debate? How does the religious perspectives on the meaning of life affect views of a women's right to control her reproductive choices? How does a commitment to a political party affect our willingness to believe hyperbolic claims and promises? On and on! It is the network of commitments of all sorts that determine the force of arguments and if argument theorists want to get serious about understanding the force of arguments, it is these underlying networks that must be addressed.

References

Amodio, D.M., & Devine, P.G. (2006). Stereotyping and evaluation in implicit race bias: Evidence for independent constructs and unique effects on behavior. Journal of Personality and Social Psychology, 91, 652-661.

Bargh, J. A., & Ferguson, M. J. (2000). Beyond behaviorism: on the automaticity of higher

mental processes. Psychological bulletin, 126(6), 925-945.

Blair, I.V., Ma, J.E., & Lenton, A.P. (2001). Imagining stereotypes away: The moderation of implicit stereotypes through mental imagery. Journal of Personality and Social Psychology, 81, 828-841.

Chalmers, D. (1996). The conscious mind. New York: Oxford University Press.

Damasio, A. (2012). Self comes to mind: Constructing the conscious brain. Random House, LLC.

Dehaene, S. (2014. Consciousness and the Brain. New York: Viking.

Dovidio, J.F. and Gaertner, S.L. (1986). Prejudice, Discrimination and Racism. Orlando: Academic Press.

Dovidio, J.F., & Gaertner, S.L. (2005). Color blind or just plain blind? The pernicious nature of contemporary racism. The Nonprofit Quarterly, 12, 40-46.

Dunham, Y., Baron, A.S., & Banaji, M.R. (2008). The development of implicit intergroup

cognition. Trends in Cognitive Sciences, 12, 248-253

Edsall, T. (2022a). How we think about politics. New York: New York Times, 8/18/2022.

Edsall, T (2022b). Why aren't you voting in your financial self-interest? New York: New York Times, 9/14/2022.

Ehrlich H.J. (1973). The Social Psychology of Prejudice New York: Wiley and Sons.

Evans, J. (2008). Dual-processing accounts of reasoning, judgment, and social cognition. Annual Review of Psychology, 59, 255-278.

Frankish, K. & Ramsey, W. (2012). The Cambridge Handbook of Cognitive Science. New York: Cambridge University Press.

Freeman, J. (2011). Argument Structure: Representation and Theory. Dordrecht: Springer.

Gardner, H. (1987). The Mind's New Science. New York: Basic Books.

Greenwald, A.G. & Krieger, L.H. (2006). Implicit bias: Scientific foundations. California Law Review. 94, 945-967.

Gutsell, J.N. & Inzlicht, M. (2010). Empathy constrained: Prejudice predicts reduced mental simulation of actions during observation of outgroups. Journal of Experimental Social Psychology. 46, 841-845.

Hamilton, D. L. (ed.) (1981). Cognitive Processes in Stereotyping and Intergroup Behavior. Hillsdale, NJ: Earlbaum Associates.

Iacoboni, M., Molnar-Szakacs, I., Gallese, V., Buccino, G., Mazziotta, J.C., & Rizzolatti, G. (2005). Grasping the intentions of others with one's own mirror neuron system. PLoS Biology, 3, (3): e79.

Jacoby, L. L., Kelley, C. M., Brown, J., & Jaseschko, J. (1989). Becoming famous overnight: Limits on the ability to avoid unconscious influences of the past. Journal of Personality and Social Psychology, 326-338.\

Krugman, P. (2020). Democrats of the living dead. New York, New York Times, 2/18/2020.

Lebrecht, S., Pierce, L.J., Tarr, M.J., & Tanaka, J.W. (2009). Perceptual other-race training reduces implicit racial bias. PLoS ONE, 4, 1-7.

Lewandowsky, S., Seifert, C., Schwarz, N., & Cook, J. (2012). Misinformation and its correction: Continued influence and successful debiasing. Psychological Science in the Public Interest, 13(3), 106-131.

Paglieri, F. & C. Castelfranchi, (2005). "Arguments as Belief Structures: Towards a Toulmin Layout of Doxastic Dynamics?" OSSA Conference Archive. Paper 43. http://scholar.uwindsor.ca/ossaarchive/OSSA6/papers/43.

Rudman, L.A. (2004). Sources of implicit attitudes. Current Directions in Psychological Science. 13, 79-82.

Strasser, C. & Seselja, D. (2011). Towards the proof-theoretic unification of Dung's argumentation framework: an adaptive logic approach. Journal of Logic and Computation, advanced access (1-24)

Thagard, P., & Aubie, B. (2008). Emotional consciousness: A neural model of how cognitive appraisal and somatic perception interact to produce qualitative experience. Consciousness and Cognition, 17(3), 811-834.

Tversky, A. and Kahneman, D. (1973.) "Availability: A Heuristic for Judging Frequency and Probability." Cognitive Psychology, 1973, 5.

Walton, D. (1996). Argumentation schemes for presumptive reasoning. Mahwah, NJ: Erlbaum.

Weinstein, M. (2002). Exemplifying an internal realist theory of truth. Philosophica, 69:(1), 11-40.

Weinstein, M. (2006). A metamathematical extension of the Toulmin agenda. In: D. Hitchcock and Verheif, J (eds.), In Arguing on the Toulmin Model: New Essays on Argument Analysis and Evaluation. Dordrecht, The Netherland: Springer, 49-69.

Weinstein, M. (2011). Arguing towards truth: The case of the periodic table. Argumentation, 25(2), 185-197.

Weinstein, M. (2013). Logic, Truth and Inquiry. London: College Publications.

Weinstein, M. (2015). Cognitive science and the model of emerging truth. In: Garssen, B. Godden, D., Mitchell, G. & Snoeck Henkemans, A. (eds.) Proceedings of the 8th International ISSA Conference, ISSA.

Weinstein, M. (2016."The Periodic Table and the Model of Emerging Truth." Foundations of Chemistry, 18:3, 195-212.

THE ROLE OF FRAMES IN PLATO'S LACHES

HARALD R. WOHLRAPP
Universität Hamburg
harald.wohlrapp@uni-hamburg.de

Abstract
Frames are highly important units in argumentative dialogues. Usually they play their role behind the curtain of awareness and, thus, can be used for all kinds of unnoticed influences on people's minds. As the concept of "frame" is, however, still rather vague the paper starts with developing a practical definition. It is an important but usually ignored fact that frames can also be latent and can then be responsible for heavy blockades in the process of understanding. In order to illustrate the subliminal force of frames a meticulous analysis of (the first half of) Plato's dialogue Laches is provided. The scenery of the two generals, Laches and Nicias, vainly trying to define the concept of bravery, is suitable to strikingly exemplify the quality of the respective difficulties.

1. The Concept of "Frame"

There is a common use of the expression 'frame' which seems to signify simply a particular way of focusing an issue. You can frame a motor car as a transportation means or as a pollution spreader or as status symbol etc. If you are the chief of a company you may have the power to frame an increment of the workload as a step in the thriving and prospering of "our great project".

In some contexts, we can nowadays observe a bickering about the correctness of a frame. If a student reckons the W.H.O. as a conspiracy club of some super-rich sponsors then this framing tends to be bowdlerized by her teacher. And this is an interesting fact: For a frame that is considered to be grossly inadequate the primary question is no more if there be some truth in thinking so, but the primary question is "as what" we are usually held to frame that issue.

This shows at least that framing is something more than an arbitrary focusing or putting an issue under a covering term. What is it?

In the first part of my paper I will propose an elaborated concept of frame which I consider of paramount importance for a subtle understanding of all kinds of thought lines, in particular of reasoning courses.

In order to provide a better access to that concept I start with sketching four theories that have served as sources for my own ideas.

1.1. The first one is Wittgenstein's "seeing as" – which was exposed in the "Philosophical Investigations"[1]. Here he shows the (today well-known) figure of the duck-rabbit-head.

For my purposes the "seeing as" is significant for the following reasons:
(A) One can see the picture in two different "frames"; and then the details can appear as different things. E.g. the beak in the duck-frame is, in the rabbit frame, a pair of ears.
(B) It is possible to see the picture under one aspect without being aware of the other possibility. E.g. we just see a picture of a duck (viz. a rabbit) and take it for the whole thing. Therefore: We might perceive things of which we perceive only certain aspects – and which might be rather different if we were able to change the usual frame.

1.2. My second source is the idea of "Reframing", originating from the Mental Research Institute in Palo Alto, CA. (Bateson, Watzlawick etc.). "Reframing" means the willful replacement of terms that are used to describe the interaction of the parts of a system (primarily a family of a mental patient) – so that the relationships between those parts could receive a new description. This could cause a change of those relationships, then a change of the problematic interactions and in the aftermath a change of the whole system including the involved persons[2].

"Reframing" adds to the "seeing as" a pragmatic and a subjective dimension. The frame is now something that determines the awareness of the social reality and its modification can be used as an instrument to rearrange problematic relationships among people.

---

[1] Wittgenstein, Ludwig (1968), Philosophical investigations, ed. G.E. M. Anscombe & R. Rush, London, II, xi

[2] Cf. "The gentle art of Reframing" in: Watzlawick, P., Weakland, J. H., & Fisch, R. (1974). *Change: Principles of Problem Formation and Problem Solution.* New York: W.W. Norton, p.115ff.

1.3 My 3rd source is Paul Feyerabend's philosophy of science which was concerned about the historical dimension of knowledge and, in particular, the "incommensurability" of consecutive theories. For the analysis of this problem Feyerabend coined the term "latent classification" which stands for a covering term that is effective in the thinking of scientists, without their being aware of it[3]. And at the hand of some striking examples he pleads that the incommensurability between old theories and new theories results from the different latent classifications at their metaphysical bottoms.

1.4 The fourth source is Charles Fillmore's "Frame Semantics"[4]. The basic idea is here, that the meaning of a word is determined not simply by its use (Wittgenstein) but rather by the bulk of intellectual content concerning the issue that the word refers to. Example: If we use the word "to buy" we understand it insofar as we are able to associate a scenery stuffed with issues like buyers, sellers, goods, money, prices, market locations and a lot of relationships between these elements of the scenery.

(Fillmore's frame concept is considered to be the origin of the above-mentioned ordinary use of the expression "frame". That use can be seen as a diluted descendant.)

There is a second remarkable feature in Fillmore's concept – and that is the hierarchical structure. Frames can provide a vertical order in the elements of the intellectual content that they embrace: We have more general and more specific layers of frames in it.

E.g. "to buy" stands below some more general frames like "to take" or "to do"; and at the same time it ranges above some more specific frames like "sale", "bargain" etc. Therefore: the immediate frame is itself an element in a hierarchical range of further frames; and we can say: The deeper our understanding of an issue, the longer that range of relevant frames.

1.5 After these preliminary clues I will now present my concept of frame. It has four dimensions, related to semantics, pragmatics, subjectivity and consciousness.

1.5.1 Semantic dimension
To frame an issue [A as B] makes the meaning of A dependent on that section of the background knowledge which is attained in the focus on B. Analogously the meaning of A can change if it is framed as a C. The perception and understanding of A tends to be limited by the respective

---

[3] Cf. Paul Feyerabend (1975), *Against Method*, London/ New York: New Left Books
[4] Fillmore, Charles J. (2006), "Frame Semantics." In *Cognitive Linguistics: Basic Readings*, edited by Dirk Geeraerts, 373-400. Berlin: De Gruyter Mouton.

frame. If a speaker frames A in a blatant idiosyncratic way he may not be understood by an audience of common people.

If A is usually framed as B then those issues and events which are outside the limits of that focus may appear alien or may even be disavowed as not existing at all.

1.5.2. Pragmatic dimension
If A is framed as B, then it plays in our activities exactly that role that fits with the knowledge content, inherent in B. If we frame [A as B], we usually act as if A really and simply *was* B – and nothing else. Depending on what A is, it appears that sooner or later A can become completely B. A motor car that was previously just framed as a status symbol may be no more than that after some time in which it is handled that way.

1.5.3. Subjective dimension
Usually the frame in which a respective issue is conceived, is not willingly chosen, but belongs to the mental furniture of a subject (person or group). Therefore, the complete information about framing should be "A is framed as B for S_1" – so that it could possibly be framed as C or as D for S_2 and S_3.

Subjects live on systems of knowledge and belief whose elements are mostly incorporated in mental and bodily reactions to the particular situation. These systems are hierarchically structured and contain framings that are features of the habits and of the life form of persons.

1.5.4. Consciousness dimension
In the basic layers of the said belief systems the frames are latent. We do not know "as what" we frame issues like e.g. life and death, war, peace, the climate change, a plague, an economic crisis, but also many circumstances and events of our private affairs. We may realize something about our basic framings if we observe ourselves very carefully in our conversations, our reactions, our fears and hopes. In grasping them one becomes conscious of additional aspects of oneself and of the world which means that one may grow in consciousness.

(Latent frames belong to what was called the "Unconsciousness": All psychoanalytic theories are mental research maps for the discovery of latent frames.)

2. The Dynamics of Frames in the central Part of the Dialogue Laches

2.1. Outline of the Dialogue

The Laches is a piece about reasonable career planning[5]. Today parents would ask a career advisor: Shall our sons study business economics, shall they learn Chinese? In this dialogue the respective question is: Shall they learn "hoplomachía", i.e. the fighting in full armour? This was a capacity very much needed for the frequent colonial wars that the city of Athens had to wage against their tributary states.

Now, that initial question about the best career of the youths is briskly and ingeniously transformed into the fundamental question of what could be done to improve the souls of the youngsters and help them to lead a life in aretè (i.e. the term for the ideal condition of the ancient aristocracy in Attica; the usual translation as "virtue" is misleading).

As the ideal of aretè is well known, but not clearly understood – one must, in order to answer the question, at first care about a real conceptualization. And as this is estimated too big a target, the five speakers (two fathers, two experts and Socrates) can agree in looking for a definition of braveness. The reason is, that braveness counts as that particular part of aretè that is related to the present topic: the fighting in armour.

With this result everything is ready for the notorious Socratic procedure. The two advisors, Laches and Nikias, are both experienced military leaders and they can now be drawn into an endeavor to define what braveness is. Of course, they both fail at the emergence of their specific limits of understanding viz. at the appearing need to amplify their understanding. The nature of their limitations is, indeed, characteristically different. Laches conceives braveness as a practical capacity which has nothing to do with theoretical knowledge; whereas Nikias (instructed by some sophist) takes it for a certain kind of knowledge. Sure, this could work as a fitting complement. But alas, Nikias is neither able to sufficiently define that kind of knowledge nor to integrate it into the practical aspects.

The dialogue itself is a large sequence of question-answer exchanges in which frames and frame-shifts amply appear. I will concentrate on three prominent episodes and show how they are pivotal steps for the deployment of the speakers' thinking.

---

[5] Cf. Plato (1964), Laches, in: *The dialogues of Plato*, 4 vols (transl: B. Jowett) 4th ed. Oxford, OUP

2.2. Three salient frame-treating episodes

2.2.1. First episode: Braveness in flight
The dialogue partners have agreed that Laces begins to determine the nature of braveness.
At once he comes out with three criteria which can be displayed like this:
x is brave ⇒ x maintains the line
 ⇒ x fights the enemy
 ⇒ x does not fly (mè feúgoi)

This approach is immediately countered by Socrates who objects that one can be fighting (and so possibly be brave) *in flight*. The objection is substantiated with two examples (the Scythians fight "flying as well as pursuing" – and the horses of Aineas who are said to "fly quickly hither and thither").

As Laches has to accept those examples he finds himself very near the contradiction that the brave does and does not fly. Still, he is able to escape it in localizing flight as a special tactic in the chariot and cavalry sections of an army – whereas his definition, so he says now, was meant for the typical situation of the fighting Greek infantry.

The next move of Socrates, however, contains a new objection which again voids the definition. It is the reference to the tactics of the Lakedemonians in the battle of Plataiai: They had fought with first flying and then turning back and attacking the Persian army so firmly that the battle was finally won.

This is a telling argument, because there is no doubt, that the Lakedemonians are Greeks – and at the times of Plataiai they were even an essential contributor in the Greek's victory over the Persian empire.

So, the most delicate element in Laches' definition is devaluated. Thus, it is obvious that his attempt operates on too small a background. What shall he do? He is not sure. Maybe he should try a completely new start?

Before I go to the second round let me please, with reference to the frame structures, explain what has happened during the first three exchanges. Laches had apparently formed his definition of braveness on the background of a distinct mental image – and this is the scenery of the Athenian way of waging war in the "phalanx", that is in long lines of armoured men, running forward side by side and chanting the "paian". Here it is essential to hold the line, move forward, combat the enemy, and, at any rate, not run away in panic – because this would crack the phalanx-formation and make it vulnerable. Therefore the "not fly" is an indispensable determinant of the brave fighter.

The situation, to which Socrates is referring, is completely different. It is a retreat which might be necessary and which can be done as well by the whole army or also by loose groups of it or even by single persons. Retreat takes place in various guises. Anyway, it is an activity which can be willingly shaped.

In the Greek text of the dialogue the word for flight is fygè. The same word, however, is also used when reference is made to the battle of Delion. That battle was lost by the Greek, the formations of the army dissolved and had to move back. And now, this battle appears under the name of "fygè Delíou" – the flight viz. retreat of Delion. (This can be said so – even if there is a more specific word for "retreat", namely "anachóresis".)

Apparently, the move back had happened in different forms: Many fled in disorder, but some also went back quietly and worthily, showing their readiness to defend themselves in case they would be attacked. Among these men were Laches and Socrates, and Laches explicitly recounts the event praising the great attitude of Socrates. Therefore, this was a splendid example of braveness in flight.

Now here comes the crucial question:
Why had Laches not taken that kind of situation into account when he tried to define braveness? I think the answer is this: The situation that a man runs away from the fighter's line – counting as the opposite of being brave – stands for him in a pragmatic frame that is clearly different from the scenery in which he himself bravely retreated (together with Socrates) after the ignominious loss of Delion.

Moreover, he is strongly focusing on the usual scenery of the awesome Athenian army so that the cases of defeat (at that time very few) are not on his record. His record is impressed by the former frame so that he finds the relevant criteria of braveness there. Sure, he is not aware that he is not just musing about braveness as an objective ideal but about his own, characteristically tailored, image of braveness.

This is a common condition in any communication: we believe to speak about the matters of fact *per se* but we can only speak about the aspects of them that we have actually in mind.

What happens now in the dialogue of Laches with Socrates? The determination of his image of braveness proves as shaped in a frame that is too narrow. Therefore, the contradictions in which he is drawn through the answers of Socrates; answers that drive his frame apart. The result is, that he is baffled. He fails to define braveness and does not know why he fails.

Yet, *this is a typical setting* when the frame in which an issue is conceived proves too narrow, but as the limitations do not bluntly appear, the frame rests unnoticed and the result is a strange confusion.

2.2.2. Second episode: military scope versus civil life
Laches had started his definition endeavor, claiming that it be not difficult; now he has deplorably failed. In order to protect him from further shortcuts Socrates first blames himself for not having asked well enough;

and then opens up the framing for a global concept of braveness. This comes in three consecutive steps. First, he says that with his question he had pointed not only to the braveness in the Greek infantry but also in cavalry and in all other sections of an army. Then he goes on and reminds Laches, that braveness is a virtue not alone in war scenery, but also in civil life with its dangers and threats of poverty, disease, pain, fear etc. The third step is then a turn from outside dangers to inside hazards – namely the demand to bravely resist being overwhelmed by feelings like desire, lust and addiction.

What happens here, that is the suspension of two major frame gaps – on the one hand the gap between the military scope and the civil life; and on the other, in civil life, the gap between being endangered by outside perils or by inside overwhelming passions. These frame suspensions are performed by Socrates, who now emphasizes the target to spread the concept of braveness over those three different areas.

In order to comprehend the difficulties one has to realize the striking differences of those three situations – a fighting soldier, a sick person in pain or a man being overwhelmed by his passions. These situations have to be compared with respect to what it means being brave there. Apparently, some quality has to be equal, but this cannot be identified by simply putting aside the differences. One must rather reflect on one's inner organization and figure out what it would mean for oneself to strive for braveness under those different circumstances.

2.2.3. Third episode: Practical attitude versus theoretical knowledge

Laches is deeply impressed by the new project. Thus, he issues his first approach rather cautiously: Braveness is a "kind of endurance of the soul". Obviously, this can only be a part of a definition – because there are several kinds of endurance and Laches has to figure out which "kind" relates to braveness.

Is it clear where to seek? Laches had initially confessed that he is not trained in conceptual work. Therefore, Socrates gives him a direction. He says that in order to choose the right kind of endurance one could remember that braveness was determined as a part of aretè and, thus, as something good.

This is, in the conceptual sphere of distinctions, inferences and frames, a move upwards – into the sphere of goodness and virtue. They are obviously more abstract than the concept of braveness, but they can help to find the correct direction for further consideration. In order to illustrate this, Socrates presents some examples of imprudent endurance (e.g. endurance in spending money); and they both agree that those are not good. After this he asks whether or not Laches would call the respective behavior "brave" – what Laches denies.

It is important, that Laches, even if he is not yet sure about the correct quality of endurance, has now a true reason to deny that question, because the following two inferences had been established before:

imprudent endurance ⇒ *not* good, *not* noble
braveness ⇒ good and noble

Thus, the kind of endurance that is sought for to complete the definition, follows logically and braveness comes out as "prudent endurance of the soul".

Up to this point no frame dynamics is noticeable. But we have seen Laches exercising in conceptual endeavors. And now, on the basis of the achieved result, Socrates attacks the third and most fundamental frame in Laches' thinking: This is the frame of practical thinking and personal experience being separated from and confined against the realm of theory and knowledge which is considered to be more or less abstract reasoning. For Laches braveness is a practical capacity to withstand difficult situations; a capacity which is independent of any theoretical understanding.

Braveness is "prudent" endurance – bare it in mind: this is not yet a unification of braveness and theoretical knowledge. However, Socrates can use it to initiate that unification. Therefore, he now suggests that Laches may examine the relationship between prudence and braveness more closely. He makes him compare two situations; in one an actor understands more about the dangers whom he is exposed to and in the second one he understands less. Then the question arises who may be more courageous. Now Laches is totally confused because the answer seems clear: The one who understands less! However, this does not fit well with the statement that imprudent endurance is not good.

This is then the end of the dialogue between Laches and Socrates. Laches utters once more, that he certainly knows what braveness is, that he has it as a distinct image in mind but that he, annoyingly, is not able to put it into an appropriate conceptual shape.

What does that mean? The fundamental frame in which for Laches the practical capacities are separated from theoretical knowledge is not enough permeable. Therefore, the coalescence of practice and theory that is certainly needed for an adequate definition of braveness cannot be achieved.

3. Final remark

This dialogue is a wonderful piece of philosophical literature because with Laches we are presented a speaker who thinks within the frame of practical capacities and who is not able to transgress into the realm of theory and knowledge; and then we are shown, in the person of Nikias, a speaker who is familiar with theory and knowledge but is not able to handle the practical content of knowledge. Thus, both fail to define braveness because of their complementary one-sidedness, which can be theorized as being limited by their habitual frames.

THE ORIGINS AND CLASSIFICATION CRITERIA OF CRITICAL QUESTIONS

SHIYANG YU
College of Philosophy, Nankai University

FRANK ZENKER
College of Philosophy, Nankai University
frank.zenker@nankai.edu.cn

Abstract
While the concept of a *critical question* (CQ) is central for argumentation theory, the field currently lacks a systematic account of their informational sources, or origins. By combining a logical, a dialectical, and a cognitive perspective, we identify four origins of CQs, namely: a (i) *presumptive*, respectively a (ii) *non-presumptive* premise that, together with other explicit arguments components, is (iii) *unnecessary*, respectively (iv) *sufficient*, to identify an argument scheme type. This proposed classification may contribute to developing a comprehensive theory of CQs.

1. Introduction

Argument schemes are variously described as "commonly adopted" (van Laar, 2011, p. 349), "conventionalized" (van Eemeren & Grootendorst, 1992, p. 96; Visser *et al.*, 2020) "forms of argument (structure of inference)" (Walton *et al.*, 2008, p. 11; Walton *et al.*, 2014, p. 89) or "general patterns of reasoning" (Bex & Verheij, 2012, p. 344). They are typically used to model externalizations of *non-monotonic* inferences—where new information can change an old conclusion—and thus map onto reasoning episodes that are "presumptive and plausibilistic in nature" (Walton, 1996, p. 13). An argument scheme's main theoretical purpose is to help *identify*, *analyze*, and *evaluate* some natural language argument, *A*. Identifying *A* entails recognizing *A*'s appearance. Analyzing *A* entails determining its structure and the scheme(s) it instantiates. Evaluating *A*, finally, entails judging its validity or goodness.

Since the work of Hastings (1962), scholars widely accept that the evaluation of an argument can leverage the *evaluative function* of critical questions (CQ) (Garssen, 2001; van Eemeren & Grootendorst, 1984; 1992;

Prakken, 2005, p. 306; Yu & Zenker, 2020; Zenker & Yu, 2022). (Hastings (1962) apparently was the first to rely on CQs to evaluate reasoning episodes (see Yu & Zenker, 2020, p. 469).) Its evaluative function is what a CQ acquires in virtue of attacking some argument scheme component, thus exerting a critical pressure that only a valid/good scheme instance can withstand.

Analysts must generally ask all CQs associated with a specific argument scheme type, which arguers or analysts themselves are in turn required to answer. To evaluate the perhaps most frequently discussed argument scheme type today, for instance, the *argument from expert opinion*, Walton (1997, p. 210, p. 223; see Walton *et al.*, 2008, p. 14) has proposed the following scheme and CQs:

The argument from expert opinion scheme
Major Premise: Source E is an expert in subject domain S containing proposition A.
Minor Premise: E asserts that proposition A (in domain S) is true (false).
Conclusion: A may plausibly be taken to be true (false).

CQs for the argument from expert opinion scheme
CQ-1: *Expertise Question*: How credible is E as an expert source?
CQ-2: *Field Question*: Is E an expert in the field that A is in?
CQ-3: *Opinion Question*: What did E assert that implies A?
CQ-4: *Trustworthiness Question*: Is E personally reliable as a source?
CQ-5: *Consistency Question*: Is A consistent with what other experts assert?
CQ-6: *Backup Evidence Question*: Is E's assertion based on evidence?

While Walton's CQ-list is widely cited today (e.g., Liao, 2021; Keppens, 2014), few scholars have asked 'why these specific CQs?' or 'is this CQ-list complete?' (e.g., Wagemans, 2011). Yet the completeness of a CQ-list is crucial because applying an *incomplete* CQ-list cannot suffice to evaluate a scheme instance as a good argument.

To evaluate whether a CQ-list is complete, it helps to clarify the informational sources, or *origins,* of a CQ. We propose four origins, namely: a (i) presumptive, respectively a (ii) non-presumptive premise that, together with other explicit argument components, is (iii) *unnecessary*, respectively (iv) *sufficient*, to identify an argument scheme type. We submit this proposed classification as a contribution to developing a comprehensive theory of CQs.

2. The logical perspective

A natural language argument must invariably be evaluated relative to some context C. Because the using of an argument scheme in C presupposes its admissibility in C, its evaluation can focus on either the argument scheme's *appropriateness* (admissibility) or its *validity* (quality) (Wagemans, 2011, p. 338; Walton, 1996). This distinction runs parallel to that between a scheme's *selection* and its *correct use* (van Eemeren & Grootendorst, 1992, p. 159). Whether CQs should only address a scheme's admissibility (e.g., Feteris, 2016) or only the instance's quality (e.g., Verheij, 2003) is a matter of theoretical preferences. Our own preference is to let CQs address the instance's *quality* (Yu & Zenker, 2020).

From a logical perspective, an argument is a set of propositions consisting of two elements: the *premises* and the *conclusion* (e.g., Ben-Ze'ev, 1995, p. 189; cf., Johnson & Blair, 2006 [¹1977], p. 10). Connecting the premises and the conclusion is a third, often implicit element: the *inference rule*. Each of these elements can be interpreted as an *origin* of a CQ, and each can be associated with a distinct *logical function*. The function of a premise-associated CQ is to evaluate the premise's *correctness*, that of an inference rule-associated CQ is to evaluate the premise's *justificatory force* for the conclusion, and that of a conclusion-associated CQ is to evaluate whether other arguments attack it (Prakken, 2010, p. 3; Verheij, 2003; 2006; Walton *et al.*, 2008, pp. 31f.).

Although these three logical functions are intuitive, the premise-associated CQ 'Is premise P_i correct?', for instance, need not be understood as serving the *sole* function of evaluating the correctness of P_i. We argue that a premise-associated CQ can also serve to *identify* the argument scheme type. A premise-associated CQ, then, can have both an evaluative and a *cognitive* function. (We return to this in Sect. 4.)

An unsolved theoretical problem is how an explicit or implicit premise can be distinguished from an inference rule. Walton & Krabbe (1995, p. 128; see Walton *et al.*, 2008, p. 32) treat the inference rule as an additional premise, one that is normally implicit. For the argument scheme from expert opinion, they consequently explicate the inference rule as a *conditional premise*.

Argument from expert opinion scheme; inference rule as explicit conditional premise (Walton et al., 2008, p. 19)
Major Premise: Source E is an expert in subject domain S containing proposition A.
Minor Premise: E asserts that proposition A (in domain S) is true (false).
Conditional Premise: If source E is an expert in subject domain S containing

proposition A, and E asserts that proposition A (in domain S) is true (false),
then A may plausibly be taken to be true (false).
Conclusion: A may plausibly be taken to be true (false).

The argument's premises and its inference rule would thus appear to constitute two distinct origins of CQs. But once the conditional premise is explicit, the argument scheme instantiates the inference rule *modus ponens* (MP). Yet the logical validity and near-universal applicability of MP make it unlikely to be challenged. Hence, a CQ that attacks MP is unlikely to cast doubt on a proponent's argument. Thus, treating the inference rule as an explicit conditional premise does not offer any immediate insights towards developing a theory of CQs.

Although this keeps from treating MP as a distinct origin of a CQ (besides the origin provided by the premises), the above conditional premise can be interpreted as the complex claim that E is honest, impartial, professional, rigorous, etc. Thus, the conditional premise conveys *more information* than the major or the minor premise, each of which are assertives rather than conditionals. We proceed to showing how a conditional premise can be transformed into CQs associated with asserted premises.

3. The dialectical perspective

Dialectically, the function of applying argument schemes and asking and answering CQs is to discharge a proponent's *burden of proof* (BOP), respectively to transfer it between proponent and opponent. For reasons given below, a proponent seeking to shift the BOP must justify only *specific* parts of an argument. This creates the opportunity to replace the conditional premise expressing the inference rule with a specific set of asserted premises.

Of course, the identity of the asserted premise(s) replacing the conditional premise may be under doubt—especially if the conclusion of an expert argument conveys a long-range scientific prediction. The BOP for the conditional premise will nevertheless count as discharged dialectically once the claim to the expert's honesty, impartiality, expertise, rigor, etc. is justified. The task of replacing the conditional premise expressing the inference rule with a specific set of asserted premises thus turns into that of identifying premises that suffice to justify the conditional premise.

From a dialectical perspective, demonstrating that a conclusion can withstand *all possible* attacks by CQs normally goes beyond the proponent's obligations. Rather, a proponent counts as having discharged their BOP for a conclusion once an opponent accepts both the premises and the inference rule. In the pragma-dialectical model, for instance, rule nine states that "[t]he protagonist has defended the standpoint at issue or a sub-standpoint conclusively by means of a complex speech act of argumentation if both the *propositional content* and its *justificatory force* called into question have been defended successfully" (van Eemeren, 2018, p. 57; *italics added*). Compared to the functions of premise- and inference rule-associated CQs, therefore, a conclusion associated CQ *lacks* a distinct dialectical function. (This CQ can nevertheless have a *cognitive* function; see Sect. 4.2.)

A proponent can discharge the BOP for a conclusion *partially* because "[t]he function of each scheme is to shift *a weigh of presumption* from one side of a dialogue to the other" (Walton, 1996, p. 46; *italics added*). Once a proponent uses a specific argument scheme, "[t]he opposing arguer [or opponent] in the dialogue can shift this weight of presumption back to the other side [...] by asking *any* of the appropriate critical questions matching that scheme" (Walton, 1996, p. 46; italics added). Among these, *some* CQs "function as starting points for finding rebuttals [and these CQs] have a BOP attached for the [opponent]" (Walton et al., 2008, p. 32; terms adapted, italics added; see Walton, 2003; 2012; 2014).

Raising these CQs can hence suffice to shift the BOP because raising them incurs a *burden of criticism* (BOC) for the opponent. After all, "[i]n some cases, it may be that a positive answer to a critical question can be assumed [, wherefore] the respondent [or opponent] has to provide an argument to show why the critical question has a negative answer" (Walton, 2003, p. 23). In leveraging that raising a CQ incurs a BOC, and that the origins of CQs can be explicit or implicit, Walton offers two criteria to distinguish *ordinary premises, assumptions*, and *exceptions*.

> "The *ordinary premises* are the ones that are *explicitly stated* in the argumentation scheme. The *assumptions* are *additional premises* that are assumed to hold, just like the ordinary premises, but if questioned by the asking of a critical question they *automatically fail to hold* unless the proponent of the argument gives some evidence to support the premise. [...] They shift the initiative back on to the proponent as soon as the question is asked. *Exceptions* are also *additional premises* except that they are assumed not to hold *unless evidence is given by the critical questioner* to show that they do hold in the case at issue. [...] They do not defeat the argument unless the questioner gives backup evidence to support the question." (Walton, 2014, p. 26; *italics added*)

The three types of premises—ordinary premises, assumptions, and exceptions—can be coordinated to three origins of CQs associated with an argument's premise(s) and its inference rule (Table 1). But this does not yield an origin of CQs associated with the conclusion. According to Walton's typology, then, arguers are only required to discharge their dialectical obligations for the premises and the inference rule.

Criterion 2 \ Criterion 1		Does the opponent incur a burden of criticism?	
		No	Yes
How is the premise formulated?	Explicitly	Ordinary premise	
	Implicitly	Assumption	Exception

Table 1: Walton's classification of the origins of CQs

Premise-associated CQs can be thought to originate with premises that support a conclusion. Walton, however, refers also to an *exception* as a 'premise'. This is counterintuitive because a premise normally *supports* the conclusion, whereas an exception *weakens* it (Govier, 1999, p. 156; Jin, 2011, p. 11; Yu & Zenker, 2019, p. 34). For the argument scheme from expert opinion, the (undercutting) exceptions are: "[w]hat E asserts is not consistent with what other experts in field F say" and "[the expert] E is not trustworthy" (Walton, 2014, p. 26). If these exceptions applied, they would weaken the argument. Conversely, the premises that support the argument's conclusion are proposition such as 'what E says is consistent' or 'E is trustworthy'. Thus, in line with the idea that premise-associated CQs originate with supporting premises, we can replace 'exception' with 'without exception' (Table 2).

Criterion 2 \ Criterion 1		Does the opponent hold a burden of criticism?	
		No	Yes
How is the premise formulated?	Explicitly	Ordinary premise	
	Implicitly	Assumption	Without exception

Table 2: Revised classification of the origins of CQs

The next section specifies Walton's criterion 1 ("Does the opponent hold a burden of criticism?") using the notion of *presumption* (Sect. 4.1) and his criterion 2 ("How is the premise formulated?") using the *cognitive* function of a CQ (Sect. 4.2).

4. Dialectical classification criteria

4.1. Presumption

In Walton's dialogical theory, a presumption is a pragmatic concept that "remove[s] potential blockage and enable[s] the dialogue to move forward [...]," thus "sav[ing] time and money and effort in communication" (Walton, 2014, p. 115). A presumption can be forwarded "'for the sake of argument' for purely practical reasons, without offering evidence to back it up" (Walton, 1996, p. xii). Normally a presumption is defeasible, indicating that hard evidence is lacking (Walton, 1996, p. 28; 2014, p. 114). The BOP for refuting a presumption (which Walton calls *burden to rebut*) lies with the opponent. It amounts to "a burden to disprove contrary evidence, should it arise in the future sequence of dialogue" (Walton, 1996, p. 27).

Thus, a presumption is "halfway between mere *supposition* and *assertion*" (Walton, 1996, p. 27; *italics added*). While a "supposition (or assumption) [...] requires only the agreement of the respondent, and carries with it no BOP on either side," an "[a]ssertion always carries with it a BOP, because assertion implies a substantive commitment to the proposition asserted" (*ibid.*; terms adapted). Hence, if an opponent objects to a presumption by supplying contrary evidence, they incur a *conditional* BOP, which turns into a *commitment* once the objection is met. Walton's criterion 1 ("Does the opponent hold a burden of criticism?"), therefore, can be specified as 'is the premise presumptive?' (Table 3).

Criterion 2 \ Criterion 1		Is the premise presumptive?	
		No	Yes
How is the premise formulated?	Explicitly	Explicit non-presumptive premise	
	Implicitly	Implicit non-presumptive premise	Implicit presumptive premise

Table 3: Criterion 1 specified as 'is the premise presumptive?'

4.2. Cognitive function

A specification of Walton's criterion 2 ("How is the premise formulated?") should answer 'why *only some* premises must be expressed explicitly?'. Dialectically, as we saw, presumptions can be likened to presumptive argument schemes because the latter correspond to an "inconclusive and defeasible argument" that has "a practical function of shifting a burden of proof in a dialogue" (Walton, 1996, p. ix; Walton *et al.*, 2008, p. 35). More precisely,

> "a presumptive inference from one proposition, *A*, to another proposition, *B*, is based on a rule of the following form: given that *A* is the case, you (the rule subject) shall proceed as if *B* were true, unless or until you have sufficient reason to believe that *B* is not the case" (Walton, 1996, p. 20; see Ulman-Margalit, 1983, p. 147).

If an arguer rejects an instance of a presumptive argument scheme, its pragmatic force can lead to "powerful negative social consequences of risking criticism, regret, reprobation, loss of esteem, or even punishment [...]" (Walton, 2014, p. 15; see Godden & Walton, 2007, p. 322; Kauffeld, 2003, p. 140). For instance, conclusions of arguments from expert opinion are presumptively acceptable in contexts where audiences trust experts. Hence, those rejecting these conclusions will expectably be criticized, a criticism that becomes stronger if good reasons for a rejection are absent.

The correspondence between a presumption and presumptive argument scheme instance thus points to the *rationality* of accepting them. Arguers can convey this correspondence—and audiences can recognize it—only in virtue of argument components that are explicit.[1] Similarly, analysts can apply a specific CQ-list only if the scheme type has been correctly identified (Walton, 2012, p. 373). Walton's criterion 2 thus points to a CQ's *cognitive* function of identifying the argument scheme type.

We distinguish a scheme type's *preliminary* identification—indicating which CQ-list is to be applied—from a *substantial* identification, confirming or falsifying the result of the preliminary identification. For example, the CQs associated with an argument's explicit components enable a substantial identification if we assume the following argument to be preliminarily identified as instantiating an argument from expert opinion:

> *Argument preliminarily identified as one from expert opinion*
> N.N. has published dozens of papers in philosophy of mind.
> N.N. asserts that humans have free will.
> So, humans have free will.

The CQs associated with the explicit premises are:

CQ-1: Has N.N. published dozens of papers in philosophy of mind?

---

[1] Implicit argument components owe their relevance to what an argument expresses explicitly. But few implicit argument components tend to be relevant to identify the argument scheme type. Indeed, the argument scheme type can normally be identified even if the argument's conclusion is implicit.

CQ-2: Did N.N. assert that humans have free will?
CQ-3: Does someone having published dozens of papers in philosophy of mind qualify as an expert?

The function of raising CQ-1 and CQ-2 is to evaluate the correctness of the first, respectively the second premise. Additionally, raising and positively answering CQ-2 contributes to identifying the argument scheme. After all, had N.N. not *asserted* but *challenged* the proposition 'human beings have free will', then the argument would not instantiate the argument from expert opinion scheme. The same holds if CQ-3 (itself not serving to evaluate one of the argument's premises) is answered negatively. The CQs associated with the explicit premises can thus be associated with two functions: to evaluate the correctness of the premises and to substantially identify the argument scheme type.

4.3. Identifying the argument scheme type

Although a conclusion-associated CQ is dialectically optional (see Sect. 3), raising it is necessary to substantially identify the argument scheme type. After all, if N.N.'s assertion is distinct from the argument's conclusion, then the argument may involve schemes *besides* the expert opinion scheme. For instance, the compound argument 'N.N. has published dozens of papers in philosophy of mind; N.N. asserts that human beings have free will; so, *Thomas* has free will' can be analyzed as:

1. Thomas has free will.
   1.1 Thomas is human.
   1.1' Humans have free will.
      1.1'.1a N.N. has published dozens of papers in philosophy of mind.
      1.1'.1b N.N. asserts that humans have free will.

From a dialectical perspective, as we saw, an arguer seeking to discharge their BOP for the conclusion (1) is obliged to merely justify the premises and the inference rule. Analysts should nevertheless raise a conclusion-associated CQ to test whether the scheme type has been correctly identified. In an argument from expert opinion, moreover, the conclusion is normally identical to the expert's assertion. This CQ, then, would *not* be formulated as "Is 'Thomas has free will' true?" (as answering this CQ goes beyond the proponent's dialectical obligations) but rather as: CQ-4: "Is the conclusion 'Thomas has free will' identical with N.N.'s assertion?"

If CQ-2 and CQ-3 are already answered positively, analysts can suppose that the argument does at least contain an argument from expert opinion. Additionally, answering CQ-4 determines whether the instance *exclusively* matches the scheme of the argument from expert opinion. If CQ-4 is *not* answered positively, the argument is plausibly a *compound argument*—comprising an argument from expert opinion and some other argument(s). (In the 'Thomas has free will'-example, the main argument pivots on the MP rule, whereas the argument from expert opinion is a sub-argument.)

A CQ-list that is appropriate for some argument scheme must be associated with an argument's explicit components (i.e., premises and conclusion) and must be sufficient to substantially identify the argument scheme type, yet without precluding that an argument simultaneously instantiates two or more argument scheme types. While this may result from having expressed an argument vaguely, the main theoretical reason is that current argument scheme classifications allow for category overlap. Although this paper cannot hope to remedy this analytical defect, we can nevertheless associate Walton's criterion with the function of *identifying* an argument scheme type (Table 4).

Criterion 2 \ Criterion 1		Is the premise presumptive?	
		No	Yes
What is the premise's role in identifying the scheme type?	Sufficient (jointly with other explicit components)	Non-presumptive premise that, together with other explicit components, is *sufficient* to identify the scheme type	
	Unnecessary	Non-presumptive premise that is *not necessary* to identify the scheme type	Presumptive premise that is *not necessary* to identify the scheme type

Table 4: Criterion 2 specified as the premise's role in identifying the argument scheme type

Absent reasons to the contrary, it becomes plausible to fill the upper left cell in Table 4 with a presumptive premise that, jointly with other explicit components, is *sufficient* to identify the argument scheme type. Thus identified, then, is a fourth and final origin of CQs.

5. Conclusion

Adopting a dialectical perspective helped to clarify the distinction between an argument's premises and its inference rule. To obtain dialectical criteria to classify the origins of CQs, we used the notions of a presumption and of a presumptive argument scheme and referred to a CQ's cognitive function. This led us to identify four origins of a CQ, namely: (i) a presumptive, respectively (ii) a non-presumptive premise that, together with other explicit arguments components, is (iii) *necessary*, respectively (iv) *sufficient*. We offer this result as a contribution to developing a comprehensive theory of CQs.

Acknowledgements

S.Y. acknowledges support from the Nankai University Development Funds for the Humanities (No. ZB22BZ0314) and from the Projects funded by China Postdoctoral Science Foundation (No. 2022T150335; No. 2020M680874). F.Z. acknowledges support from European Network for Public Policy Argumentation (COST Action CA17132).

References

Ben-Ze'ev, A. 1995. Emotions and argumentation. *Informal Logic* 17: 189-200.
Bex, F., and Verheij, B. 2012. Solving a murder case by asking critical questions: An approach to fact-finding in terms of argumentation and story schemes. *Argumentation* 26: 325-353.
Feteris, E. 2016. Prototypical argumentative patterns in a legal context: The role of pragmatic argumentation in the justification of judicial decisions. *Argumentation* 30: 61-79.
Garssen, B. 2001. Argument schemes. In F.H. van Eemeren (ed.), *Crucial Concepts in Argumentation Theory*, pp. 81–99. Amsterdam: Amsterdam University Press.
Godden, D. M., and Walton, D. 2007. A theory of presumption for everyday argumentation. *Pragmatics and Cognition* 15: 313-346.
Govier, T. 1999. *The Philosophy of Argument*. Newport News, VA: Vale Press.
Hastings, A.C. 1962. *A Reformulation of the Modes of Reasoning in Argumentation*. Dissertation. Northwestern University, Evanston, IL.
Jin, R. 2011. The structure of pro and con arguments: A survey of the theories. In J.A. Blair, and R.H. Johnson (eds.), *Conductive Argument: An Overlooked Type of Defeasible Reasoning*, pp. 10-30. London: College Publications.
Johnson, R. H., and Blair, J. A. 2006 [1977 1st]. *Logical Self-defense*. New York: International Debate Education Association.
Liao, Y. 2021. The legitimacy crisis of arguments from expert opinion: Can't we trust experts? *Argumentation* 35: 265-286.

Kauffeld, F. 2003. The ordinary practice of presuming and presumption with special attention to veracity and burden of proof. In van Eemeren, F.H. (ed.), *Anyone Who Has a View: Theoretical Contributions to the Study of Argumentation*, pp. 136-146. Dordrecht: Kluwer.

Keppens, J. 2014. On modelling non-probabilistic uncertainty in the likelihood ratio approach to reasoning. *Artificial Intelligence and Law* 22: 239-290.

Prakken, H. 2010. On the nature of argument schemes. In Dialectics, dialogue and argumentation. An examination of Douglas Walton's theories of reasoning and argument, ed. C.A. Reed and C. Tindale, 167–185. London: College Publications. http://www.cs.uu.nl/groups/IS/archive/henry/schemes10.pdf. Accessed 18 Feb 2020.

Prakken, H. 2005. AI & law, logic and argument schemes. *Argumentation* 19: 303-320.

Ulman-Margalit, E. 1983. On presumption. *Journal of Philosophy* 80: 143-163.

van Eemeren, F.H. 2018. *Argumentation Theory: A Pragma-dialectical Perspective*. Cham: Springer International Publishing AG.

van Eemeren, F.H., and R. Grootendorst. 1992. *Argumentation, Communication, and Fallacies*. Hillsdale: Lawrence Erlbaum.

van Eemeren, F.H., and Grootendorst, R. 1984. Speech Acts in Argumentative Discussions. Dordrecht: Foris Publications.

Verheij, B. 2006. Evaluating arguments based on Toulmin's scheme. In D. Hitchcock and B. Verheij (eds.), *Arguing on the Toulmin Model. New Essays in Argument Analysis and Evaluation*, pp. 181–202. Netherlands: Springer.

Verheij, B. 2003. Dialectical argumentation with argumentation schemes: An approach to legal logic. *Artificial Intelligence and Law* 11: 167–195.

Visser, J., Lawrence, J., Wagemans, J., and Walton, D. 2020. Annotating argument schemes. *Argumentation*. https://doi.org/10.1007/s10503-020-09519-x.

Walton, D., Tindale, C., and Gordon, T. 2014. Applying recent argumentation methods to some ancient examples of plausible reasoning. *Argumentation* 28: 85-119.

Walton, D. 2012. Building a system for finding objections to an argument. *Argumentation* 26: 369-391.

Walton, D., C. Reed, and F. Macagno. 2008. *Argumentation Schemes*. Cambridge: Cambridge University Press.

Walton, D. 2003. Is there a burden of questioning? *Artificial Intelligence and Law* 11: 1-43.

Walton, D. 1997. *Appeal to Expert Opinion*. University Park: Pennsylvania State University.

Walton, D. 1996. *Argumentation Schemes for Presumptive Reasoning*. Mahwah: Lawrence Erlbaum.

Walton, D., and Krabbe, E. C. W. 1995. *Commitment in Dialogue: Basic Concepts of Interpersonal Reasoning*. Albany, NY: State University of New York Press.

Wagemans, J. 2011. The assessment of argumentation from expert opinion. *Argumentation* 25: 329-339.

Yu, S., and Zenker, F. 2019. A dialectical view on conduction: Reasons, warrants, and normal suasory inclinations. *Informal Logic* 39: 32-69.

Yu, S., and Zenker, F. 2020. Schemes, critical questions, and complete argument evaluation. *Argumentation* 34: 469-498.

Zenker, F., and Yu, S. 2022. Authority argument schemes, types, and critical questions. *Argumentation* 37: 25–51.

MULTI-CONTEXT POLYLOGUES: ANALYSIS OF INTERTWINED LEGAL & PUBLIC ARGUMENTS

GÁBOR Á. ZEMPLÉN
Department of Argumentation Theory and Marketing,
Eötvös Loránd University-Faculty of Economics, ELTE,
Budapest
zemplen@gtk.elte.hu

JÁNOS TANÁCS
Department of Argumentation Theory and Marketing, ELTE,
Budapest

Abstract

Argumentative exchanges in specific public or legal contexts contain moves that interact with arguments in other contexts. Locating the argumentative threads that run across the various public and institutional contexts is complex challenge. Expert-testimony can reasonably contradict judicial justification and legal arguments (e.g. press-rectification) can regulate public arguments.

Section 1 introduces the tools used to provide an analysis of selected constituents of multi-party, multi-position and multi-context polylogues. Section 2 sketches a simple multi-context polylogue and reconstructs the controversy that followed the Hungarian Government's 2016 billboard message: "Did you know? Brussels wants to settle a whole city's worth of illegal immigrants in Hungary". Increasing the complexity, Section 3 looks at polylogues that unfolded after two series of articles targeted the reputation of Viktor Orbán and other well-known and powerful public figures in the Fidesz-MPP party.

The developed tables of the multi-context polylogues can map Agents/Players who initiate dialogues in a primary Context and can also map the uptake of moves in the different contexts. The framework helps the understanding and analysis of potentially illicit moves, conflicting interpretations of phrases and appraisals of argumentative moves in significant lawsuits and public debates.

1. The interaction of public and legal arguments

Some multi-party argumentative discussions — argumentative polylogues (Lewiński, & Aakhus 2014) —are played out in part in specific normative institutional (e.g. legal) contexts, and also in the sphere of public argumentation. Press rectification, for example, is an argumentative legal discourse, where messages in the media are subjected to legal investigations, and legal interpretation classifies and declares certain moves in evidentiary processes as legitimate or not.

Such multi-context polylogues are potentially high stake, can have an impact on the public image of the actors, can jeopardize not just a publicly held standpoint, but the reputation or the perceived rationality of actors. They are also often so intricate, that taking them to be composites of standard dyadic interactions is probably unwise. A Judge can order the printing of a retraction message in a journal, so judicial decision can impact and regulate public argument. The press can ridicule the legal arguments of the Judge and evaluate the judicial justification. In some cases, public discourse over rulings impact expert discourse or induce changes in the legal handling of public argument.

In our communicative approach institutional and public argument interact, and if we can learn to see communication as a tapestry, into which the threads of argument are interwoven (Jacobs and Jackson 1992: 174), the investigated polylogues can be portrayed as unfolding on multiple layers, where initiating a dialogue in one context can influence several dialogues and various contexts. Threads of argumentative discourse in one layer of legal discourse link up with elements from other layers.

The type of complex argumentative exchange we are examining with this framework is not uncommon, and in the long run these polylogues shape public trust in institutions and influence deliberative cultures. Although important foci of argument analysis, polylogues with both social and judicial relevance inherently pose challenges of analysis and evaluation.

To handle reconstructive issues of polylogues with multiple players, with distinct, and potentially incompatible positions that are expressed *across a variety of places* (Aakhus and Lewiński 2017) we propose a minimalist approach to address *differences in places*. The temporally unfolding argumentative exchanges with cross-context interactions make it difficult to provide precise formulations of an issue/standpoint, and evaluation at the level of both quality and effectiveness become complex in entangled nets of legal and public argument.

In order to study how moves are played out in the significantly different contexts in the polylogues that nevertheless interact, the analysis presumes as opposed to argues for speech-act pluralism and assumes that a move can have several valence values in various contexts. The analysis is sensitive to the contextual differences and details (phrases, propositions, argument-structure), but to keep the overview simple, the positions are reconstructed elliptically. We provide sufficient detail to ground a (contestable) interpretation and evaluation of the selected moves in multi-party and multi-position polylogues, as we reduce the argumentative polylogues to sets of truncated dialogues.

Multi-context polylogues play out in contexts with significantly different institutionalized procedures. As various players (we call them Agents) can initiate an argumentative exchange in a polylogue, for the constituent analysis we concentrate on Agents, who target a specific primary audience in a primary Context with a Dialogue-start that comes with burden of proof (Protagonist roles). A Dialogue-start is usually initiated in the intended, primary context acknowledging some field-specific normative rules, but arguments supporting the position can surface in other contexts, where speech-acts may be subjects to unintended burden of proof. Some Agents in certain contexts can use declaratives to give 'binding' interpretation of other elements of other discourses in other contexts of the multi-layered fabric.

The partial mapping of the polylogue focuses on selected moves of the argumentative exchanges, including moves with contradictory evaluations in various contexts. The next section reconstructs a social debate in Hungary and introduces the notation used in the examples of intertwined legal and public arguments.

2. A simple multi-context polylogue on immigrants

In this first argumentative polylogue the first Dialogue-start is a short message. In the 2016 campaign on a referendum the Government littered the country with billboards. The campaign consisted of various statements after the phrase "Did you know?", including $\alpha(1)$:

> α Government billboard announcing: "Did you know? Brussels wants to settle a whole city's worth of illegal immigrants in Hungary." (1)

When taking up a Protagonist role in a specific context, the Agent of the dialogue-start $\alpha(1)$ has an intended Audience (the Public). In the upcoming popular vote the question is: "Do you want the European Union to be able to impose compulsory resettlement of non-Hungarian citizens in Hungary without the consent of Parliament?" The message of $\alpha(1)$ informs the electorate, and persuades the public to vote 'No'.

The first dialogue-start (α) aims to persuade the electorate, but many viewed the billboard messages inappropriate. A petitioner contested $\alpha(1)$ in a legal argument, as a false statement not fit to inform the electorate.

In the second dialogue-start β initiated an investigation within a rather elaborate legal-argumentative context. This procedural dialogue-start with a request/petition to trigger an advantageous decision in relation to the contested text published by the Government is imbued with argumentation, but before a judgement is reached, it is unclear which parts of it bear relevance for the reconstruction, so let us describe the move elliptically as

> β( ) An individual requests institutional rejection of **(1)**, claiming that α made an illegitimate (fallacious) move.

In reaction to β( ), a third Agent is triggered to take up a Protagonist role. In this example, the National Election Commission (NEC) had to *respond to β( ) on the merits of α(1)*, and resolve the second dialogue-start by providing an argumentative judicial decision within the given timeframe. Instead of a decision on the veracity of **α(1),** only the function of the message is disambiguated in the NEC declarative:

> γ The National Election Commission (NEC) classifies the message "[The function of the billboard-messages is to] inform them [members of the electorate] of their content in order to facilitate their decision" (NEC decision 53/2016) (2)
> The NEC concludes that judicial decision on the veracity of the statement is excluded, because "it [the statement] falls within the scope of the protection of political expression" (NEC decision 53/216). (3)

The institutional interpretation **γ(2,3)** does not reject the claim that **α(1)** is false, yet rejects the position of **β( )**, and approves of the message **α(1)** as permissible *in an election period*. As **α** is related to the subject of the upcoming referendum, the Government is directly addressing the electorate with the posters published. *For this time-period* the Government is communicating **α(1)** in good faith when it communicates its political opinion (a false statement?) on the subject of a referendum.

Both **β(_)** and **γ(2,3)** are engaging with the temporally precedent **α(1)** from the context of public argumentation. Although **β(_)** and **γ(2,3)** are both in a legal context, the argumentative contexts are significantly different, as the latter Protagonist can declare **α(1)** as true or false, as a legitimate or illegitimate move, and also **β(_)** as successful or failed.

The polylogue does not end with the verdict, as an article interpreted and evaluated the decision which was upheld by the Curia in yet another dialogue-start by a journalist (Fábián 2017):

> δAn online article evaluates the argumentative support in **γ(2,3)**: "In an amazingly twisted reasoning, the Supreme Court has ruled why the government's referendum posters are untrue. The bottom line:

information that appears in the government's "Did you know?" information campaign - such as Brussels wanting to settle a city's worth of illegal immigrants in Hungary - cannot be considered facts in ordinary terms, and therefore the government cannot be accused of making untrue statements." (4)

The elliptically introduced simple multi-context polylogue with four Agents and Dialogue-starts can be charted as in **Table I**.

Table I. Multi-context polylogue "Did you know?" with 4 dialogue-starts.

Agent & Dialogue-start	Context	Aim of Dialogue-start	Success in Context Primary / Secondary
α(1) Hungarian Government billboard "Did you know?..."	Public 1	Persuade the electorate to vote "No"	? / Yes
β( ) An individual's legal complaint	Legal 1	Persuade/convince the NEC to rule against α(1)	No / ?
γ (2, 3) National Election Commission decision	Legal 2	Provide a justified ruling that withstands an appeal	Yes / No
δ (4) Media-representative's evaluation	Public 2	Persuade the readers not to go to vote	Yes / n.a.

In this polylogue we can speak of challenges and contrary positions (several are implicit), but it would be an oversimplification to reduce the differences as contradictory or easily resolvable. It is probably too reductive to propose two sides to the polylogue (odd dialogue starts are pro Government and even ones against) or to assume that moves are in direct (and exclusive) dialectical relationship to one another, as in an analytical overview used in pragma-dialectics.

Table I. maps the Dialogue-starts by Agents in specific Contexts, and the last column tabulates success for the dialogue-starts in their primary and secondary contexts. In cases where evaluation depends on the perspective, we put "?" to describe success or failure.

Was α(1) successful in the original or primary context? In line with the Government's original aim 98% of the votes on the 2. Oct 2016 was 'No' on the referendum. However, *as a referendum*, it was *invalid* as only 44,08% of voters voted. In the secondary, legal context **α** was, however, successful (unlike **β**).

Apart from presenting problems, the mapping offers some insights into the polylogue, and helps appreciate the complexity of intertwined legal and public arguments. The dialogue-start **β** is terminated by decision **γ** of the NEC, and so is unsuccessful as a dialogue-start in the primary legal context. The decision **γ** is upheld by the Curia, and so is successful in the primary legal context. The twisted reasoning of **γ** is ridiculed in the dialogue start **δ** in the public context. In **δ** the argumentative oddity of **γ** is used in support of the failed petition **β**.

In the later public discourse on the referendum, including the argumentative exchanges on the voting strategy public arguments utilized **β** and **γ** to counter **α** and successfully convince many to *abstain* from voting. The polylogue contributed to the referendum ending up being invalid, showing that winning an argument in one context can have unintended (potentially negative) consequences for the Agent.

3. Press-rectification as multi-context polylogue

After the first example of a multi-context polylogue, where in the various contexts the analyzed moves had conflicting (argumentative) evaluation, we now turn to more complex cases. The press-correction lawsuits investigated in this section were triggered by investigative journalism pieces published by the weekly *ÉS (Life and Literature)*. The "watchdog of democracy" directly targeted - among others - the reputation of Viktor Orbán, Prime Minister in 1999 and former Prime Minister by 2005.

The methods introduced in the previous section are adapted: we focus on a short explication of agents and dialogue-starts in primary contexts, but to highlight crucial elements of the polylogue only cite linguistically small constituents for some of the agents. Section 3.1. sets the scene and gives an overview of the agents in **Table II.**, and in two sections we follow up on two of the threads of the multi-layered canvas.

3.2 discusses lawsuits against $α_1$ highlighting the conflicting interpretations among (linguistic and judicial) experts on a semantic issue.

3.3 discusses conflicting (1st and 2nd degree) judicial interpretations of a phrase used in $α_2$, highlighting the strategic advantage of picking the right Claimant for the legal context.

The editorial staff reflected on the trials, and the landmark cases are well documented in thick volumes edited by Kovács & Tarnói (2000, 2005). We therefore minimize the references and note that a more detailed exposition discussing the secondary literature and the legal background is available in Hungarian (Tanács & Zemplén 2017).

3.1. Investigative journalism on the Orbán family businesses

The series of articles we now turn to linked politicians with mining companies, vineyards, as well as with ethically and legally problematic activities. A reputable weekly broadsheet, *Élet és Irodalom* (henceforth ÉS) published α1 and α2:

a_1 Investigative journalism articles "Boys in the Mine" focused on the mining contracts of the father of the prime minister, and companies near the party Fidesz-MPP (Ószabó & Vajda, 1999). The papers discussed whether "the fact that one of the members of the Orbán family has become one of the leading figures in Hungarian political life over the past ten years, and who has held the most powerful post in the Republic of Hungary since last June, has played a role in the Orbán family's gentrification, or whether it is just a typical business-success story". (5)

a_2 Investigative journalism article-series "Tokaj Wine-Battles" discussed politicians interacting with corporations, stating among others that on certain days "the former Prime Minister attended meetings of the members of Szárhegydűlő-Sárazsadány-Tokajhegyalja Kft., which was partly owned by his wife, where the acquisition of state subsidies and real estate was also discussed" and "actually given advice to his wife's company". (6)[1]

Following the two article-series, some of the content was challenged in legal contexts, and the judicial decisions were appealed, resulting in intricate polylogues. Actors included Judicial experts and Parliamentary Committees, and some verdicts met with reasoned opposition even from legal experts.

The penetration of the articles was high nationwide (for a_2 in a poll 79 knew of the allegations, 56% assumed a_2 to be true). In the public context a primary target of both a_1 and a_2 was Viktor Orbán, and the reputation loss for him was significant, with an 8% drop in one month after a_2 was published (hvg.hu 2005).

Interestingly in the legal context the winning cases were not filed by the politicians who were the targets of the articles, but organizations (we will return to the strategic significance of Claimant-choice). And by the end of the polylogue not only the reputation of the politicians involved was damaged, but also that of the journal, the authors, investigative journalism in general, as well as that of the judiciary branch.

In **Table II.** we overview the selected Agents, with a short description of the dialogue-start in a primary context. The last two columns list the contexts, where the dialogue-start of an agent can be considered successful or failed (primary contexts are in bold). The following two subsections will introduce the agents in the polylogue.

---

[1] Attila Rajnai was the author (in ÉS 2005/47, repr. Kovács & Tarnói, 2005). The paper's position (6) is translated from (Bod 2007).

Table II. "Boys in the Mine" & "Tokaj Wine-Battles" Polylogue-overview.

Agent	Context	Dialogue-start	Success	Failure
$α_1$ (5) "Boys in the Mine" (ÉS, 1999)	Public 1	Inform the readers / electorate	**P1**, P2	L1, L3, L5
$α_2$ (6) "Tokaj Wine-Battles" (ÉS, 2005)	Public 1	Inform the readers / electorate	**P1**, P2, L1	L3, L5
$β_{1a}$ ( ) Fidesz, a party organization	Legal 1	Challenge $α_1$ on ownership-detail	L1, L3, L5	P1
	Legal 1	Challenge $α_1$ on enrichment	-	L1
$β_{1b}$ ( ) L. Kövér & 10 party members	Legal 1	Challenge $α_1$ on fiscal issues	-	L1
$β_{1c}$ ( ) J. Áder & Fidesz party director				
$γ_{1a}$ ( ) Judicial decision	Legal 2	Provide a justified ruling	L2	P1
	Legal 2	Provide a justified ruling	L2, P1	
$γ_{1b}$ ( ) Judicial decision	Legal 2	Dismissing the case	L2, P1	
$γ_{1c}$ ( ) Judicial decision				
$α_{1a}$ (7) ÉS expert counterargument	Public 2	Inform / persuade the readers to reject $γ_{1a}$	P1	
$δ_{1a}$ ( ) ÉS appeal with expert testimony	Legal 3	Persuade/convince the Court to reject $γ_{1a}$	P1	L3
$ε_{1a}$ ( ) Judicial decision 2nd degree	Legal 4	Provide a justified ruling	L4, L5	P1
$ζ_{1a}$ ( ) Judicial decision 3rd degree	Legal 5	Provide a justified ruling	L5	
$β_{2a}$ (8) An organization (SzSzT ltd.)	Legal 1	Challenge $α_2$ on "member's meeting"	L3	L1, L2
$γ_{2a}$ (9) Judicial decision	Legal 2	Provide a justified ruling	P1, P2	L2
$ω_{2a}$ ( ) Judicial expert	Public 2	Approve $γ_{2a}$	?, P1	L4
$δ_{2a}$ ( ) SzSzT Ltd. appeal	Legal 3	Persuade/convince the Court to reject $γ_{2a}$	L3	P1
$ε_{2a}$ (10) Judicial decision 2nd degree	Legal 4	Provide a justified ruling	L4	P1

3.2. Semantic disambiguation of (ir)relevant details

Soon after communicating $α_1$, three lawsuits were started. We will investigate the first ($β_{1a}$) in more detail later and summarize the two unsuccessful dialogue-starts.

$β_{1a}$ was launched by Fidesz-MPP as a party, and the subject was the ownership of certain companies mentioned in $α_1$: are they owned by Fidesz?

$β_{1b}$ challenged $α_1$ on the claim that party leaders' enrichment had been aided by the money from the sale of the party-headquarters. The court of second instance dismissed the action by $β_{1b}$ as a matter of law, saying that $α_1$ reported an existing, well-founded opinion, and that this could not be the subject of a lawsuit for press-rectification. (Similar to the *stasis* in the NEC decision on the "Did you know..." campaign examined in Section 2.).

$β_{1c}$ was filed against the newspaper by the party's director and János Áder (later President of Hungary from 2012 to 2022). $β_{1c}$ alleged that ÉS falsely claimed in $α_1$ that Fidesz had not accounted for its finances to the party's membership. ÉS made no such claim, its allegation was that Fidesz had not accounted to its membership for the money it had received from the sale of its headquarters. The plaintiffs withdrew this claim at the first hearing, withdrew the lawsuit, and the court dismissed it.

In the winning case $β_{1a}$ the veracity of five statements in $α_1$ was contested (ÉS 19/20 Aug. 1999, Kovács & Tarnói 2000, 109-111). In three of the five allegations directly and in one case indirectly, the crucial issue became the meaning of the phrase "Fidesz-cég" (Fidesz-company).

$β_{1a}$ claims that sentences from $α_1$ are inadmissible as they use "Fidesz-cég".

If the phrase "Fidesz-cég" does not mean "a company near Fidesz", but means a "company of Fidesz", the sentences are false. The court of first instance resolved the linguistic issue and the weekly was ordered to publish a rectification (Perekről, 2000).

Judicial decision $γ_{1a}$ ruled in favour of the plaintiff on four counts, and, concerning the fifth contested claim, it ruled in favour of the plaintiff in part and the defendant in part.

The judicial interpretation $γ_{1a}$ was contested after the court ruling by many. It was challenged in the journal ÉS in an opinion-piece by the linguist Péter Hajdú, who published a counterargument $α_{1a}$:

Expert $α_{1a}$ asserted that *if* the journal thought that these companies belong to the political party, another phrase would have been used: "My question is whether the term "Fidesz-cég" means that the company in question is owned by Fidesz. Not necessarily, and most likely not. I.e.:

If the newspaper were to describe Centum, for example, as a Fidesz-owned company, it would write "a Fidesz cége." (7)

The judge's decision γ_{1a} was also contested in court, as the defendant journal ÉS appealed the decision. In appeal δ_{1a} the expert (linguist László Kálmán) highlighted in his testimony the ambiguity of the term "Fidesz-cég".

At odds with the expert interpretations in the public and legal contexts (α_{1a} and δ_{1a}) the earlier decision was upheld in ε_{1a} and ζ_{1a} (Kovács & Tarnói, 2000: 323-331).

Table III. Polylogue positions on the permissible use of a key phrase with a short description of the use in the primary context, and success in intended / unintended contexts.

Agent & dialogue start	Is "Fidesz-cég" permissible?	Use in context	Success
α_1 (5) "Boys in the Mine" (ÉS, 1999)	Yes	An abbreviated form of referring to corporations introduced in α_1	Yes /
β_{1a} ( ) Fidesz, a party organization	No	Challenge legal use in α_1 by an appeal to legal definition	Yes / No
γ_{1a} ( ) Judicial decision	No	Apply legal definition to rule against α_1 in favour of β_{1a}	Yes/ No
α_{1a} (7) ÉS linguist expert	Yes	Expert opinion to defend α_1 with counterargument to γ_{1a} (phrase can mean *Near-Fidesz*)	Yes / -
δ_{1a} ( ) Expert for ÉS appeal	Yes	Challenge γ_{1a} in court, because meaning in non-technical domains is ambiguous	No / Yes
ε_{1a} ( ) Judicial decision 2nd degree	No	Reject the challenge by δ_{1a}, and support interpretation of the phrase by γ_{1a}.	Yes / -

In such a simple analysis (see Table III.) we can pinpoint expert testimonies α_{1a} and δ_{1a} and contrast it with judicial position γ_{1a}. A notable feature of γ_{1a} is that it is only supported by *one* possible meaning of "Fidesz-cég" and does not handle the ambiguous use observed in natural language. The Press retracted a part of the message, but not in accord with the rational conviction of the journalists. Various organizations kept follow up record on mining contracts in later years (Pethő & Zöldi 2017).

3.3. The impact of Claimant-choice on a polylogue

Due to the increased publicity and the resulting concentrated media attention, the polylogue triggered by α_2 was in the limelight of public discourse, and the public dialogue co-developed with processing of the Claimant's move β_{2a}. Even a Parliamentary committee was set up to study the relevance of advice given by Viktor Orbán and the financial moves of his wife, and discrepancies in the politician's testimony were found, strongly supporting the Press's findings in α₂.

Although α₂ targeted the former Prime Minister, the veracity of the report was not challenged by him in the legal context. The managing director of a Ltd. company considered the phrasing of α₂ to be damaging to its good reputation. As part of the rectification the company (SzSzT, in short for Szárhegy-dűlő-Sárazsadány-Tokajhegyalja Kft.) requested the following text to be published:

> β₂ₐ is challenging α₂, requesting the journal to publish: "Correction - In the article entitled Tokaj Wine Battles II. in the 11 March 2005 issue of our newspaper, we falsely stated - or rather rumored - that Szárhegy-dűlő-Sárazsadány-Tokajhegyalja Ltd. held a members' meeting on 6 September 2000, 4 June 2001 and 2 December 2001, and therefore no minutes could be taken of such members' meetings." (8)

The editorial staff of the weekly refused to publish the rectification, published an editorial statement. Repeating the request in a legal context β₂ₐ filed a press-rectification. The judge rejected the argument,

> The judicial decision γ₂ₐ stated that "it is not in doubt that the members of the plaintiff company met occasionally - and also on the dates in question - and discussed the company's affairs in substance, and the veracity of the information provided about the content of the meetings has not yet been challenged on the merits, to the court's knowledge."   (9)

Table IV. Polylogue positions on the key phrase to describe the meeting of the members of the company ("taggyűlés").

Agent & Dialogue-start	Use of "taggyűlés"	Support for use in Dialogue-start	Notes on Use
α₂ (6) "Tokaj Wine-Battles" (ÉS, 2005)	Permissible	Based on oral interviews and some documents	The methodology and data collection is explicated
β₂ₐ (8) An organization (SzSzT ltd.)	Not permissible	Appeal to legal definition in Economic Act	The type of move cannot be initiated by the targeted politicians, only by organizations named in α₂
γ₂ₐ (9) Judicial decision	Permissible	The veracity of the information in α₂ about the content of the meetings could not be challenged on the merits	Protection of the text under investigation trumps Claimant-protection
δ₂ₐ ( ) SzSzT appeal	Not permissible	γ₂ₐ failed to follow the guidelines of the Economic Act on	Arguments repeated and slightly modified to attack α₂ & γ₂ₐ

		formal definition of member's meeting	
ω_{2a} ( ) Judicial expert	Permissible	γ_{2a} is applauded not falling for the trap set by β_{2a}	Contributes to extended expert-discourse on legal argumentation
ε_{2a} (10) 2nd degree Judicial decision	Not permissible	Statements in question are not proved to be true beyond reasonable doubt	The protection of the Claimant trumps text-protection

Within the legal community several experts commented on, and in general applauded the decision (ω_{2a}( ) stands for a judicial expert in support of γ_{2a}). The company SzSzT Ltd. criticized γ_{2a}, appealed and repeatedly stressed that the meetings referred to do not satisfy the formal definition of 'a company's member's meeting' (according to the Economic Act), and supported the claim with a verified signed document, in which the owners of the company state that no member's meeting took place on the given days. Appeal δ_{2a} of the company initiated a second degree court procedure:

ε_{2a} in the second order Court reverted the decision, stating that without "the authentic minutes of the general meeting of members obtained from the Court of Registration" the defendant does "not prove beyond reasonable doubt the truth of the allegations in the proceedings" Consequently, the statements made by α_2 "had to be regarded as untrue and untrue statements of fact, which in themselves constitute an infringement." (10)

The polylogue, again, resulted in reputational loss both for the targeted politicians, for the emblematic organ of Hungarian investigative journalism ('The fact-finder is naked' run the headline of *Magyar Nemzet* following the appeal decision, Kovács & Tarnói 2005, 394), and for the judiciary. The second order decision was criticized by many – including legal experts.

When initiating a legal discourse, it seems clear that the choice of claimant has significant influence on the available moves. In the last polylogue the success in the legal context was limited for the Agents affected the most by α_2, as only a certain type of claimant (a company) could include the moves β_{2a} and δ_{2a}. In the legal context this choice of Claimant yielded a winning strategy, as a narrow technical definition of a phrase could be used by a company to argue against the veracity of statements in α_2.

4. Can polylogue-analysis respond to complexity and urgency?

Actors in multi-position and multi-context discussions can involve Parliamentary Committees, Legal Actors (the plaintiff, the defendant, the judge, etc.), just as firms, political parties, and other organizations. The contexts brought into play can be various. In most detail we looked at landmark Hungarian press-rectification cases and judicial decisions that have also triggered social debates before the emergence of an illiberal democracy. Our more recent, first example came from a political campaign and exemplify state communication of contemporary Hungary.

Argumentative exchanges in hybrid settings, where legal and public argument are intertwined are common not only in hybrid regimes; similar dynamics can unfold concerning advertisements, or in the wake of crises and catastrophes. Complex translation takes place when messages in the media get scrutinized in the context of legal argumentation, and when judicial decisions are scrutinized in the media. Can valid, or at least contestable reconstructions be given of such polylogues?

The current primary tools of knowledge-transmission, the journal article format, and the conference proceedings naturally impose challenges due to the sheer extension and intricacy of some of the debates. The increasingly complex examples introduced in the paper highlight how determining the meaning of key locutions of the polylogues varies across and within contexts, as does truth or falsity of a statement, or (non-)admissibility of a move. Our examples are interlinked, yet the dialogue starts are initiated in different contexts with differences in regulations and in the level of burden of proof.

References

Aakhus, M., & Lewiński, M. (2017). Advancing polylogical analysis of large-scale argumentation: Disagreement management in the fracking controversy. *Argumentation*, 31, 179-207.
Bod, T. (2007, March 30). Fiúk a szőlőben. *Élet és Irodalom*. https://www.es.hu/cikk/2007-04-01/bod-tamas/fiuk-a-szoloben.html
Fábián, T. (2016, Aug 29). Kúria: A kormány tájékoztató kampánya nem hazug, mivel nem is állít tényeket. *Index*.
hvg.hu (2005, Apr 13). Medián: a kormánypártok utolérték a Fideszt. *Heti Világgazdaság*. http://hvg.hu/kozvelemeny.median/20050413median/2
Jacobs, S., & Jackson, S. (1992). Relevance and digressions in argumentative discussion: A pragmatic approach. *Argumentation*, 6, 161-176.
Kovács, Z. & Tarnói, G. (eds.) (2000). *Fiúk a bányában. Fidesz-perek az ÉS ellen*. Budapest, Irodalom Kft.,
Kovács, Z. & Tarnói, G. (eds.) (2005). *Fiúk a szőlőben. A Tokaji borcsaták per- és sajtóanyaga*. Budapest, Élet és Irodalom.

Kúria: A kormány tájékoztató kampánya nem hazug, mivel nem is állít tényeket. *Index*, http://bit.ly/2lcxg7n

Lewiński, M., & Aakhus, M. (2014). Argumentative polylogues in a dialectical framework: A methodological inquiry. *Argumentation*, 28(2), 161-185.

Ószabó, A. & Vajda, É. (1999, Aug 6). Fiúk a bányában. Az Orbán család vállalkozásai. *Élet és Irodalom*. https://www.es.hu/cikk/1999-08-06/oszabo-attila-vajda-eva/fiuk-a-banyaban.html

Perekről. (2000, Jan 7). *Élet és Irodalom*. https://www.es.hu/old/0001/visszhang.htm

Pethő, A. & Zöldi, B. (2017, May 9). Rejtett állami munkákból is jött pénz az Orbán-család gyorsan szerzett milliárdjaihoz. *Direkt36*. https://www.direkt36.hu/rejtett-allami-munkakbol-is-jott-penz-az-orban-csalad-gyorsan-szerzett-milliardjaihoz/

Tanács, J., & Zemplén, G. (2017). Press-Rectification and Strategic Maneuvering: How Picking the Right Claimant Affects Judicial Interpretations. *Iustum Aequum Salutare*

LIST OF CONTRIBUTORS TO ALL VOLUMES
(IN ALPHABETICAL ORDER BY SURNAME)

SCOTT AIKIN
Vanderbilt University

JOSÉ ALHAMBRA
Autonomous University of Madrid

JOSE M. ALONSO-MORAL
Universidade de Santiago de Compostela, Spain

KATIE ATKINSON
University of Liverpool

SHARON BAILIN
Faculty of Education, Simon Fraser University, Vancouver Canada

MARK BATTERSBY
Department of Philosophy, Capilano University, Vancouver Canada

TREVOR BENCH-CAPON
University of Liverpool

SARAH BIGI
University of the Sacred Heart, Milan, Italy

PETAR BODLOVIĆ
NOVA Institute of Philosophy (ArgLab),
FCSH, Nova University of Lisbon

MIEKE BOON
University of Twente

ELENA CABRIO
Université CôteD'Azur, CNRS, Inria, I3S

JOHN CASEY
Northeastern Illinois University

ALEJANDRO CATALA
Universidade de Santiago de Compostela, Spain

DORIANA CIMMINO
Independent researcher

LIST OF CONTRIBUTORS

DANIEL COHEN
Colby College

FEDERICA COMINETTI
Università dell'Aquila

CLAUDIA COPPOLA
Università Roma Tre, La Sapienza Università di Roma

MARÍA INÉS CORBALÁN
ArgLab-IFILNOVA, NOVA Universidade de Lisboa

HÉDI VIRÁG CSORDÁS
Assistant Lecturer at Budapest University of Technology and Economics

GIULIA D'AGOSTINO
Università della Svizzera italiana (USI)

DANIEL DE OLIVEIRA FERNANDES
University of Fribourg, Switzerland

EMMANUELLE DIETZ
Airbus Central R&T, Germany

ÁLVARO DOMÍNGUEZ-ARMAS
NOVA Institute of Philosophy, NOVA University of Lisbon

GONEN DORI-HACOHEN
University of Massachusetts Amherst

IOVAN DREHE
Technical University of Cluj-Napoca

LUCIJA DUDA
University of Manchester

MICHEL DUFOUR
University Sorbonne-Nouvelle

ALINA DURRANI
University of Massachusetts Amherst

CATARINA DUTILH NOVAES
Department of Philosophy, Vrije Universiteit Amsterdam

List of Contributors

ISABELA FAIRCLOUGH
University of Central Lancashire

LOGAN FIELDS
Advancing Machine and Human Reasoning (AMHR) Lab, University of South Florida

JOSÉ ÁNGEL GASCÓN
Departamento de Filosofía, Universidad de Murcia

GIULIA GIUNTA
University of Neuchâtel

GEOFFREY C. GODDU
University of Richmond

SARA GRECO
USI-Università della Svizzera italiana

MARCELLO GUARINI
University of Windsor, Canada

PASCAL GYGAX
University of Fribourg, Switzerland

DALE HAMPLE
Western Illinois University

AMALIA HARO MARCHAL
University of Granada

ANNETTE HAUTLI-JANISZ
University of Passau

BITA HESHMATI
University of Groningen

MIKA HIETANEN
Lund University

MICHAEL J. HOPPMANN
Northeastern University, Boston

BROOKE HUBSCH
The Pennsylvania State University

LIST OF CONTRIBUTORS

BETH INNOCENTI
University of Kansas

CHIARA JERMINI-MARTINEZ SORIA
Università della Svizzera italiana

ANTONIS KAKAS
Dept. Computer Science, University of Cyprus, Cyprus

ALEXANDRA KARAKAS
Assistant Lecturer at Budapest University of Technology and Economics

IRYNA KHOMENKO
Taras Shevchenko National University of Kyiv

ZLATA KIKTEVA
University of Passau

KONRAD KILJAN
University of Warsaw
Laboratory of The New Ethos, Warsaw University of Technology

GABRIJELA KIŠIČEK
University of Zagreb

MARCIN KOSZOWY
Laboratory of The New Ethos, Warsaw University of Technology

ADAMOS KOUMI
Dept. Computer Science, University of Cyprus, Cyprus

MANFRED KRAUS
University of Tübingen

LEONARD KUPŚ
Faculty of Psychology and Cognitive Science
Adam Mickiewicz University, Poznań, Poland

NIILO LAHTI
The University of Eastern Finland, School of Theology

JOHN LAWRENCE
Centre for Argument Technology, University of Dundee, UK

List of Contributors

LAWRENCE LENGBEYER
United Stated Naval Academy

MARCIN LEWIŃSKI
Nova Institute of Philosophy
Nova University Lisbon, Portugal

JIAXING LI
Nankai University

YAN-LIN LIAO
Department of Philosophy, Sun Yat-sen University, China

JOHN LICATO
Advancing Machine and Human Reasoning (AMHR) Lab, University of South Florida

DAVIDE LIGA
University of Luxembourg

EDOARDO LOMBARDI VALLAURI
Università Roma Tre

COSTANZA LUCCHINI
Università della Svizzera italiana (USI)

CHRISTOPH LUMER
University of Siena, Italy

GIORGIA MANNAIOLI
Università Roma Tre, La Sapienza Università di Roma

MAURIZIO MANZIN
University of Trento

ZAID MARJI
Advancing Machine and Human Reasoning (AMHR) Lab, University of South Florida

HUBERT MARRAUD
Universidad Autónoma de Madrid (Spain)

SANTIAGO MARRO
Université CôteD'Azur, CNRS, Inria, I3S

LIST OF CONTRIBUTORS

VIVIANA MASIA
Università Roma Tre

DAVIDE MAZZI
University of Modena and Reggio Emilia (Italy)

GUIDO MELCHIOR
University of Graz

CHIARA MERCURI
Università della Svizzera Italiana

DIMA MOHAMMED
Institute of Philosophy, Faculty of Social and Human Sciences, NOVA University of Lisbon, Portugal

ELENA MUSI
University of Liverpool

HENRI MÜTSCHELE
Heinrich- Heinrich Heine University Düsseldorf, Germany

ZI-HAN NIU
Department of Philosophy, Sun Yat-sen University, China

PAULA OLMOS
Universidad Autónoma de Madrid

MARIANA OROZCO
University of Twente

RAHMI ORUÇ
Ibn Haldun University, Comparative Literature, ArguMunazara Research Center

STEVE OSWALD
University of Fribourg, Switzerland

WENQI OUYANG
Department of Philosophy, Sun Yat-sen University, China

FABIO PAGLIERI
Istituto di Scienze e Tecnologie della Cognizione, Consiglio Nazionale delle Ricerche (ISTC-CNR), Italy

ROOSMARYN PILGRAM
Leiden University Centre for Linguistics

FEDERICO PUPPO
University of Trento

MENNO H. REIJVEN
University of Amsterdam

THÉOPHILE ROBINEAU
Université Paris Cité

ANDREA ROCCI
Università della Svizzera italiana (USI)

MARIA GRAZIA ROSSI
Institute of Philosophy, Faculty of Social and Human Sciences, NOVA University of Lisbon, Portugal

LUCIA SALVATO
Università Cattolica del Sacro Cuore

CRISTIÁN SANTIBÁNEZ,
Universidad Católica de la Santísima Concepción

MENASHE SCHWED
Ashkelon Academic College, Israel

BLAKE D. SCOTT
Institute of Philosophy, KU Leuven

HARVEY SIEGEL
University of Miami

ILIA STEPIN
Universidade de Santiago de Compostela, Spain

JÁNOS TANÁCS
Department of Argumentation Theory and Marketing, ELTE, Budapest

GIULIA TERZIAN
ArgLab-IFILNOVA, NOVA Universidade de Lisboa

CHRISTOPHER W. TINDALE
Department of Philosophy, University of Windsor, Ontario, Canada

LIST OF CONTRIBUTORS

SERENA TOMASI
University of Trento

MARIUSZ URBAŃSKI
Faculty of Psychology and Cognitive Science
Adam Mickiewicz University, Poznań, Poland

MEHMET ALÌ ÜZELGÜN
IFILNOVA, Universidade Nova de Lisboa
CIES-ISCTE, Instituto Universitário de Lisboa

CHARLOTTE VAN DER VOORT
Leiden University Centre for Linguistics
JAN ALBERT VAN LAAR
University of Groningen

LOTTE VAN POPPEL
Center for Language and Cognition Groningen

SERENA VILLATA
Université CôteD'Azur, CNRS, Inria, I3S

JACKY VISSER
Centre for Argument Technology, University of Dundee, UK

JEAN H.M. WAGEMANS
University of Amsterdam

HAILONG WANG
Wuhan University

JIANFENG WANG
Fujian Normal University, China

MARK WEINSTEIN
Montclair State University

HARALD R. WOHLRAPP
Universität Hamburg

MING-HUI XIONG
Guanghua Law School, Zhejiang University, China

OLENA YASKORSKA-SHAH

Università della Svizzera italiana (USI)

SHIYANG YU
College of Philosophy, Nankai University

GÁBOR Á. ZEMPLÉN
Eötvös Loránd University-Faculty of Economics, ELTE, Budapest

FRANK ZENKER
College of Philosophy, Nankai University, Tianjin, China

www.ingramcontent.com/pod-product-compliance
Lightning Source LLC
Chambersburg PA
CBHW071311150426
43191CB00007B/583